ADAMS'S NEW ARITHMETIC—REVISED EDITION

ARITHMETIC,

IN WHICH THE

PRINCIPLES OF OPERATING BY NUMBERS

ARE

ANALYTICALLY EXPLAINED

AND

SYNTHETICALLY APPLIED.

ILLUSTRATED BY COPIOUS EXAMPLES

DESIGNED FOR THE USE OF SCHOOLS AND ACADEMIES

BY DANIEL ADAMS, M. D.,
AUTHOR OF THE SCHOLAR'S ARITHMETIC, SCHOOL GEOGRAPHY, ETC

BOSTON:
PHILLIPS, SAMPSON, AND COMPANY.
NEW YORK: COLLINS AND BROTHER
KEENE, N. H.: J. H. SPALTER.

PREFACE.

THE "Scholar's Arithmetic," by the author of the present work, was first published in 1801. The great favor with which it was received is an evidence that it was adapted to the wants of schools at the time.

At a subsequent period the analytic method of instruction was applied to arithmetic, with much ingenuity and success, by our late lamented countryman, WARREN COLBURN. This was the great improvement in the modern method of teaching arithmetic. The author then yielded to the solicitations of numerous friends of education, and prepared a work combining the analytic with the synthetic method, which was published in 1827, with the title of "Adams' New Arithmetic."

Few works ever issued from the American press have acquired so great popularity as the "New Arithmetic." It is almost the only work on arithmetic used in extensive sections of New England. It has been re-published in Canada, and adapted to the currency of that province. It has been translated into the language of Greece, and published in that country. It has found its way into every part of the United States. In the state of New York, for example, it is the text-book in ninety-three of the one hundred and fifty-five academies, which reported to the regents of the University in 1847. And, let it be remarked, it has secured this extensive circulation solely by its merits. Teachers, superintendents, and committees have adopted it because they have found it fitted to its purpose, not because hired agents have made unfair representations of its merits, and, of the defects of other works, seconding their arguments by liberal pecuniary offers — a course of dealing recently introduced, as unfair as it is injurious to the cause of education. The merits of the "New Arithmetic" have sustained it very successfully against such exertions. Instances are indeed known, in which it has been thrown out of schools on account of the "liberal offers" of those interested in other works, but has subsequently been readopted without any efforts from its publishers or author.

The "New Arithmetic" was the pioneer in the field which it has occupied. It is not strange, then, that teachers should find defects and deficiencies in it which they would desire to see removed, though they might not think that they would be profited by exchanging it for any other work. The repeated calls of such have induced the author to undertake a revision, in which labor he would present acknowledgments to numerous friends for important and valuable suggestions. Mr. J. HOMER FRENCH, of Phelps, N. Y., well known as a teacher, has been engaged with the author in this revision, and has rendered important aid. Mr. W. B. BUNNELL, also, principal of Yates Academy, N. Y., formerly principal of an academy in Vermont, has assisted throughout the work, having prepared many of the articles. The revision after Percentage is mostly his work.

The characteristics of the "New Arithmetic," which have given the work so great popularity, are too well known to require any notice here These, it is believed, will be found in the new work in an improved form.

One of the peculiar characteristics of the new work is a more natural and philosophical arrangement. After the consideration of simple whole numbers, that of simple fractional numbers should evidently be introduced, since a part of a thing needs to be considered quite as frequently as a whole thing. Again; since the money unit of the federal currency is divided decimally, Federal Money certainly ought not to precede Decimal Fractions. It has been thought best to consider it in connection with decimals. Then follow Compound Numbers, both integral and fractional, the reductions preceding the other operations, as they necessarily must. Percentage is made a general subject, under which are embraced many particulars. The articles on Proportion, Alligation, and the Progressions will be found well calculated to make pupils thoroughly acquainted with these interesting but difficult subjects.

Care has been taken to avoid an arbitrary arrangement, whereby the processes will be purely mechanical to the learner. If, for instance, all the reductions in common fractions precede the other operations, the pupil will have occasion to divide one fraction by another long before he shall have learned the method of doing it, and must proceed by a rule, to himself perfectly unintelligible. The studied aim has been throughout the entire work to enable the ordinary pupil to understand every thing as he advances. The author is yet to be convinced that mental discipline will be promoted, or any desirable end be subserved, by conducting the pupil through blind, mechanical processes. Just so far as he can understand, and no farther, is there prospect of benefit. No good results from presenting things, however excellent in themselves, if they are beyond the comprehension of the learner.

Those teachers who prefer to examine their classes by questions, will find that little will escape the pupil's attention, who shall correctly answer all those in the present work, while teachers who practise the far superior method of recitation by analysis, will find the work admirably adapted to their purpose.

The examples, it is hoped, will require very full applications of the principles.

Many antiquated things, which it has been fashionable to copy in arithmetics, from time immemorial, have been omitted or improved, while new and practical matter has been introduced. A Key to this revision is in progress.

With these remarks, the work is submitted to the candid examination of the public, by THE AUTHOR.

Keene, N. H., February, 1848.

SUGGESTIONS TO TEACHERS.

The writer complies with the request of the venerable author of "Adams' Arithmetic," to preface the new work with a few suggestions to his associates in the work of instruction. Though he has been engaged for sometime past in assisting to make the work better fitted to accomplish its design, he is perfectly satisfied that improvement in school education is rather to be sought in improved use of the books which we now have, than in making better books. Better arithmeticians would be made by the book as it was before the present revision, using it as it might be used, than will probably be made in most cases with the new work, even though the former were very defective, the latter perfect. Exertion, then, to bring teachers to a higher standard, will be more effective in improving school education, than any efforts at improving school *books* can possibly be. It is here where the great improvement must be sought. Without the cooperation of competent teachers, the greatest excellences in any book will remain unnoticed, and unimproved. Pupils will frequently complain that they have never found one that could explain some particular thing, of which a full explanation is given in the book which they have ever used, and their attention only needed to have been called to the explanation.

Then let teachers make themselves, in the first place, thoroughly acquainted with arithmetic. The idea that they can, "study and keep ahead of their classes," is an absurd one. They must have surveyed the whole field in order to conduct inquirers over any part, or there will be liability to ruinous misdirection. Young teachers are little aware of their deficiencies in knowledge, and still less aware of the injurious effects which these deficiencies exert upon pupils, who are often disgusted with school education, because they are made to see in it so little that is meaning.

In the next place, let no previous familiarity with the subject excuse teachers from carefully preparing each lesson before meeting their classes. Thereby alone will they feel that freshness of interest, which will awaken a kindred interest among their pupils; and if on any occasion they are compelled to omit such preparation, they will discover a declining of interest with their classes. Teachers who are obliged to have their books open, and watch the page while their classes recite, are unfit for their work.

Pupils should be taught how to study. That, after all, is the great object of educating. The facilities for merely acquiring knowledge are abundant, if persons know how to improve them. The members of classes will often fail in recitation, not because they have not tried, but have not known how to get their lesson. They neglect trying, because they can do so little to advantage. They may read over a statement in their book a dozen times they say, but cannot remember it, —*because they do not understand it.* An hour spent with each pupil individually in questioning him on the meaning of each sentence, which he may be required to read, will be of incalculable advantage.

When pupils shall have been taught how to study, let them be required *to get their lessons, and recite them.* If the present book is not thought by teachers to contain a sufficient description, and a sufficient explanation of everything, let them try to find one that does, for if pupils present themselves before the blackboard at the time of recitation, with the expectation that the teacher is to explain to the class, and help them through with what they cannot go through themselves, they will not feel that they must have studied themselves; and the paltry oralizing of the teacher will not be listened to, or if heard, will not be understood, or at best, not retained in memory. Pupils may be made to see things for the moment, while no abiding impression will remain on their minds. They will often proceed, in such a manner, through a book, and, with the mistaken idea that they understand its contents, the evil of superficialism may be perpetuated by them, perhaps, as teachers. Pupils will never have a sufficient understanding of a subject till they shall have studied it carefully themselves, and mastered each part by severe personal application.

Recitation by analysis will be found more conducive to thorough scholarship than adherence to any written questions. Let the class, or any member of the class, be able to commence at the beginning and go through with the entire lesson without any suggestion from the teacher, — a thing that is perfectly practicable and easily attainable. Let pupils be called on, at the pleasure of the teacher, in any part of the class, to go on with the recitation, even to proceed with it in the midst of a subject, the topic in no case ever being named by the teacher. They will thereby become accustomed to give their attention to the recitation, and they will be profited from it, besides securing *habits* of attention, which will be of incalculable value.

In fine, let arithmetic be studied properly, and more valuable mental discipline will be acquired from it, than is often attained from the whole course in mathematics usually assigned by college faculties. It is not the extent, but the value of acquisitions in mathematics, which is desirable.

W. B. B.

INDEX.

SIMPLE NUMBERS.

COMMON FRACTIONS.

DECIMAL FRACTIONS AND FEDERAL MONEY.

COMPOUND NUMBERS.

MEASURE OF CAPACITY.

PERCENTAGE.

ARITHMETIC

NOTATION AND NUMERATION.

¶ **1.** A *single* thing, as a dollar, a horse, a man, &c., is called a *unit*, or *one*. One and one more are called *two*, two and one more are called *three*, and so on. Words expressing how many (as one, two, three, &c.) *are called numbers.*

This way of expressing numbers by *words* would be very slow and tedious in doing business. Hence two shorter methods have been devised. Of these, one is called the *Roman** *method, by letters;* thus, I represents one; V, five; X, ten, &c., as shown in the note at the bottom of the page.

The other is called the *Arabic method, by certain characters,* called *figures.* This is that in general use.

---

* In the Roman method, by letters, I represents *one;* V, *five;* X, *ten;* L, *fifty;* C, *one hundred;* D, *five hundred;* and M, *one thousand.*

As often as any letter is repeated, so many times its value is repeated, unless it be a letter representing a *less* number placed before one representing a *greater;* then the less number is taken from the greater; thus, IV represents *four;* IX, *nine,* &c., as will be seen in the following

TABLE.

One	I.	Ninety	LXXXX. or XC
Two	II.	One hundred	C.
Three	III.	Two hundred	CC.
Four	IIII. or IV.	Three hundred	CCC.
Five	V.	Four hundred	CCCC.
Six	VI.	Five hundred	D. or IↃ.*
Seven	VII.	Six hundred	DC.
Eight	VIII.	Seven hundred	DCC.
Nine	VIIII. or IX.	Eight hundred	DCCC.
Ten	X.	Nine hundred	DCCCC.
Twenty	XX.	One thousand	M. or CIↃ.†
Thirty	XXX.	Five thousand	IↃↃ. or $\overline{\text{V}}$.‡
Forty	XXXX. or XL.	Ten thousand	CCIↃↃ. or $\overline{\text{X}}$.
Fifty	L.	Fifty thousand	IↃↃↃ.
Sixty	LX.	Hundred thousand	CCCIↃↃↃ. or $\overline{\text{C}}$
Seventy	LXX.	One million	$\overline{\text{M}}$.
Eighty	LXXX.	Two million	$\overline{\text{MM}}$.

* IↃ is used instead of D to represent five hundred, and for every additional Ↄ annexed at the right hand, the number is increased *ten times.*

† CIↃ is used to represent one thousand, and for every C and Ↄ put at each end, the number is increased *ten times.*

‡ A line over any number increases its value *one thousand times.*

In the Arabic method the first nine numbers have each a separate character to represent it; thus,

¶ 2. A *unit*, or single thing, is represented by this character, ...

	Units.
A *unit*, or single thing, is represented by this character,	1.
Two units, by this character,	2.
Three units, by this character,	3.
Four units, by this character,	4.
Five units, by this character,	5.
Six units, by this character,	6.
Seven units, by this character,	7.
Eight units, by this character,	8.
Nine units, by this character,	9.

Note 1. These nine characters are called *significant figures*, because they each represent some number. Sometimes, also, they are called *digits*.

Note 2. The value of these figures, as here shown, is called their *simple value*. It is their value always when single.

Nine is the largest number which can be expressed by a single figure. There is another character, 0; it is called a *cipher*, *naught*, or *nothing*, because it denotes the *absence* of a thing. Still it is of frequent use in expressing numbers.

By these ten characters, variously combined, *any* number may be expressed.

The unit 1 is but a *single* one, and in this sense it is called a unit of the *first* order. All numbers expressed by one figure are units of the first order.

¶ 3. Ten has no appropriate character to represent it, but it is considered as forming a unit of a *second* or higher order, consisting of *tens*. It is represented by the same unit figure 1 as is a *single* thing, but it is written at the left hand of a cipher; thus, 10, ten. The 0 fills the *first* place, at the *right* hand, which is the place of *units*, and the 1 the *second* place from the right hand, which is the place of *tens*. Being put in a new place, it has a new value, which is ten times its *simple* value, and this is what is called a *local* value.

Questions. — **¶ 1.** What is a single thing called? What is a number? Give some examples. How many ways of expressing numbers shorter than writing them out in words? What are they called? Which is the method in general use? In the Arabic method, how many numbers have each a separate character?

¶ 2. How is one represented? Make the characters to nine. What are these nine characters called? Why? What is the simple value of figures? What is the largest number which can be represented by a single figure? What other character is frequently used? Why is it called naught? How many are the Arabic characters? What are numbers expressing single things called?

There may be one, two, or more tens, just as there are one, two, or more units, or single things; it takes ten cents to make one ten-cent piece; just so it takes ten single things to make one ten. All figures in the *second* place express units of the 2d order, that is, units of *tens.*

	Tens. Units.	
One ten is . . .	10	ten.
Two tens are . .	20	twenty
Three tens " .	30	thirty
Four tens " . .	40	forty.
Five tens " . .	50	fifty.
Six tens " . .	60	sixty.
Seven tens " . .	70	seventy.
Eight tens " . .	80	eighty.
Nine tens " . .	90	ninety.
One ten, one unit,	11	eleven.
One ten, two units,	12	twelve.

Note. Twenty, thirty, &c., are contractions for two tens, three tens, &c.

One ten and one unit, 11, are called eleven; one ten and two units, 12, twelve, &c. In this way the units of the 1st order are united with the *tens*, that is, with the units of the 2d order, to form the numbers from 10 to 20, from 20 to 30, to 40, and so on to 99, which is the *largest* number that can be represented by two figures.

The weeks in a year are 5 tens and 2 units, (5 of the second order and 2 of the first order now described,) and are expressed thus, 52, (fifty-two.) In the same manner express on your slate, or on the blackboard, the two orders united, so as to form all the numbers from 10 to 99.

¶ 4. Ten tens are called one hundred, which forms a unit of a still higher, or 3d order, and is expressed by writing two ciphers at the right hand of the unit 1, . . . thus,

Hund. Tens. Units.	
100	one hundred.
200	two hundred.
300	three hundred,
	&c.

Note. When there are no units or tens, we write ciphers in their places, which denote the absence of a thing, (¶ 2.)

Questions.—¶ **3.** How is ten represented? What is it considered as forming? Consisting of what? What place does the cipher fill? The one? Where is unit's place, and where ten's place, counting from the right? How much is the value of a figure increased by being removed from a lower to a higher place? In which place does it retain its simple value? In ten's place, what is its value called? What is 1 ten and 1 unit called? 1 ten and 2 units? How are the numbers from 10 to 99 expressed? Of what is the number forty made up? *Ans.* 4 tens and no units. Sixty? What do you unite, to form the number twenty-three? thirty-seven? seventy-five? &c. Of what are twenty, thirty, &c., contractions? What is the *largest*, and what the *least*, number you can express by one figure? by *two* figures?

Three hundred sixty-five, the days in a year, are expressed thus, 365; 3 being in the place of hundreds, 6 in the place of tens, and 5 in the place of units.

After the same manner, the pupil may be required to unite the *three* orders, and express any number from 99 to 999.

¶ 5. We have seen that figures have *two* values, viz., *simple* and *local.*

The *simple* value of a figure is its value when *standing alone;* thus, the simple value of 7 is seven.

The *local* value of a figure is its value *according to its distance from the place of units;* thus, the *local* value of 7, in the number 75, is 7 tens, or seventy, while its *simple* value is seven; in the number 756, its local value is seven hundred.

Note. From the fact that 10 is 1 more than 9, it follows, as may be found by trial, that the local value of every figure at the left of units, except 9, exceeds a certain number of nines by the simple value of the figure. Take the number 623: 2 (tens) is 2 more than 2 nines, and 6, (hundreds,) 6 more than a certain number of nines. On this principle is founded a method of proof in the subsequent rules, by casting out the nines.

¶ 6. Ten hundred make one thousand, which is called a unit of the next higher, or 4th order, consisting of *thousands,* and is expressed by writing *three* ciphers at the right hand of the unit 1, giving it a new local value; thus, 1000, one thousand.

To thousands succeed tens and hundreds of thousands, forming units of the 5th and 6th orders.

Questions.—**¶ 4.** What are 10 tens called? What do they form? How many places are required to express hundreds? How much does 1 cipher, placed at the right hand of 1, increase it? 2 ciphers? How do you express two hundred? &c. What are 4 hundreds, 9 tens, and 5 units called? How is one hundred ninety-three expressed? What place does the 3 occupy? the 9? the 1? How do you express the absence of an order? How is the number of days in a year expressed?

¶ 5. How many values have figures? What are they? What is the simple value? local value? What is the value of 5 in 59? Is it its simple, or a local value? Is the value of 8, in 874, simple or local? of the 7? of the 4?

¶ 6. How do you express one thousand? seven thousand? A thousand is a unit of what order? How many thousands are 30 hundreds? What after thousands, and of what order? The 6th order is what? In writing nine hundred and two thousand and nine, where do you place ciphers? Why?

¶ 7. In this table of the six orders now described, you see the unit 1 moving from right to left, and at each removal forming the unit of a higher order. There are other orders yet undescribed, to form which the unit moves onward still towards the left, its value being increased *ten times* by each removal.

Note 1. The *Ordinal* numbers, 1st, 2d, 3d, &c., may be called *indices* of their respective orders.

Note 2. *Various Readings.* In the number 546873, the left hand figure 5 expresses 5 units of the 6th order, or it may be rendered in the next lower order with the 4, and together they may be read 54 units of the 5th order, (ten thousands,) and connecting with the 6, they may be read, 546 units of the 4th order, or 546000. Hence, *units of any higher order may be rendered in units of any lower order.*

TABLE.

6th order. Hundreds of Thousands.	5th order. Tens of Thousands.	4th order. Thousands	3d order. Hundreds.	2d order. Tens.	1st order. Units
					1
				1	0
			1	0	0
		1	0	0	0
	1	0	0	0	0
1	0	0	0	0	0
,	,	,	,	,	,
9	9	9	9	9	9

To hundreds of thousands succeed units, tens, and hundreds of millions.

¶ 8. To millions succeed billions, trillions, quadrillions, quintillions, sextillions, septillions, octillions, nonillions, decillions, undecillions, duodecillions, tredecillions, &c., to each of which, as to units, to thousands, and to millions, are assigned *three* places, viz., *units, tens, hundreds,* as in the following examples:

Questions. — ¶ **7.** How is the unit 1 of the 1st order made a unit of the 2d order? of the 3d order, &c., to the 6th order? What may the ordinal numbers, 1st, 2d, 3d, &c., be called? 7 units of the 6th order are how many units of the 4th order? *The teacher will multiply such questions.* What is the least, and what the largest, number which can be expressed by 2 places? 3 places? &c. What after hundreds of thousands? Of what order will millions be? tens of millions? hundreds of millions?

¶ **8.** What after millions? How many places are allotted to billions? to trillions? &c. Give the names of the orders after trillions In reading large numbers, what is frequently done? Why? The 1st period at the right is the period of what? the 2d? the 3d? the 4th &c

2

	of Septillions	of Sextillions.	of Quintillions.	of Quadrillions.	of Trillions.	of Billions.	of Millions.	of Thousands.	of Units.
	Units	Hundreds Tens Units	Hundreds Tens Units	Hundreds Tens Units	Hundreds Tens Units	Hundreds Tens Units	Hundreds Tens Units	Hundreds Tens Units	Hundreds Tens Units
	3	0 8 2	7 1 5	2 0 3	1 7 4	5 9 2	8 3 7	4 6 3	5 1 2
	3,	0 8 2,	7 1 5,	2 0 3,	1 7 4,	5 9 2,	8 3 7,	4 6 3,	5 1 2
	9th Period.	8th Period.	7th Period.	6th Period.	5th Period.	4th Period.	3d Period.	2d Period.	1st Period.

To facilitate the reading of large numbers, we may point them off into periods of *three figures each*, as in the 2d example. The names and the order of the periods being known, this division enables us to read numbers consisting of many figures as easily as we can read those of only three figures. Thus, in looking at the above examples, we find the first period at the left hand to contain one figure only, viz., 3. By looking under it, we see that it stands in the 9th period from units, which is the period of septillions; therefore we read it 3 septillions, and so on, 82 sextillions, 715 quintillions, 203 quadrillions, 174 trillions, 592 billions, 837 millions, 463 thousands, 512.

¶ 9. From the foregoing we deduce the following principles:—

Numbers increase from right to left, and decrease from left to right, in a ten-fold ratio; and it is

A FUNDAMENTAL LAW OF THE ARABIC NOTATION; that,

Questions. — ¶ 9. How do numbers increase? how decrease, and in what proportion? To what is 1 ten equal? 1 hundred? 1 thousand? &c. To what are 10 units equal? 10 hundreds? &c. What is a fundamental law of the Arabic notation? What is notation? numeration? How do you write numbers? read numbers? If you were to write a number containing units, tens, hundreds, and millions, but no thousands, how would you express it?

I. Removing any figure one place towards the *left*, increases its value *ten times*, and

II. Removing any figure one place towards the *right*, decreases its value *ten times*.

The expressing of numbers as now shown is called *Notation*. The reading of any number set down in figures is called *Numeration*.

To write numbers.—Begin at the left hand, and write in their respective places the units of each order mentioned in the number. If any of the intermediate orders of units be omitted in the number mentioned, supply their respective places with ciphers.

To read numbers.—Point them off into periods of three figures each, beginning at the right hand; then, beginning at the left hand, read each period separately.

Let the pupil write down and read the following numbers:

Two million, eighty thousand, seven hundred and five.
One hundred million, one hundred thousand and one.
Fifty-two million, sixty thousand, seven hundred and three.
One hundred thirty-two billion, twenty-seven million.
Five trillion, sixty billion, twenty-seven million.
Seven hundred trillion, eighty-six billion, and nine.
Twenty-six thousand, five hundred and fifty men.
Two million, four hundred thousand dollars.
Ninety-four billion, eighty thousand minutes.
Sixty trillion, nine hundred thousand miles.
Eighty-four quintillion, seven quadrillion, one hundred million grains of sand.

¶ 10. Numbers are employed to express quantity.

Quantity is anything which can be measured. Thus, *Time* is quantity, as we can measure a portion of it by days, hours, &c. *Distance* is quantity, as it can be measured by miles, rods, &c.

By the aid of numbers quantities may either be added together, or one quantity may be taken from another.

Arithmetic is the art of making calculations upon quantities by means of numbers.

Questions.—¶ 10. Numbers are employed to express what? What is quantity? By what is a quantity of grain measured? a quantity of cloth? What is arithmetic? What is an abstract number? a denominate number? What is the unit of a number? What is the unit value of 8 bushels? of 16 yards? of 20 pounds of sugar? of 3 quarts of milk? of 9 dozen of buttons? of 18 tons of hay? of 16 hogsheads of molasses?

A number applied to no kind of thing, as 5, 10, 18, 36, is called an *abstract* number.

A number applied to some kind of thing, as 7 horses, 25 dollars, 250 men, is called a *denominate* number.

The *unit*, or *unit value* of a number, is one of the kind which the number expresses; thus, the unit of 99 days is 1 day; the unit of 7 dollars is 1 dollar; the unit of 15 acres is 1 acre. In like manner the unit of 9 tens may be said to be 1 ten; the unit of 8 hundred to be 1 hundred; the unit of 6 thousand to be 1 thousand, &c.

ADDITION OF SIMPLE NUMBERS.

¶ 11. 1. James had 5 peaches, his mother gave him 3 more; how many had he then? *Ans.* 8.

Why? *Ans.* Because 5 and 3 are 8.

2. Henry, in one week, got 17 merit marks for perfect lessons, and 6 for good behavior; how many merit marks did he get? *Ans.* ——. Why?

3. Peter bought a wagon for 36 cents, and sold it so as to gain 9 cents; how many cents did he get for it?

4. Frank gave 15 walnuts to one boy, 8 to another, and had 7 left; how many walnuts had he at first?

5. A man bought a chaise for 54 dollars; he expended 8 dollars in repairs, and then sold it so as to gain 5 dollars; how many dollars did he get for the chaise?

The putting together of two or more numbers, (as in the foregoing examples,) so as to make one *whole number*, is called *Addition*, and the whole number is called the *Sum*, or *Amount*.

6. One man owes me 5 dollars, another owes me 6 dollars, another 8 dollars, another 14 dollars, and another 3 dollars what is the sum due to me?

7 What is the amount of 4, 3, 7, 2, 8, and 9 dollars?

8. In a certain school, 9 study grammar, 15 study arithmetic, 20 attend to writing, and 12 study geography; what is the whole number of scholars?

Questions.—¶ **11.** What is addition? What is the answer, or number sought, called? What is the sign of addition? What does it show? How is it sometimes read? Whence the word *plus*, and what is its signification? What is the sign of equality, and what does it show?

SIGNS.—A cross, +, one line horizontal and the other perpendicular, is the sign of *Addition*. It shows that numbers with this sign between them are to be added together; thus, $4+7+14+16$ denote that 4, 7, 14, and 16 are to be added together. It is sometimes read *plus*, which is a Latin word signifying *more*.

Two parallel, horizontal lines, =, are the sign of *Equality*. It signifies that the number *before* it is equal to the number *after* it; thus, $5+3=8$ is read 5 and 3 are 8; or, 5 plus 3 are equal to 8.

In this manner let the pupil be instructed to commit the following

ADDITION TABLE.

2 + 0 = 2	3 + 0 = 3	4 + 0 = 4	5 + 0 = 5
2 + 1 = 3	3 + 1 = 4	4 + 1 = 5	5 + 1 = 6
2 + 2 = 4	3 + 2 = 5	4 + 2 = 6	5 + 2 = 7
2 + 3 = 5	3 + 3 = 6	4 + 3 = 7	5 + 3 = 8
2 + 4 = 6	3 + 4 = 7	4 + 4 = 8	5 + 4 = 9
2 + 5 = 7	3 + 5 = 8	4 + 5 = 9	5 + 5 = 10
2 + 6 = 8	3 + 6 = 9	4 + 6 = 10	5 + 6 = 11
2 + 7 = 9	3 + 7 = 10	4 + 7 = 11	5 + 7 = 12
2 + 8 = 10	3 + 8 = 11	4 + 8 = 12	5 + 8 = 13
2 + 9 = 11	3 + 9 = 12	4 + 9 = 13	5 + 9 = 14
6 + 0 = 6	7 + 0 = 7	8 + 0 = 8	9 + 0 = 9
6 + 1 = 7	7 + 1 = 8	8 + 1 = 9	9 + 1 = 10
6 + 2 = 8	7 + 2 = 9	8 + 2 = 10	9 + 2 = 11
6 + 3 = 9	7 + 3 = 10	8 + 3 = 11	9 + 3 = 12
6 + 4 = 10	7 + 4 = 11	8 + 4 = 12	9 + 4 = 13
6 + 5 = 11	7 + 5 = 12	8 + 5 = 13	9 + 5 = 14
6 + 6 = 12	7 + 6 = 13	8 + 6 = 14	9 + 6 = 15
6 + 7 = 13	7 + 7 = 14	8 + 7 = 15	9 + 7 = 16
6 + 8 = 14	7 + 8 = 15	8 + 8 = 16	9 + 8 = 17
6 + 9 = 15	7 + 9 = 16	8 + 9 = 17	9 + 9 = 18

$5+9=$ how many?
$8+7=$ how many?
$4+3+2=$ how many?
$6+4+5=$ how many?
$2+0+4+6=$ how many?
$7+1+0+8=$ how many?
$3+0+9+5=$ how many?

$9 + 2 + 6 + 4 + 5 =$ how many?
$1 + 3 + 5 + 7 + 8 =$ how many?
$1 + 2 + 3 + 4 + 5 + 6 =$ how many?
$8 + 9 + 0 + 2 + 4 + 5 =$ how many?
$6 + 2 + 5 + 0 + 8 + 3 =$ how many?

¶ 12. When the numbers to be added are *small*, the addition is readily performed in the *mind*, and this is called *mental arithmetic;* but it will frequently be more convenient and even necessary, when the numbers are *large*, to write them down before adding them, and this is called *written arithmetic.*

1. Harry had 43 cents, his father gave him 25 cents more: how many cents had he then?

SOLUTION.—One of these numbers contains 4 tens and 3 units. The other number contains 2 tens and 5 units. To unite these two numbers together into one, write them down one under the other, placing the *units* of one number directly under *units* of the other, and the *tens* of one number directly under *tens* of the other, and draw a line underneath.

```
43 cents.
25 cents.
—
```

Beginning at the column of units, we add each column separately; thus, 5 units and 3 units are 8 units, which we set down in units' place.

```
43 cents.
25 cents.
—
 8
```

We then proceed to the column of tens, and say, 2 tens and 4 tens are 6 tens, or 60, which we set down directly under the column in tens' place, and the work is done.

```
     43 cents.
     25 cents.
     —
Ans. 68 cents.
```

It now appears that Harry's whole number of cents is 6 tens and 8 units, or 68 cents; that is, $43 + 25 = 68$.

Units are written under units, tens under tens, &c.; because none but figures of the same unit value can be added to each other; for 5 units and 3 tens will make neither 8 tens nor 8 units, just as 5 cows and 3 sheep will make neither 8 cows nor 8 sheep.

Questions.—¶ 12. What distinction do you make between *mental* and *written* arithmetic? How do you write numbers for addition? Where do you begin to add? and where do you set the amount? How do you proceed? Why do you write units under units, tens under tens, &c.?

2. A farmer bought a chaise for 210 dollars, a horse for 70 dollars, and a saddle for 9 dollars; what was the whole amount?

Write the numbers as before directed, with units under units, tens under tens, &c.

OPERATION.

Chaise, 210 *dollars*.
Horse, 70 *dollars*.
Saddle, 9 *dollars*.

Answer, 289 *dollars*.

Add as before. The units will be 9, the tens 8, and the hundreds 2; that is, 210 + 70 + 9 = 289.

After the same manner are performed the following examples, *in which the amount of no column exceeds nine*.

3. A man had 15 sheep in one pasture, 20 in another pasture, and 143 in another; how many sheep had he in the three pastures? 15 + 20 + 143 = how many?

4. A man has three farms, one containing 500 acres, another 213 acres, and another 76 acres; how many acres in the three farms? 500 + 213 + 76 = how many?

5. Bought a farm for 2316 dollars, and afterwards sold it so as to gain 550 dollars; what did I sell the farm for? 2316 + 550 = how many?

6. A chair-maker sold, in one week, 30 Windsor chairs, 36 cottage, 102 fancy, and 21 Grecian chairs; how many chairs did he sell? 30 + 36 + 102 + 21 = how many?

7. A farmer, after selling 500 bushels of wheat to a commission merchant, 320 to a miller, and sowing 117 bushels, found he had 62 bushels left; how many bushels had he at first? 500 + 320 + 117 + 62 = how many?

8. A dairyman carried to market at one time 231 pounds of butter, at another time 124, at another 302, at another 20, and at another 12; how many pounds did he carry in all?
Ans. 689 pounds.

9. A box contains 115 arithmetics, 240 grammars, 311 geographies, 200 reading books, and 133 spelling books; how many books are there in the box? *Ans*. 999.

¶ 13. Hitherto the amount of any one column, when added up, has not exceeded 9, and consequently has been expressed by a *single* figure. But it will frequently happen that the amount of a single column will *exceed* 9, requiring *two* or *more figures* to express it.

1. There are three bags of money. The first contains 876

dollars, the second 653 dollars, the third 426 dollars; what is the amount contained in all the bags?

OPERATION.

First bag,	876	*dollars.*
Second "	653	"
Third "	426	"
	1955	"

SOLUTION. — Writing the numbers as already described, we add the units, and find them to be 15, equal to 5 units, which we write in units' place, adding the 1 ten with the tens; which being added together are 15 tens, equal to 5 tens, to be written in tens' place, and 1 hundred, to be added to the hundreds. The hundreds being added are 19, equal to 9 hundreds, to be written in hundreds' place, and 1 thousand, to be written in thousands' place.

Ans. 1955 dollars.

PROOF.—We may reverse the order, and, beginning at the top, add the figures downwards. If the two results are alike, the work may be supposed to be right, for it is not likely that the same mistake will be made twice, when the figures are added in a different order.

NOTE. — *Proof by the excess of nines.* If the work be right, there will be just as many of any small number, as 9, with the same remainder, in the amount, as in the several numbers taken together. Hence,

OPERATION.

876	3
653	5
426	3
1955	2

In the upper number, 8 (hundreds) is 8 more than a certain number of nines, (¶ 5) 7 (tens) is 7 more. Adding the 8 and 7, and the 6 units together, the sum is 21 = 2 nines and 3 remainder, which we set down at the right hand, as the excess of nines in this number. In the same manner, 5 is found to be the excess of nines in the second number, and 3 in the third number. These several excesses being added together, make 1 nine and an excess of 2, which is the same as the excess of nines in the general amount, found in the same manner. This method will detect every mistake, except it be 9, or an exact number of nines.

To find what will be the excess after casting the nines out of any number, begin at the left hand, *and add together the figures which express the number;* thus, to cast the nines out of 892, we say 8 (passing over 9) + 2 (dropping 9 from the sum) = 1.

From the examples and illustrations now given, we derive the following

RULE.

I. Write the numbers to be added, one under another, plac-

Questions. — ¶ **13.** If the amount of the column does not exceed 9, what do you do? What when it exceeds 9? How do you add each column? What do you do with the amount of the left column? For what number do you carry? If the amount of a column be 36, what would you set down, and how many would you carry? On what principle do you do this? How is addition proved? Why? Repeat the rule for addition.

ing units under units, tens under tens, &c., and draw a line underneath.

II. Begin at the unit column, and add together all the figures contained in it; if the amount does not exceed 9, write it under the column; but if it exceed 9, write the units in units' place, and carry the tens to the column of tens.

III. Add each succeeding column in the same manner, and set down the whole amount of the last column.

EXAMPLES FOR PRACTICE.

2.	3.
2 8 6 3 7 0 5 4 2 1 0 6 1	4 3 6 7 5 8 3 0 2 1 4 6 3
3 1 0 7 4 2 9 3 1 5 6 3 8	1 7 5 2 3 4 9 7 1 3 6 2 0
6 2 5 3 0 3 4 7 9 2	6 0 8 1 2 7 5 3 0 6 2 1 7
2 4 7 1 3 5	5 6 5 2 1 7 4 6 3 0 1 2 8
8 6 7 3	8 7 0 3 2 6 3 4 7 2 0 1 3

4. Add together 587, 9658, 67, 431, 28670, 85, 100000, 6300, and 1. *Amount*, 145799.

5. What is the amount of 8635, 7, 2194, 16, 7421. 93, 5063, 135, 2196, 89, and 1225? *Ans.* 27074.

6. A man being asked his age, answered that he left England when he was 12 years old, and that he had afterwards spent 5 years in Holland, 17 years in Germany, 9 years in France, whence he sailed for the United States in the year 1827, where he had lived 22 years; what was his age?

Ans. 65 years.

7. A company contract to build six warehouses; for the first they receive 36850 dolls.; for the second, 43476 dolls.; for the third, 18964 dolls.; for the fourth, 62840 dolls.; for the fifth, 71500 dolls.; for the sixth, as much as for the first three; to what do these contracts amount? *Ans.* 332920 dollars.

8. James had 7 marbles, Peter had 4 marbles more than James, and John had 5 more than Peter; how many marbles in all? *Ans.* 34.

9. There are seven men; the first man is worth 67850 dollars; the second man is worth 2500 dolls. more than the first man; the third, 3168 dolls. more than the second; the fourth, 16973 dolls. more than the third; the fifth, 40600 dolls. more than the fourth; the sixth, 19888 dolls. more than the fifth; and the seventh, 49676 dolls. more than the sixth; how many dollars are they all worth? *Ans.* 784934 dollars.

10. What is the interval in years between a transaction

that happened 275 years ago, and one that will happen 125 years hence? *Ans.* 400 years.

11. What is the amount of 46723, 6742, and 986 dollars?

12. A man has three orchards; in the first there are 140 trees that bear apples, and 64 trees that bear cherries; in the second, 234 trees bear apples, and 73 bear cherries; in the third, 47 trees bear plums, 36 bear pears, and 25 bear cherries: how many trees in all the orchards, and how many of each kind?

Ans. 619 trees; 374 bear apples; 162 cherries; 47 plums; and 36 pears.

13. A gentleman purchased a farm for 7854 dollars; he paid 194 dollars for having it drained and fenced, and 300 dollars for having a barn built upon it; how much did it cost him, and for how much must he sell it, to gain 273 dollars?

Ans. { It cost him 8348 dollars.
He must sell it for 8621 dollars.

¶ 14. Review of Numeration and Addition

Questions. — What are numbers? What are the methods of expressing numbers? What is numeration? notation? fundamental law in the Arabic notation? How does the Arabic differ from the Roman method? What is understood by units of different orders? What is quantity? Arithmetic? What is understood by the simple value of figures? the local value? the unit value of a number? Explain the difference between an abstract and a denominate number. What is addition? the rule? proof? For what number do you carry, and why?

EXERCISES.

1. Washington was born in the year of our Lord 1732; he was 67 years old when he died; in what year did he die? *Ans.* 1799.

2. The invasion of Greece by Xerxes took place 481 years before Christ; how long ago is that this current year?

3. There are two numbers; the less is 8671, the difference between the numbers is 597; what is the greater number? *Ans.* 9268.

4. A man borrowed a sum of money, and paid in part 684 dollars; the sum left unpaid was 876 dollars; what was the sum borrowed?

5. There are four numbers; the first 317, the second 812 the third 1350, and the fourth as much as the other three; what is the sum of them all? *Ans.* 4958.

6. A gentleman left his daughter 16 thousand 16 hundred

and 16 dollars; he left his son 1800 more than his daughter what was his son's portion, and what was the amount of the whole estate?

Ans. { Son's portion, 19416. Whole estate, 37032.

7. A man, at his death, left his estate to his four children who, after paying debts to the amount of 1476 dollars, received 4768 dollars each; how much was the whole estate? *Ans.* 20548.

8. A man bought four hogs, each weighing 375 pounds; how much did they all weigh? *Ans.* 1500.

9. The fore quarters of an ox weigh one hundred and eight pounds each, the hind quarters weigh one hundred and twenty-four pounds each, the hide seventy-six pounds, and the tallow sixty pounds; what is the whole weight of the ox? *Ans.* 600.

10. The imports into the several States in 1842 were as follows: Me. 606864 dollars, N. H. 60481, Vt. 209868, Mass. 17986433, R. I. 323692, Ct. 335707, N. Y. 57875604, N. J. 145, Pa. 7385858, Del. 3557, Md. 4417078, D. C. 29056, Va. 316705, N. C. 187404, S. C. 1359465, Ga. 341764, Al. 363871, La. 8033590, O. 13051, Ky. 17306, Tenn. 5687, Mich. 80784, Mo. 31137, Fa. 176980 dollars; what was the entire amount? *Ans.* 100162087.

SUBTRACTION OF SIMPLE NUMBERS.

¶ 15. 1. Charles, having 18 cents, bought a book, for which he gave 6 cents; how many cents had he left?

2. John had 12 apples; he gave 5 of them to his brother; how many had he left?

3. Peter played at marbles; he had 23 when he began, but when he had done he had only 12; how many did he lose?

4. A man bought a cow for 17 dollars, and sold her again for 22 dollars; how many dollars did he gain?

5. Charles is 9 years old, and Andrew is 13; what is the difference in their ages?

6. A man borrowed 50 dollars, and paid all but 18; how many dollars did he pay? that is, take 18 from 50, and how many would there be left?

The taking of a less number from a greater (as in the foregoing examples) is called *Subtraction*. The greater number is called the *Minuend*, the less number the *Subtrahend*, and what is left after subtraction is called the *Difference*, or *Remainder*.

7. If the minuend be 8, and the subtrahend 3, what is the difference or remainder? *Ans.* 5.

8. If the subtrahend be 4, and the minuend 16, what is the remainder?

SIGN. — A short horizontal line, —, is the sign of subtraction. It is usually read *minus*, which is a Latin word signifying *less*. It shows that the number *after* it is to be taken from the number *before* it. Thus, 8 — 3 = 5, is read 8 minus or less 3 is equal to 5; or, 3 from 8 leaves 5. The latter expression is to be used by the pupil in committing the following

SUBTRACTION TABLE.

2 — 2 = 0	6 — 3 = 3	5 — 5 = 0	7 — 7 = 0
3 — 2 = 1	7 — 3 = 4	6 — 5 = 1	8 — 7 = 1
4 — 2 = 2	8 — 3 = 5	7 — 5 = 2	9 — 7 = 2
5 — 2 = 3	9 — 3 = 6	8 — 5 = 3	10 — 7 = 3
6 — 2 = 4	10 — 3 = 7	9 — 5 = 4	8 — 8 = 0
7 — 2 = 5	4 — 4 = 0	10 — 5 = 5	9 — 8 = 1
8 — 2 = 6	5 — 4 = 1	6 — 6 = 0	10 — 8 = 2
9 — 2 = 7	6 — 4 = 2	7 — 6 = 1	9 — 9 = 0
10 — 2 = 8	7 — 4 = 3	8 — 6 = 2	10 — 9 = 1
3 — 3 = 0	8 — 4 = 4	9 — 6 = 3	
4 — 3 = 1	9 — 4 = 5	10 — 6 = 4	
5 — 3 = 2	10 — 4 = 6		

7 — 3 = how many?	18 — 7 = how many?
8 — 5 = how many?	28 — 7 = how many?
9 — 4 = how many?	22 — 13 = how many?
12 — 3 = how many?	33 — 5 = how many?
13 — 4 = how many?	41 — 15 = how many?

¶ 16. When the numbers are *small*, as in the foregoing examples, the taking of a less number from a greater is readily done in the *mind;* but when the numbers are *large*,

Questions. — ¶ **15.** What is subtraction? What is the greater number called? the less number? that which is left after subtraction? What is the sign of subtraction? How is it usually read? What does minus mean? What does the sign of subtraction show?

the operation is most easily performed part at a time, and therefore it is necessary to *write* the numbers down before performing the operation

1. A farmer, having a flock of 237 sheep, lost 114 of them by disease; how many had he left?

Here we have 4 units to be taken from 7 units, 1 ten to be taken from 3 tens, and 1 hundred to be taken from 2 hundreds. It will therefore be most convenient to write the less number under the greater, observing, as in addition, to place units under units, tens under tens, &c., thus:

OPERATION.

From 237 *the minuend,*
Take 114 *the subtrahend.*
———
123 *the remainder.*

SOLUTION. — We begin with the units, saying, 4 (units) from 7, (units,) and there remain 3, (units,) which we set down directly under the column in units' place. Then proceeding to the next column, we say, 1 (ten) from 3, (tens,) and there remain 2, (tens,) which we set down in *tens'* place. Proceeding to the next column we say, 1 (hundred) from 2, (hundreds,) and there remains 1, (hundred,) which we set down in *hundreds'* place, and the work is done. It now appears that the number of sheep left was 123; that is, 237 — 114 = 123, *Ans.*

NOTE. — We write units under units, tens under tens, &c., that those of the same unit value may be subtracted from each other; for we can no more take 3 tens from 7 units than we can take 3 cows from 7 sheep.

Examples in which each figure in the subtrahend is less than the figure above it.

2. There are two farms; one is valued at 3750, and the other at 1500 dollars; what is the difference in the value of the two farms? *Ans.* 2250.

3. A man's property is worth 8560 dollars, but he has debts to the amount of 3500 dollars; what will remain after paying his debts? *Ans.* 5060.

4. From 746 subtract 435. *Rem.* 311.

5. From 4983 subtract 2351. *Rem.* 2632.

6. From 658495 subtract 336244. *Rem.* 322251.

7. From 8764292 subtract 7653181. *Rem.* 1111111.

Questions. — ¶ **16.** When the numbers are small, how may the operation be performed? When they are large, what is more convenient? How are the two numbers to be written? Where do you begin the subtraction?

¶ 17. 1. James, having 15 cents, bought a pen-knife for which he gave 7 cents, how many cents had he left?

OPERATION.

15 *cents.*	A difficulty presents itself here; for we cannot take 7 from 5; but we can take 7 from 15, and there will remain 8.
7 *cents.*	
—	
8 *cents left.*	

2. A man bought a horse for 85 dollars, and a cow for 27 dollars; what did the horse cost him more than the cow?

OPERATION.
85
27
—

SOLUTION. — The same difficulty presents itself here as in the last example, that is, the unit figure in the subtrahend is greater than the unit figure in the minuend. To obviate this difficulty, we may take 1 (ten) from the 8 (tens) in the minuend, which will leave 7 (tens,) and add it to the 5 units, making 15 units, (7 tens + 15 units = 85,) thus,

TENS.	UNITS.
7	15
2	7
5	8

We now take 7 units from 15 units, and 2 tens from 7 tens, and have 5 tens and 8 units, or 58 remainder; that is 85 — 27 = 58 dollars more for the horse than for the cow.

The operation may be shortened as follows.

OPERATION.

Horse,	85	*dollars.*
Cow,	27	"
	—	
Diff.	58	"

We have 8 tens and 5 units in the minuend, and 2 tens and 7 units in the subtrahend. We can now, *in the mind,* suppose 1 ten taken from the 8 tens, which would leave 7 tens, and this 1 ten we can suppose joined to the 5 units, making 15. We can now take 7 from 15, as before, and there will remain 8, which we set down. The taking of 1 ten out of 8 tens, and joining it with the 5 units, is called *borrowing ten.* Proceeding to the next higher order, or tens, we must consider the upper figure, 8, from which we borrowed, 1 less, calling it 7; then, taking 2 (tens) from 7, (tens,) there will remain 5, (tens,) which we set down, making the difference 58 dollars, *Ans.*

Questions. — ¶ 17. In subtracting 7 from 15, what difficulty presents itself? How do you obviate it? In taking 27 from 85, instead of taking 7 from 5 what do you take it from? Whence the 15? From what do you subtract the 2 tens? Why not from 8 tens instead of 7 tens? What is this operation called? Explain how the operation is performed in example 3. There is another method, often practised, erroneously called *borrowing ten,* — explain the principle on which it is done. When we subtract units from units, of what name will the remainder be? tens from tens, what? hundreds from hundreds, what?

NOTE. — It has been usual to perform subtraction, where the figure in the subtrahend is larger than the figure above it, on another principle. If to two unequal numbers the same number be added, the difference between them will remain the same. Thus, the difference between 17 and 8 is 9, and the difference between 27 and 18, each being increased by 10, is also 9. Take the last example.

TENS.	UNITS.
8	15
3	7
5	8

Adding 10 units to 5 units in the minuend, and 1 ten to 2 tens in the subtrahend, we have increased both by the same number, and the remainder is not altered, being 58.

This method, which has been erroneously called *borrowing ten*, may be practised by those who prefer, though the former is more simple and equally convenient.

3. From 10000 subtract 9.

OPERATION.
10000
9
―――
9991

SOLUTION. — In this example we have 0 units from which to subtract 9 units, and going to tens of the minuend, we have 0 tens, nor hundreds, nor thousands, but we have 1 ten thousand from which, borrowing 10 units, we have 9990, that is, 9 thousands, 9 hundreds and 9 tens left. Taking 9 units from 10 units, we have 1 unit, then no tens in the subtrahend from 9 tens in the minuend leave 9 tens, no hundreds from 9 hundreds leave 9 hundreds, no thousands from 9 thousands leave 9 thousands.

4. A man borrowed 713 dollars and paid 475 dollars; how much did he then owe? *Ans.* 238 dollars.

5. From 1402003 take 681404. *Rem.* 720599.

6. What is the difference between 36070324301 and 28040373315? *Ans.* 8029950986.

7. From 81324036521 take 2546057867. *Rem.* 78777978654.

¶ 18. TO PROVE ADDITION AND SUBTRACTION. — Addition and subtraction are the reverse of each other. Addition is putting together; subtraction is taking asunder.

1. A man bought 40 sheep and sold 18 of them; how many had he left?
40 — 18 = 22 sheep left. *Ans.*

2. A man sold 18 sheep and had 22 left; how many had he at first?
18 + 22 = 40 sheep at first. *Ans.*

Hence, subtraction may be proved by *addition*, and addition by *subtraction*.

To prove subtraction, add the remainder to the *subtrahend*, and, if the work is right, the *amount* will be equal to the *minuend*.

To prove addition, subtract, successively, from the amount the *several numbers* which were added to produce it, and, if the work is right, there will be no *remainder*. Thus $7+8+6=21$; *proof*, $21-6=15$, and $15-8=7$, and $7-7=0$.

NOTE. — *Proof by excess of nines.* We may cast out the nines in the remainder and subtrahend; if the excess equals the excess found by casting out the nines from the minuend, the work is presumed to be right.

From the remarks and illustrations now given, we deduce the following

RULE.

I. Write down the numbers, the less under the greater, placing units under units, tens under tens, &c., and draw a line under them.

II. Beginning with units, take successively each figure in the *lower* number from the figure *over* it, and write the remainder directly below.

III. When a figure of the subtrahend exceeds the figure of the minuend over it, borrow 1 from the next left hand figure of the minuend; and add it to this upper figure as 10, in which case the left hand figure of the minuend must be considered one less.

NOTE. — Or when the lower figure is greater than the one above it we may add 10 to the upper figure, and 1 to the next lower figure

EXAMPLES FOR PRACTICE.

1. If a farm and the buildings on it be valued at 10000, and the buildings alone be valued at 4567 dollars, what is the value of the land? *Ans.* 5433 dollars.

2. The population of New York in 1830 was 1,918,608; in 1840 it was 2,428,921; what was the increase in ten years? *Ans.* 510,313.

3. George Washington was born in the year 1732, and died in the year 1799; to what age did he live? *Ans.* 67 years.

4. The Declaration of Independence was published July 4th, 1776; how many years to July 4th the present year?

Questions. — ¶ **18.** Addition is the reverse of what? Subtraction, of what? How will you show that they are so? How do you prove subtraction? How can you prove addition by subtraction? Repeat the rule for subtraction.

5. The Rocky Mountains, in N. A., are 12,500 feet above the level of the ocean, and the Andes, in S. A., are 21,440 feet; how many feet higher are the Andes than the Rocky Mountains? *Ans.* 8,940 feet.

NOTE. — Let the pupil be required to prove the following examples.

6. What is the difference between 7,648,203 and 928,671? *Ans.* 6,719,532.

7. How much must you add to 358,642 to make 1,487,945? *Ans.* 1,129,303.

8. A man bought an estate for 13,682 dollars, and sold it again for 15,293 dollars; did he gain or lose by it? and how much? *Ans.* 1,611 dollars

9. From 364,710,825,193 take 27,940,386,574.

10. From 831,025,403,270 take 651,308,604,782.

11. From 127,368,047,216,843 take 978,654,827,352.

¶ 19. Review of Subtraction.

Questions. — What is subtraction? What is the rule? What is understood by borrowing ten? Of what is subtraction the reverse? How is subtraction proved? How is addition proved by subtraction?

EXERCISES.

1. How long from the discovery of America by Columbus, in 1492, to this present year?

2. Supposing a man to have been born in the year 1773, how old was he in 1847? *Ans.* 74.

3. Supposing a man to have been 80 years old in the year 1846, in what year was he born? *Ans.* 1766.

4. There are two numbers, whose difference is 8764; the greater number is 15687; I demand the less. *Ans.* 6923.

5. What number is that which, taken from 3794, leaves 865? *Ans.* 2929.

6. What number is that to which if you add 789, it will become 6350? *Ans.* 5561.

7. A man possessing an estate of twelve thousand dollars, gave two thousand five hundred dollars to each of his two daughters, and the remainder to his son; what was his son's share? *Ans.* 7000 dollars.

8. From seventeen million take fifty-six thousand, and what will remain? *Ans.* 16,944,000.

9. What number, together with these three, viz., 1301 2561, and 3120, will make ten thousand? *Ans.* 3018.

10. A man bought a horse for one hundred and fourteen dollars, and a chaise for one hundred and eighty-seven dollars; how much more did he give for the chaise than for the horse?

11. A man borrows 7 ten dollar bills and 3 one dollar bills, and pays at one time 4 ten dollar bills and 5 one dollar bills; how many ten dollar bills and one dollar bills must he afterwards pay to cancel the debt?

Ans. 2 ten doll. bills and 8 one doll.

12. The greater of two numbers is 24, and the less is 16; what is their difference?

13. The greater of two numbers is 24, and their difference 8; what is the less number?

14. The sum of two numbers is 40, the less is 16; what is the greater?

EXERCISES IN ADDITION AND SUBTRACTION.

15. A man carried his produce to market; he sold his pork for 45 dollars, his cheese for 38 dollars, and his butter for 29 dollars; he received, in pay, salt to the value of 17 dollars, 10 dollars' worth of sugar, 5 dollars' worth of molasses, and the rest in money; how much money did he receive?

Ans. 80 dollars.

16. A boy bought a sled for 28 cents, and gave 14 cents to have it repaired; he sold it for 40 cents; did he gain or lose by the bargain? and how much? *Ans.* He lost 2 cents.

17. One man travels 67 miles in a day, another man follows at the rate of 42 miles a day; if they both start from the same place at the same time, how far will they be apart at the close of the first day? —— of the second? —— of the third? —— of the fourth? *Ans.* To the last, 100 miles.

18. One man starts from Boston Monday morning, and travels at the rate of 40 miles a day; another starts from the same place Tuesday morning, and follows on at the rate of 70 miles a day; how far are they apart Tuesday night?

Ans. 10 miles.

19. A man, owing 379 dollars, paid at one time 47 dollars at another time 84 dollars, at another time 23 dollars, and at another time 143 dollars; how much did he then owe?

Ans. 82 dollars.

20. Four men bought a lot of land for 482 dollars; the first man paid 274 dollars, the second man 194 dollars less than

the first, and the third man 20 dollars less than the second, how much did the second, the third, and the fourth man pay?

Ans. { The second paid 80.
The third paid 60.
The fourth paid 68. }

21. Four men bought a horse; the first man paid 21 dollars, the second 18 dollars, the third 13 dollars, and the fourth as much as the other three, wanting 16 dollars; how much did the fourth man pay? and what did the horse cost?

Ans. Fourth man paid — dolls.; horse cost 88 dolls.

22. From 1,000,000 take 1, and what remains? (*See* ¶ 17 *Ex.* 3.)

MULTIPLICATION OF SIMPLE NUMBERS.

¶ 20. 1. If one orange costs 5 cents, how many cents must I give for 2 oranges? —— how many cents for 3 oranges? —— for 4 oranges?

2. One bushel of apples cost 20 cents; how many cents must I give for 2 bushels? —— for 3 bushels?

3. One gallon contains 4 quarts; how many quarts in 2 gallons? —— in 3 gallons? —— in 4 gallons?

4. Three men bought a horse; each man paid 23 dollars for his share; how many dollars did the horse cost them?

5. In one dollar there are one hundred cents; how many cents in 5 dollars?

6. How much will 4 pairs of shoes cost at 2 dollars a pair?

7. How much will two pounds of tea cost at 43 cents a pound?

8. There are 24 hours in one day; how many hours in 2 days? —— in 3 days? —— in 4 days? —— in 7 days?

9. Six boys met a beggar, and gave him 15 cents each; how many cents did the beggar receive?

In this example we have 15 cents (the number which each boy gave the beggar) to be repeated 6 times, (as many times as there were boys.)

When questions occur where the same number is to be repeated several times, the operation may be shortened by a rule called *Multiplication.*

In multiplication the number to be repeated is called the *Multiplicand.*

The number which shows *how many times* the multiplicand is to be repeated, is called the *Multiplier.*

The result, or answer, is called the *Product.*

The multiplicand and multiplier taken together are called *Factors*, or producers, because when multiplied together they produce the product.

10. There is an orchard in which are 5 rows of trees, and 27 trees in each row; how many trees in the orchard?

In the first row, . . . 27 *trees.*
" *second,* . . . . 27 "
" *third,* . . . . . 27 "
" *fourth,* . . . . 27 "
" *fifth,* . . . . . 27 "

In the whole orchard, 135 *trees.*

Solution.—The whole number of trees will be equal to the amount of *five* 27's added together.

In adding, we find that 7 taken five times amounts to 35. We write down the five units, and reserve the three tens; the amount of 2 taken five times is 10, and the 3, which we reserved, makes 13, which, written at the left of units, makes the whole number of trees 135.

If we have learned that 7 taken 5 times amounts to 35, and that 2 taken 5 times amounts to 10, it is plain we need write the number 27 but *once*, and then, setting the multiplier under it, we may say, 5 times 7 are 35, writing down the 5, and reserving the 3 (tens) as in addition. Again, 5 times 2 (tens) are 10, (tens,) and 3, (tens,) which we reserved, make 13, (tens,) as before.

Multiplicand, 27 *trees in each row.*
Multiplier, . 5 *rows.*

Product, . 135 *trees, Ans.*

From the above example, it appears that multiplication is a short way of performing many additions, and it may be defined,—*The method of repeating one of two numbers as many times as there are units in the other.*

Sign.—Two short lines, crossing each other in the form of the letter X, are the sign of multiplication. When placed between numbers it shows that they are to be multiplied together; thus, $3 \times 4 = 12$, signifies that 3 times 4 are equal to 12, or 4 times 3 are equal to 12; and thus, $4 \times 2 \times 7 = 56$, signifies that 4 multiplied by 2, and this product by 7, equals 56.

Questions.—¶ **20.** When questions occur in which the same number is to be repeated several times, how may the operation be shortened? In multiplication, what is the multiplicand? the multiplier? the product? What are factors? Why? What is multiplication? Illustrate by the two methods of performing the 10th example. How do you define multiplication? What is the sign? Repeat the table.

NOTE. — Before any progress can be made in this rule, the following table must be committed perfectly to memory.

MULTIPLICATION TABLE.

2 times 0 are 0
2 × 1 = 2
2 × 2 = 4
2 × 3 = 6
2 × 4 = 8
2 × 5 = 10
2 × 6 = 12
2 × 7 = 14
2 × 8 = 16
2 × 9 = 18
2 × 10 = 20
2 × 11 = 22
2 × 12 = 24

3 × 0 = 0
3 × 1 = 3
3 × 2 = 6
3 × 3 = 9
3 × 4 = 12
3 × 5 = 15
3 × 6 = 18
3 × 7 = 21
3 × 8 = 24
3 × 9 = 27
3 × 10 = 30
3 × 11 = 33
3 × 12 = 36

4 × 0 = 0
4 × 1 = 4
4 × 2 = 8
4 × 3 = 12
4 × 4 = 16
4 × 5 = 20
4 × 6 = 24
4 × 7 = 28
4 × 8 = 32
4 × 9 = 36
4 × 10 = 40
4 × 11 = 44
4 × 12 = 48

5 × 0 = 0
5 × 1 = 5
5 × 2 = 10
5 × 3 = 15
5 × 4 = 20
5 × 5 = 25
5 × 6 = 30
5 × 7 = 35
5 × 8 = 40
5 × 9 = 45
5 × 10 = 50
5 × 11 = 55
5 × 12 = 60

6 × 0 = 0
6 × 1 = 6
6 × 2 = 12
6 × 3 = 18
6 × 4 = 24
6 × 5 = 30
6 × 6 = 36
6 × 7 = 42
6 × 8 = 48
6 × 9 = 54
6 × 10 = 60
6 × 11 = 66
6 × 12 = 72

7 × 0 = 0
7 × 1 = 7
7 × 2 = 14
7 × 3 = 21
7 × 4 = 28
7 × 5 = 35
7 × 6 = 42
7 × 7 = 49
7 × 8 = 56
7 × 9 = 63
7 × 10 = 70
7 × 11 = 77
7 × 12 = 84

8 × 0 = 0
8 × 1 = 8
8 × 2 = 16
8 × 3 = 24
8 × 4 = 32
8 × 5 = 40
8 × 6 = 48
8 × 7 = 56
8 × 8 = 64
8 × 9 = 72
8 × 10 = 80
8 × 11 = 88
8 × 12 = 96

9 × 0 = 0
9 × 1 = 9
9 × 2 = 18
9 × 3 = 27
9 × 4 = 36
9 × 5 = 45
9 × 6 = 54
9 × 7 = 63
9 × 8 = 72
9 × 9 = 81
9 × 10 = 90
9 × 11 = 99
9 × 12 = 108

10 × 0 = 0
10 × 1 = 10
10 × 2 = 20
10 × 3 = 30
10 × 4 = 40
10 × 5 = 50
10 × 6 = 60
10 × 7 = 70
10 × 8 = 80
10 × 9 = 90
10 × 10 = 100
10 × 11 = 110
10 × 12 = 120

11 × 0 = 0
11 × 1 = 11
11 × 2 = 22
11 × 3 = 33
11 × 4 = 44
11 × 5 = 55
11 × 6 = 66
11 × 7 = 77
11 × 8 = 88
11 × 9 = 99
11 × 10 = 110
11 × 11 = [illegible]
11 × 12 = 132

12 × 0 = 0
12 × 1 = 12
12 × 2 = 24
12 × 3 = 36
12 × 4 = 48
12 × 5 = 60
12 × 6 = 72
12 × 7 = 84
12 × 8 = 96
12 × 9 = 108
12 × 10 = 120
12 × 11 = 132
12 × 12 = 144

$9 \times 2 =$ how many?	$4 \times 3 \times 2 = 24.$
$4 \times 6 =$ how many?	$3 \times 2 \times 5 =$ how many?
$8 \times 9 =$ how many?	$7 \times 1 \times 2 =$ how many?
$3 \times 7 =$ how many?	$8 \times 3 \times 2 =$ how many?
$5 \times 5 =$ how many?	$3 \times 2 \times 4 \times 5 =$ how many?

¶ 21. 1. There are on a board, 3 rows of stars, and 4 stars in a row; how many stars on the board?

DIAGRAM OF STARS.

* * * *

* * * *

* * * *

A slight inspection of the diagram will show that the number of stars may be found by considering that there are either 3 rows of 4 stars each, (3 times 4 are 12,) or 4 rows of 3 stars each, (4 times 3 are 12;) therefore, we may use *either* of the given numbers for a multiplier, as best suits our convenience. We generally write the numbers as in subtraction, the larger uppermost, with units under units, tens under tens, &c. Thus,

Multiplicand, 4 *stars.*
Multiplier, 3 *rows.*
—
Product, 12 *stars, Ans.*

NOTE.—4 and 3 are the *factors,* which produce the product 12.

This diagram of stars is commended to the particular attention of the pupil, as it is intended to make use of it hereafter in illustrating operations in multiplication and also in division.

First, you will notice the *terms of the diagram,* and their application.

TERMS OF THE DIAGRAM.

Stars in a row.	Using this term as a representation or symbol of the *multiplicand* and one factor of the product.
Number of rows.	Using this term as a symbol of the *multiplier* and the other factor of the product.
Number of stars	Using this term as a symbol of the *product,* for when the stars in a row are taken as many times as there are *rows of stars,* then the product will be the whole number of stars contained in the diagram.

As the stars in a row symbolize the multiplicand, it follows that the *multiplier* (number of rows) in reality simply expresses the *number of times* the multiplicand (stars in a row) is to be taken. Hence, to multiply by 1, (1 row of stars,) is to take the multiplicand (stars in a row) 1 time; to multiply by 2 (2 rows,) is to take the multiplicand (stars in a row) 2 times; to multiply by 3, (3 rows,) is to take the multiplicand (stars in a row) 3 times, and so on.

Illustration.—What cost 7 yards of cloth at 3 dollars a yard? (7 rows, 3 stars in a row.) The two numbers as given in the question are both *denominate;* but how are they to be considered in the *operation?* The price of 7 yards will evidently be 7 times the price of 1 yard, that is, 7 times 3 *dollars; dollars* (number of stars) is the thing sought by the question; and hence, 3 *dollars* being of the same name as the *thing* or answer sought, is the *true* multiplicand. That number which was yards in putting the question, being taken for the *multiplier*, in this relation is not to be considered yards, but *times* of taking the multiplicand; and hence, in the operation, it must always be considered an *abstract* number For multiplication is a short way of performing many additions, and to talk of adding 3 dollars to itself 7 yards times is nonsense. But we can repeat 3 dollars as many times as 1 yard is repeated to make 7 yards.

There is then a *true* multiplicand and a *true* multiplier. The true multiplicand is that number which is of the same name as the answer sought; the *true* multiplier is that number which indicates the *times* the true multiplicand is to be repeated, or taken; but as it respects the *operation*, it has been shown above that we may use either of the given numbers as the multiplier; that is, the multiplicand and multiplier may change places; still, the *product* will always be of the same name as the true multiplicand.

This application of the terms of the diagram to the terms of the question we shall call *symbolizing* the question.

Questions.—¶ **21.** You have in your book a diagram of stars; what is the first use made of it? What is the difference between 4 times 7, and 7 times 4? Which of the given numbers may be used for the multiplier? What are the terms of the diagram? What do these terms symbolize? 6 times 7 are 42,—which of these numbers is the multiplicand? the multiplier? the product? and what, in the diagram, is a symbol of each? What does the multiplier express? Show by the diagram what it is to multiply by 1, by 2, by 3, &c. What must the multiplier always be considered? What do you understand by the *true* multiplicand? by the true multiplier? What will the product always be? Give an example to show that you understand these principles What do you understand by symbolizing a question?

NOTE. — Let the teacher see to it that these principles are *well* understood by the pupil before he proceeds.

As the pupil advances, the teacher should, from time to time, refer him back to a review of these principles.

¶ 22. 1. What will 84 barrels of flour cost, at 7 dollars a barrel?

SOLUTION. — The price of 84 barrels will evidently be 84 times the price of 1 barrel, 7 stars in one row × by 84, number of rows, 7 dollars is the *true* multiplicand, &c.; but as it will be more convenient, the multiplicand and multiplier may change places, and we may consider it 7 rows of 84 stars in a row, and multiply the number of barrels, 84, by the price of 1 barrel, thus —

OPERATION.

Multiplicand, 84
Multiplier, 7
———
Product, 588 *dolls.*

Writing the larger number uppermost, as in subtraction, (¶ 16,) and the multiplier under units of the multiplicand, we begin at the right hand and say, 7 × 4 (units) = 28 (units) = 2 tens and 8 units; we set down the 8 units in units' place, as in addition, and reserving the 2 tens, we say, 7 × 8 (tens) = 56 (tens,) and 2 (tens) which we reserved, make 58 (tens,) or five hundred and 8 tens, which we set down at the left of the 8 units, and the whole make 588 dollars, the cost of 84 barrels of flour, at 7 dollars a barrel, *Ans.*

2. A merchant bought 273 hats, (stars in a row,) at 8 dollars each, (number of rows;) what did they cost (number of stars)? *Ans.* 2184 dollars.

3. How many inches are there in 253 feet, (stars in a row,) every foot being 12 inches (number of rows)?

OPERATION.

253
12
———
Ans. 3036

SOLUTION. — The product of 12, with each of the significant figures or digits, having been committed to memory from the multiplication table, it is just as easy to multiply by 12 as by a single figure. Thus, 12 times 3 are 36, &c.

4. What will 476 barrels of fish cost, at 11 dollars a barrel? *Ans.* 5236 dollars.

From these examples we deduce the following

Questions. — ¶ 22. How will you explain the first example? When you multiply units by units, what is your product? When tens by units, what? How can you multiply by 12? How do you write down numbers for multiplication? How do you perform multiplication when the multiplier does not exceed 12?

RULE.

I. *To set down numbers for multiplication.* Write down the multiplicand, under which write the multiplier, setting units under units, tens under tens, &c.

II. *To perform multiplication when the multiplier does not exceed* 12. Begin at the right hand, and multiply each figure in the multiplicand by the multiplier, setting down and carrying as in addition.

EXAMPLES FOR PRACTICE.

5. A farmer sold 29 bags of wheat, each bag containing 3 bushels; how many bushels did he sell? 29 × 3 = how many?

6. A farmer, who had two farms, raised 361 bushels of wheat on one, and 5 times as much on the other; how many bushels did he raise on both? *Ans.* 2166 bushels.

7. A miller sold 42 loads of flour, each load containing 9 barrels, at 7 dollars a barrel; how many barrels of flour did he sell, and what did the whole cost?

Ans. He sold —— barrels; cost, 2646 dollars.

¶ 23. 1. A piece of valuable land, containing 33 acres, (number of rows,) was sold for 246 dollars an acre, (stars in a row;) what did the whole cost?

Note 1. — When the multiplier exceeds 12, it is more convenient to multiply by each figure separately: —

FIRST OPERATION.

Multiplicand, 246 *dollars, price of* 1 *acre.*
Multiplier, 33 *number of acres.*

1*st product,* 738 *dollars, price of* 3 *acres.*

Solution. — In this example, the multiplier consists of 3 tens and 3 units. First, multiplying by the 3 units, gives us 738 dollars, the price of 3 acres.

Having found the price of 3 acres, our next step is to get the price of 30 acres.

SECOND OPERATION.

246 *dollars, price of* 1 *acre.*
33 *number of acres.*

738 *dollars, price of* 3 *acres.*
738 (*tens*) *price of* 30 *acres.*

8118 *dollars, price of* 33 *acres.*

To do this, we multiply by the 3 tens, (thirty,) and write the first figure of the product (8) *in tens'* place, that is, *directly under the figure* by which we multiply. For the price of 30 acres being 10 times the price of 3 acres, it will consist of the same figures each being removed 1 place towards the left, by which its value is increased 10 times. Then adding together the price of 3 acres, and the price of 30 acres, we have the price of 33 acres.

NOTE. — The correctness of the above operation results from the fact that when units (1st order) are multiplied by units, (1st order,) the product is units, 1st order. Tens (2d order) × units, (1st order,) the product is units of the 2d order. Hundreds (3d order) × tens, (2d order,) the product is units of the 4th order.

And universally, —

If a figure of any order be multiplied by some figure of another order, the product will be units of that order indicated by *the sum of their indices, minus* 1. Thus, 7 of the 5th order, (70000,) multiplied by 4 of the 3d order, (400,) their indices being $5 + 3 = 8$, and $8 - 1 = 7$, their product will be 28 units of the 7th order, that is, 28 millions.

2. How many yards in 23 pieces of broadcloth, each piece containing 67 yards?

OPERATION.

```
   7 yards in each piece.
  23 number of pieces.
 ---
 201 yards in  3 pieces.
134   "    "  20 pieces.
----
1541  "    "  23 pieces.
```

SOLUTION. — Multiplying 67 yards by 3, we get 201 yards in 3 pieces; and multiplying 67 by 2 tens, we get 134 tens, = 1340 yards in 20 pieces. Add the two products together, and we get 1540 yards (No. of stars) in 23 pieces
Ans. 1540 yards.

Hence, — *To perform multiplication when the multiplier exceeds* 12, —

RULE.

I. Multiply the multiplicand by each figure in the multiplier separately, first by the units, then by the tens, &c., remembering always to place the first figure of each product directly under its multiplier.

II. Having multiplied in this manner by each figure in the multiplier, add these several products together, and their *sum* will be the answer.

PROOF. — Take the multiplicand for the multiplier, and the multiplier for the multiplicand, and if the product be the same as at first, the work may be supposed to be right.

Questions. — **¶ 23.** How do you multiply when the multiplier exceeds 12? Where do you write the first figure of each product? Why? What do you do with the several products? Repeat the rule What is the method of proof? A figure of any one order multiplied by some figure of another order, the product will be what?

EXAMPLES FOR PRACTICE.

3. There are 320 rods in a mile; how many rods are there in 57 miles? $320 \times 57 =$ how many?

4. It is 436 miles from Boston to the city of Washington; how many rods is it?

5. What will 784 chests of tea cost, at 69 dollars a chest? $784 \times 69 =$ how many?

6. If 1851 men receive 758 dollars apiece, how many dollars will they all receive? *Ans.* 1403058 dollars.

NOTE. —*Proof by the excess of nines.* Casting out the nines in the multiplicand, we have an excess of 6, which we write before the sign of multiplication. Also, we find the excess in the multiplier to be 2, which we write after the sign. The product of the nines cast out from each factor will be an exact number of nines, since every nine multiplied by nine produces an exact number of nines. Hence, if there is an excess of nines in the entire product, it must be from an excess in the product of the excesses, 6 and 2, found in the factors. Multiplying 6 by 2, and casting out nine from the product, we write the excess, 3, over the sign; and casting out the nines from the product of the factors, we find the excess will be the same number 3, which we write under the sign, and presume that the work is right.

```
   1851
    758
-------
1403058
-------
 PROOF.
   3
 6 × 2
   3
```

7. There are 24 hours in a day; if a ship sail 7 miles in an hour, how many miles will she sail in 1 day, at that rate? how many miles in 36 days? how many miles in 1 year, or 365 days? *Ans.* 61320 miles in 1 year.

8. A merchant bought 13 pieces of cloth, each piece containing 28 yards, at 6 dollars a yard; how many yards were there, and what was the whole cost?

Ans. to the last, 2184 dollars.

9. Multiply 37864 by 235. *Product,* 8898040.

10. " 29831 " 952. " 28399112.

11. " 93956 " 8704. " 817793024.

12. The factors of a certain number are 25 and 87; what is the number? *Ans.* 2175.

13. A hatter sold 15 cases of hats, each containing 24 hats worth 8 dollars apiece; how many hats did he sell, and to how many dollars did they amount?

Ans. to the last, 2880 dollars.

14. A grazier sold 23 head of cattle every year for 6 years, at an average price of 17 dollars a head; how many head of cattle did he sell, to how much did they amount each year, and to how much did they amount in 6 years?

Ans. to the last, 2346 dollars.

Contractions in Multiplication.

¶ 24. I. *When the multiplier is a composite number.*

Any number which can be produced by multiplying two or more numbers together, is called a *Composite* number, and

The numbers which are multiplied together to produce it are called its *Component* parts, or *Factors;* thus, 15 can be produced by multiplying together 3 and 5, and is, therefore, a composite number, and the numbers 3 and 5 are its component parts.

So, also, 24 is a composite number. Its component parts may be 2 and 12, (2 × 12 = 24,) or 3 and 8, (3 × 8 = 24,) or 4 and 6, (4 × 6 = 24,) or 2, 3, and 4, (2 × 3 × 4 = 24,) or 2, 2, 2, 3, (2 × 2 × 2 × 3 = 24.)

1. What will 18 yards of cloth cost, at 4 dollars a yard? 3 × 6 = 18. It follows, therefore, that 3 and 6 are component parts of 18.

OPERATION.

4 *dollars, cost of* 1 *yard.*
3 *yards.*

12 *dollars, cost of* 3 *yards.*
6 (3 × 6 =) 18 *yards.*

72 *dollars, cost of* 18 *yards.*

SOLUTION.—If 1 yard cost 4 dollars, 3 yards will cost 3 times 4 dollars, = 12 dollars; and, if 3 yards cost 12 dollars, 18 yards (3 × 6 = 18) will cost 6 times as much as 3 yards, that is, 6 times 12 dollars = 72 dollars. Hence,

To perform multiplication when the multiplier exceeds 12, and is a composite number,—

RULE.

I. Separate the multiplier into two or more component parts, or factors.

II. Multiply the multiplicand by one of the component parts, and the product thus obtained by the other, and so on, if the component parts be more than two, till you have multiplied by each one of them. The last product will be the product required.

Questions.—¶ 24. What is a composite number? What are the component parts, or factors? Why is 15 a composite number? How many factors may a composite number have? Is 11 a composite number? Why not? Explain the 1st example. How do you multiply by a composite number? Does it matter by which factor you multiply first? Have you performed the 3 operations, (Ex 2.) and compared their products?

EXAMPLES FOR PRACTICE.

2. What will 136 tons of potash cost, at 96 dollars per ton?

Let the pupil make 3 operations, and multiply, 1st, by 12 and 8; 2dly, by 4, 4, and 6; 3dly, by 96, and compare the operations; he will find the results to be the same in each case. *Ans.* 13056 dollars.

3. Supposing 342 men to be employed in a certain piece of work, for which they are to receive 112 dollars each; how much will they all receive?

$8 \times 7 \times 2 = 112$. *Ans.* 38304 dollars.

4. How many acres of land in 48 farms, each containing 367 acres? *Ans.* 17616 acres.

5. Supposing 168 persons to be employed in a woollen factory, at an average price of 274 dollars each per year; how much will they all receive?

$8 \times 7 \times 3 = 168$. *Ans.* 46,032 dollars.

6. Multiply	853	by 56.	*Product,*	47,768.
7. "	18109	" 35.	"	633,815.
8. "	1947271	" 81.	"	157,728,951.

¶ 25. II. *When the multiplier is* 10, 100, 1000, *&c.*

1. What will 10 acres of land cost, at 25 dollars per acre?

SOLUTION. — The price of 10 acres will be 10 times the price of 1 acre, or 10 times 25 dollars.

OPERATION.

25 *dollars, price of* 1 *acre.*
250 *dollars, price of* 10 *acres.*

Now if we annex a cipher to 25, the 5 units are made 5 tens, and the 2 tens are made 2 hundreds. Each figure, then, is increased ten-fold, and consequently the whole number is multiplied by 10.

It is also evident that if 2 ciphers were annexed to 25, the 5 units would be made 5 hundreds, and 2 tens would be made 2 thousands each figure being increased a hundred fold, or multiplied by 100. If 3 ciphers were annexed, each figure would be multiplied by 1000, &c Hence,

When the multiplier is 10, 100, 1000, or 1, *with any number of ciphers annexed,—*

RULE.

Annex as many ciphers to the multiplicand as there are

Questions. — **¶ 25.** How are the figures of a number affected, by placing one cipher at the right hand? two ciphers? three ciphers? &c. How, then, do you multiply by 1, with any number of ciphers annexed?

ciphers in the multiplier, and the multiplicand, so increased, will be the product required.

EXAMPLES FOR PRACTICE.

2. What will 76 barrels of flour cost, at 10 dollars a barrel? *Ans.* 760 dollars.

3. If 100 men receive 126 dollars each, how many dollars will they all receive? *Ans.* 12600 dollars.

4. What will 1000 pieces of broadcloth cost, estimating each piece at 312 dollars? *Ans.* 312000 dollars.

5. Multiply 5682 by 10000.

6. " 82134 " 100000

¶ 26. III. *When there are ciphers on the right hand of the multiplicand, multiplier, either or both.*

1. What will 40 acres of land cost, at 27 dollars per acre?

OPERATION.

27 *dollars, price of* 1 *acre.*
4
———
108 *dollars, price of* 4 *acres.*
1080 *dollars, price of* 40 *acres.*

SOLUTION. — The price of 40 acres will be 40 times the price of 1 acre. But 40 being a composite number, ($4 \times 10 = 40$,) we multiply by 4, one component part, to get the price of 4 acres, and then to multiply the price of 4 acres by 10, the other component part, we annex a cipher to get the price of 40 acres.

2. What will 200 acres of land cost, at 400 dollars an acre?

FIRST OPERATION.

400 *dollars, price of* 1 *acre.*
200
———
000
000
800
———
80000 *dollars, price of* 200 *acres.*

SOLUTION. — The 200 acres will cost 200 times the price of 1 acre. We see in the operation that the product is 8 with 4 ciphers at the right hand, the same number as in the multiplicand and multiplier counted together. We may then shorten the operation, as follows: —

SECOND OPERATION.

400
200
———
80000

Multiplying the significant figures together, we place their product, 8, under the 2. Then we annex to this product 4 ciphers, the number in both factors. Hence,

To perform multiplication when there are ciphers on the right hand of either or both the factors, —

RULE.

I. Set the significant figures under each other, placing the ciphers at the right hand.

II. Multiply the significant figures together.

III. Annex as many ciphers to the product as there are on the right hand of both the factors.

EXAMPLES FOR PRACTICE.

3. If 1300 men receive 460 dollars apiece, how many dollars will they all receive? *Ans.* 598000 dollars.

4. It takes 200 shingles to lay 1 course on the roof of a barn, and there are 60 courses on each of the two sides; how many shingles will it take to cover the barn? *Ans.* 24000.

5. A certain storehouse contains 30 bins for storing wheat, and each bin will hold 400 bushels; how many bushels of wheat can be stored in it? *Ans.* 12000 bushels.

¶ **27.** IV. *When there are ciphers between the significant figures of the multiplier.*

1. What is the product of 378, multiplied by 204?

FIRST OPERATION.

```
  378
  204
 ----
 1512
 000
756
-----
77112
```

Multiplying by a cipher produces nothing. Therefore, in the multiplication, we may omit the cipher, and multiply by the significant figures only, as in the second operation.

SECOND OPERATION

```
  378
  204
 ----
 1512
756
-----
77112
```

Hence, to perform multiplication,

When there are ciphers between the significant figures of the multiplier, —

RULE.

Omit the ciphers, and multiply by the significant figures only, remembering to place the 1st figure of each product directly under its multiplier.

Questions. — ¶ **26.** How do you set down numbers for multiplication, when there are ciphers on the right hand of the multiplicand, multiplier, either or both? How do you multiply? How many ciphers do you annex to the product? If there were 2 ciphers on the right hand of your multiplicand, and 5 on the right hand of your multiplier, how many would you annex to the product?

¶ **27.** When there are ciphers between the significant figures of the multiplier how do you multiply? Where do you set the 1st figure of the product?

EXAMPLES FOR PRACTICE.

2. Multiply 154326 by 3007. *Product*, 464058282.
3. Multiply 543 by 206. *Product*, 111858.
4. Multiply 1620 by 2103. *Product*, 3406860
5. Multiply 36243 by 32004. *Product*, 1159920972.
6 Multiply 101,010,101 by 1,001,001.
Product, 101,111,212,111,101.

¶ 28. Other Methods of Contraction.

I. *When the multiplier is* 9, 99, *or any number of* 9's,—

Annex as many ciphers to the multiplicand as there are *nines* in the multiplier, and from the number thus produced, subtract the multiplicand; the remainder will be the product. Thus,

Multiply 6547 by 999.

OPERATION.

6547000
6547
6540453 *Ans.*

Let the pupil prove the operation by actual multiplication.

II. *When the multiplier is* 13, 14, 15, 16, 17, 18, *or* 19,—

Multiply 32046375 by 14.

OPERATION.

32046375 × 14
128185500
448649250 *Ans.*

Place the multiplier at the right of the multiplicand, with the sign of multiplication between them; multiply the multiplicand by the unit figure of the multiplier, and set the product one place to the *right* of the multiplicand. This product, *added* to the multiplicand, makes the *true* product.

NOTE.—*If the multiplier be* 101, 102, *&c., to* 109,—

Multiply 72530486 by 103.

OPERATION.

72530486 × 103
217591458
7470640058 *Ans.*

Multiply as above, and set the product *two* places to the right of the multiplicand, and add them together for the true product.

Questions.—¶ **28.** When the multiplier is 9, 99, &c., why does the contraction, as above directed, give the true product? *Ans.* Multiplying by 9 repeats the multiplicand 9 times; annexing a cipher repeats or increases it 10 times, which is 1 time too many: hence the rule, subtract it 1 time, &c. When the multiplier is 13, 14, &c., why? When 101, 102, &c., why? When 21, 31, &c., why?

III. *When the multiplier is* 21, 31, *and so on to* 91,—

Multiply 83107694 by 31.

OPERATION.

83107694 × 31
249323082

2576338514 *Ans.*

Multiply by the *tens' figure* only of the multiplicand, and set the unit figure of the product under the place of the *tens*, and so on; then add them together for the true product.

¶ 29. Review of Multiplication.

Questions. — What is multiplication, and how defined? Explain the use of the diagram of stars, and show its application to Ex. 1, ¶ 22. What must the true multiplier always be? the product? Why can the factors exchange places? How do you multiply by 12, or less? by a number greater than 12? by a composite number? by 1 with ciphers annexed? by any number with ciphers annexed? when there are ciphers between the significant figures? When units of different orders are multiplied together, of what order is the product?

EXERCISES.

1. An army of 10700 men, having plundered a city, took so much money, that, when it was shared among them, each man received 46 dollars; what was the sum of money taken? *Ans.* 492200 dollars.

2. Supposing the number of houses in a certain town to be 145, each house, on an average, containing two families, and each family 6 members, what would be the number of inhabitants in that town? *Ans.* 1740.

3. If 46 men can do a piece of work in 60 days, how many men will it take to do it in one day? *Ans.* 2760.

4. Two men depart from the same place, and travel in opposite directions, one at the rate of 27 miles a day, the other 31 miles a day; how far apart will they be at the end of 6 days? *Ans.* 348 miles.

5. What number is that, the factors of which are 4, 7, 6 and 20? *Ans.* 3360.

6. If 18 men can do a piece of work in 90 days, how long will it take one man to do the same? *Ans.* 1620 days.

7. What sum of money must be divided between 27 men, so that each man may receive 115 dollars?

8. There is a certain number, the factors of which are 89 and 265; what is that number?

9. What is that number, of which 9, 12, and 14 are factors?

10. If a carriage wheel turn round 346 times in running

1 mile, how many times will it turn round in the distance from New York to Philadelphia, it being 95 miles?

Ans. 32870.

11. In one minute are 60 seconds, how many seconds in 4 minutes? —— in 5 minutes? —— in 20 minutes? —— in 40 minutes? *Ans.* to the last, 2400 seconds.

12. In one hour are 60 minutes; how many *seconds* in an hour? —— in two hours? how many seconds from nine o'clock in the morning till noon?

Ans. to the last, 10800 seconds.

13. Multiply 275827 by 19725. *Product*, 5440687575.

14. Two men, A and B, start from the same place at the same time, and travel the same way; A travels 52 miles a day, and B 44 miles a day; how far apart will they be at the end of 10 days? *Ans.* 80 miles.

15. A farmer sold 468 pounds of pork at 6 cents a pound, and 48 pounds of cheese at 7 cents a pound, and received in payment 42 pounds of sugar at 9 cents a pound, 100 pounds of nails at 6 cents a pound, 108 yards of sheeting at 10 cents a yard, and 12 pounds of tea at 95 cents a pound; how many cents did he owe? *Ans.* 54 cents.

16. A boy bought 10 oranges; he kept 7 of them, and sold the others for 5 cents apiece; how many cents did he receive?

Ans. 15 cents.

17. The component parts of a certain number are 4, 5, 7, 6, 9, 8, and 3; what is the number? *Ans.* 181440.

18. In 1 hogshead are 63 gallons; how many gallons in 8 hogsheads? In 1 gallon are 4 quarts; how many quarts in 8 hogsheads? In 1 quart are 2 pints; how many pints in 8 hogsheads? *Ans.* to the last, 4032 pints.

19. The component parts of a multiplier are 5, 3 and 5, and the multiplicand is 118; what is the multiplier? what the product? *Ans.* to the last—the product is 8850.

20. An army consists of 5 divisions, each division of 8 brigades, each brigade of 4 regiments, each regiment of 9 companies, and each company of 77 men, rank and file; the number of officers, &c., to the whole army is 42, the number belonging peculiarly to each division is 19, to each brigade 25, to each regiment 11, and to each company 14; how many men in the army? *Ans* 133937

DIVISION OF SIMPLE NUMBERS.

¶ 30. 1. James has 12 apples in a basket, which he distributes equally among several boys, giving them 4 apples each; how many boys receive them?

Solution. — He can give the apples to as many boys as the times he can take 4 apples out of the basket, which is 3 times. *Ans.* 3 boys.

2. If a man travel 4 miles in an hour, in how many hours will he travel 24 miles?

Solution. — It will take him as many hours as 4 is contained times in 24. *Ans.* 6 hours

3. James divided 28 apples equally among 3 of his companions; how many did he give to each?

Solution. — The 28 apples are to be divided into 3 equal parts, and one part given to each boy, who will thus receive 9 apples. It will require 27 apples to give 3 boys 9 apples each, since $9 \times 3 = 27$. There will be one apple left, which must be cut into 3 equal parts, and 1 part given to each boy.

Note. — If a unit, or whole thing, be divided into 2 equal parts, *one* of those parts is called *one half;* if into 3 equal parts, 1 part is called 1 *third; two parts* are called 2 *thirds*, &c. If divided into 4 *equal parts*, one part is called 1 *fourth*, or *one* quarter; 2 parts are called 2 *fourths*, or 2 *quarters;* 3 parts, 3 fourths, or 3 quarters, &c. If divided into 5 equal parts, the parts are called *fifths*. If into 6 equal parts, the parts are called *sixths*, &c.

4. Seven men bought a barrel of flour, each man paying an equal share; for what part of the barrel did 1 man pay? —— 2 men? —— 3 men? —— 4 men? —— 5 men? —— 6 men?

5. Twelve men built a steamboat, each man doing an equal share of the work; how much of the work did 3 men do? —— 5 men? —— 7 men? —— 9 men? —— 11 men?

6. A boy had two apples, and gave one half an apple to each of his companions; how many were his companions? *Ans.* 4.

7. A boy divided four apples among his companions, by giving them one third of an apple each; among how many did he divide his apples? *Ans.* 12.

Questions. — ¶ 30. What do you understand by 1 half of any thing or number? 1 third? 2 thirds? 1 seventh? 4 sevenths? 6 fifteenths? 8 tenths? 5 twentieths? 9 twelfths? How many halves make a whole one? How many thirds? fourths? sevenths? ninths? twelfths? fifteenths? twentieths? &c. How many thirds make three whole ones?

8. How many quarters in 5 oranges?

Solution.—In 1 orange there are 4 quarters, and in 5 oranges there are $5 \times 4 = 20$ quarters. *Ans.*

9. How many oranges would it take to give 12 boys one quarter of an orange each? *Ans.* 3 or.

10. How much is one half of 12 apples? *Ans.* 6 ap.

11. How much is one third of 12?

12. How much is one fourth of 12? *Ans.* 3.

13. A man had 30 sheep, and sold one fifth of them; how many of them did he sell? *Ans.* 6.

14. A man purchased sheep for 7 dollars apiece, and paid for them all 63 dollars; what was their number? *Ans.* 9.

¶ 31. 1. How many oranges, at 3 cents each, may be bought for 12 cents?

Solution.—As many times as 3 cents can be taken from 12 cents so many oranges may be bought; the object, therefore, is to find how many times three is contained in 12.

	12 *cents.*
First orange,	3 *cents.*
	9
Second orange,	3 *cents.*
	6
Third orange,	3 *cents.*
	3
Fourth orange,	3 *cents.*
	0

We see, in this example, that 12 contains 3 four times, for we subtract 3 from 12 four times, after which there is no remainder; consequently, *subtraction* alone is sufficient for the operation; but we may come to the same result by a much shorter process, called *division*

Ans. 4 oranges.

We see from the above, that one number will be contained in another as many times as it can be subtracted from it, and hence, that

Division is a short way of performing many subtractions of the same number. The *minuend* is called the *dividend*, the number which is subtracted at one time is called the *divisor*, and the number which indicates the number of times the subtraction is performed is called the *quotient.*

The cost of one orange, (3 cents,) multiplied by the number of oranges, (4,) is equal to the cost of all the oranges, (12 cents;) 12 is, therefore, a *product*, and 3 *one* of its factors; and to find how many times 3 is contained in 12, is to find the *other* factor, which, multiplied into 3, will produce 12. Hence, the process of division consists in finding one factor of a product when the other is known.

2. A man would divide 12 cents equally among 3 children; how many would each child receive?

SOLUTION.— The numbers in this are the same as in the former example, but the object is different. In the former example, the object was to see how many times 3 cents are contained in 12 cents; in this, to divide 12 cents into 3 equal parts. Still the object is to find a number, which, multiplied into 3, will produce 12. This, as in the former example, is, *Ans.* 4 cents.

Hence *Division* may be defined—

I. The method of finding how many times one number is contained in another of the same kind. (Ex. 1.) Or,

II. The method of dividing a number into a certain number of equal parts. (Ex. 2.)

The *Dividend* is the number to be divided, and answers to the product in multiplication.

The *Divisor* is the number by which we divide, and answers to one of the factors.

The *Quotient* is the result or answer, and is the other factor. When anything is left, it is called the *Remainder*.

NOTE.—In the first use of division, the divisor and dividend must be of the same kind, for it would be absurd to ask how many times a number of pounds of butter is contained in a number of gallons of molasses. In the second use, the quotient is of the same kind with the dividend, for if a number of acres of land should be divided into several parts, each part will still be acres of land.

SIGN.— The sign of division is a short horizontal line between two dots, thus $\div$. This shows that the number before it is to be divided by the number after it; thus $27 \div 9 = 3$, is read, 27 divided by 9 is equal to 3; or, to shorten the expression, 9 in 27 3 times. Or the dividend may be written in place of the upper dot, and the divisor in place of the lower dot; thus $\frac{27}{9}$ shows that 27 is to be divided by 9 as before.

Questions.—¶ **31.** In what way is the first example performed? How might the operation be shortened? How often is one number contained in another? What, then, is division? Show its relation to multiplication. What is the object in the first, and what in the second, example? Define division. What is the dividend? to what does it answer in subtraction, and to what in multiplication? What the divisor, and to what does it answer in subtraction and multiplication? What the quotient, and to what does it answer? Explain the divisor and dividend in the first use of division. The dividend and quotient in the second

DIVISION TABLE.

NOTE. — The expression used by the pupil in reciting the table may be, 2 in 2 one time, 2 in 4 two times, 4 in 12 three times, &c

$\frac{2}{2}=1$	$\frac{3}{3}=1$	$\frac{4}{4}=1$	$\frac{5}{5}=1$	$\frac{6}{6}=1$	$\frac{7}{7}=1$
$\frac{4}{2}=2$	$\frac{6}{3}=2$	$\frac{8}{4}=2$	$\frac{10}{5}=2$	$\frac{12}{6}=2$	$\frac{14}{7}=2$
$\frac{6}{2}=3$	$\frac{9}{3}=3$	$\frac{12}{4}=3$	$\frac{15}{5}=3$	$\frac{18}{6}=3$	$\frac{21}{7}=3$
$\frac{8}{2}=4$	$\frac{12}{3}=4$	$\frac{16}{4}=4$	$\frac{20}{5}=4$	$\frac{24}{6}=4$	$\frac{28}{7}=4$
$\frac{10}{2}=5$	$\frac{15}{3}=5$	$\frac{20}{4}=5$	$\frac{25}{5}=5$	$\frac{30}{6}=5$	$\frac{35}{7}=5$
$\frac{12}{2}=6$	$\frac{18}{3}=6$	$\frac{24}{4}=6$	$\frac{30}{5}=6$	$\frac{36}{6}=6$	$\frac{42}{7}=6$
$\frac{14}{2}=7$	$\frac{21}{3}=7$	$\frac{28}{4}=7$	$\frac{35}{5}=7$	$\frac{42}{6}=7$	$\frac{49}{7}=7$
$\frac{16}{2}=8$	$\frac{24}{3}=8$	$\frac{32}{4}=8$	$\frac{40}{5}=8$	$\frac{48}{6}=8$	$\frac{56}{7}=8$
$\frac{18}{2}=9$	$\frac{27}{3}=9$	$\frac{36}{4}=9$	$\frac{45}{5}=9$	$\frac{54}{6}=9$	$\frac{63}{7}=9$

$\frac{8}{8}=1$	$\frac{9}{9}=1$	$\frac{10}{10}=1$	$\frac{11}{11}=1$	$\frac{12}{12}=1$
$\frac{16}{8}=2$	$\frac{18}{9}=2$	$\frac{20}{10}=2$	$\frac{22}{11}=2$	$\frac{24}{12}=2$
$\frac{24}{8}=3$	$\frac{27}{9}=3$	$\frac{30}{10}=3$	$\frac{33}{11}=3$	$\frac{36}{12}=3$
$\frac{32}{8}=4$	$\frac{36}{9}=4$	$\frac{40}{10}=4$	$\frac{44}{11}=4$	$\frac{48}{12}=4$
$\frac{40}{8}=5$	$\frac{45}{9}=5$	$\frac{50}{10}=5$	$\frac{55}{11}=5$	$\frac{60}{12}=5$
$\frac{48}{8}=6$	$\frac{54}{9}=6$	$\frac{60}{10}=6$	$\frac{66}{11}=6$	$\frac{72}{12}=6$
$\frac{56}{8}=7$	$\frac{63}{9}=7$	$\frac{70}{10}=7$	$\frac{77}{11}=7$	$\frac{84}{12}=7$
$\frac{64}{8}=8$	$\frac{72}{9}=8$	$\frac{80}{10}=8$	$\frac{88}{11}=8$	$\frac{96}{12}=8$
$\frac{72}{8}=9$	$\frac{81}{9}=9$	$\frac{90}{10}=9$	$\frac{99}{11}=9$	$\frac{108}{12}=9$

$28 \div 7$, or $\frac{28}{7}$ = how many?	$49 \div 7$, or $\frac{49}{7}$ = how many?
$42 \div 6$, or $\frac{42}{6}$ = how many?	$32 \div 4$, or $\frac{32}{4}$ = how many?
$54 \div 9$, or $\frac{54}{9}$ = how many?	$99 \div 11$, or $\frac{99}{11}$ = how many?
$32 \div 8$, or $\frac{32}{8}$ = how many?	$84 \div 12$, or $\frac{84}{12}$ = how many?
$33 \div 11$, or $\frac{33}{11}$ = how many?	$108 \div 12$, or $\frac{108}{12}$ = how many?

NOTE. — The pupil should be *thoroughly exercised* in the foregoing table.

¶ 32. The principles of division will be made more plain to the pupil by turning his attention to the same diagram to which it was directed while illustrating the principles of multiplication, since division is the reverse of multiplication.

DIAGRAM OF STARS.

* * * *
* * * *
* * * *

In multiplication, we call the whole number of stars a symbol of the product. In division, a symbol of the dividend.

In multiplication, *stars in a row*, and *number of rows*, are symbols of the multiplicand and multiplier, which are factors of the product.

In division, *stars in a row*, and *rows of stars*, are symbols of the divisor and quotient, which are factors of the dividend.

¶ 33. When the object in division is to find how many times one number, or quantity, is contained in another number, or quantity, the divisor must be of the same kind as the dividend, (stars in a row,) and the quotient will be a *number* telling *how many times* (rows of stars.)

On the other hand—when the object is to divide a number or quantity, into a given number of equal parts, the *quotient* will be of the same name or kind as the dividend, (stars in a row.) If we divide 35 apples into 5 parts, the quotient, 7 apples, will be one part, (stars in a row,) and the divisor, 5, will be a *number*, that is, the number of parts, (rows of stars.)

¶ 34. It has been remarked that *division is a short way of performing many subtractions.* How often can 3 be subtracted from 963? *Ans.* 321 times. To set down 963 and subtract 3 from it 321 times would be a long and tedious process; but by division we may decompose the number 963 thus; $963 = 900 + 60 + 3$, and say 3 is contained in 9(00,) 3(00) times, in 6(0,) 2(0) times, and in 3, 1 time = 321 times, which brings us to the same result in a much shorter way.

¶ 35. 1. How many yards of cloth, at 3 dollars a yard, can be bought for 936 dollars?

SOLUTION. — As many yards as 3 dollars are contained times in 936

Questions. — **¶ 32.** What, in the diagram of stars, may be taken as a symbol of the dividend in division? *Stars in a row* and *rows of stars* are taken as symbols of what, in multiplication? of what in division? If the dividend be 108, the divisor 12, and the quotient 9, how would you make a diagram to correspond?

¶ 33. When the object is to find how many times one number or quantity is contained in another, the divisor will be of what name or kind? the quotient will be what? When the object is to divide a number or quantity into a given number of parts, the quotient will be of what name or kind? the divisor will be what?

¶ 34. How can you make it appear from the diagram that division is a shorter way of performing many subtractions? Find on the blackboard the quotient of 963 divided by 3 How would this example be performed by subtraction?

dollars, or as many as 3 can be subtracted times from 936; 936 dollars is the dividend, (number of stars,) 3 dollars (stars in a row) the divisor.

PREPARATION.

Dividend,
Divisor, 3) 936

Write the divisor on the left of the dividend, separate them by a curved line, and draw a line underneath.

OPERATION.

3) 936
Quotient, 312

We may decompose the dividend thus, 936 = 900 + 30 + 6, and divide each part separately.

Beginning at the left hand, we say, 3 in 9, 3 times. This quotient 3 is 3 hundred, because the 9 which we divided is hundreds; therefore we write it under the 9 in the place of hundreds.

Proceeding to the next figure, we say, 3 in 3, 1 time, which, being 1 ten, we write it in tens' place. Lastly, 3 in 6, 2 times, which, being units, we write the 2 in units' place, and the work is done. The quotient (number of rows) is 3 hundred, (300,) 1 ten, (10,) and 2 units, or 312 yards, *Ans.*

NOTE. — The quotient figure will always be of the same order of units as the figure divided to obtain it.

2. 2846 ÷ 2 = how many? 840 ÷ 4 = how many? 500 ÷ 5 = how many?

3. If you give 856 dollars to 4 men, how many dollars will you give to each?

OPERATION.

Divisor, Dividend,
4 *men,*) 856 *dollars.*
Quotient, 214 *dollars.*

SOLUTION. — Write down the numbers as before. Divide the first figure, 8, (hundreds,) in the dividend as before. Proceeding to the next figure, 5, (tens,) 4 is contained 1 (ten) time in 4 of the tens, or 40, and there is 1 ten left, which, added to the 6 units, will make 16, and 4 in 16 units, 4 (units) times. *Ans.* 214 dollars.

Here again we see that the 856 is taken in three parts, 800, 40, and 16, and each part is divided separately. When this decomposing into parts can be done in the mind, as in these examples, the process is called *Short Division.* It can always be done when the divisor does not exceed 12.

4. Answer the following questions after the same manner, viz., 650 ÷ 5 = how many? 8490 ÷ 6; or, what expresses the same thing, $\frac{8490}{6}$ = how many? $\frac{91840}{7}$ = how many? $\frac{10920}{8}$ = how many?

5. What is the quotient of 14371 divided by 7?

OPERATION.

7)14371
Quotient, 2053

There are two other things to be learned in this operation. First, the divisor, 7, is not contained in 1, the first figure of the dividend then take two figures, or so many as shall contain the divisor, and say, 7 in 14, 2 times; we write 2 in the quotient, in thousands' place, because we divided 14 thousands.

Then, again, proceeding to the next figure, 3 *in the dividend will not contain the divisor*, 7; to obviate this difficulty, *we place a cipher in the quotient*, joining the 3 to the 7 tens, calling it 37 tens, and so proceed. *Ans.* 2053.

Hence, *for Short Division*, this general

RULE.

I. Write the divisor at the left hand of the dividend, separate them by a line, and draw a line under the dividend, to separate it from the quotient.

II. Find how many times the divisor is contained in the first left hand figure or figures of the dividend, and place the result directly under the last figure of the dividend taken, for the first figure of the quotient.

III. If there be no remainder, divide the next figure in the dividend in the same way; but, if there be a remainder, join it to the *next* figure of the dividend as so many tens, and then find how many times the divisor is contained in *this* amount, and set down the result as before.

IV. Proceed in this manner till all the figures in the dividend are divided.

EXAMPLES FOR PRACTICE.

6. A man has 256 hours' work to do; how many days will it take him, if he work 8 hours each day?

Ans. 32 days.

7. $\frac{2370247}{11}$ = how many? *Ans.* 215477.

8. In 1 gallon are 4 quarts; how many gallons in 2784 quarts? *Ans.* 696 gallons.

9. Seven men undertake to build a barn, for which they are to receive 602 dollars; into how many equal parts must the money be divided? How much will 1 part be? —— 3 parts? —— 5 parts? (See ¶ 30.)

Ans. to the last, 430 dollars.

Questions. — ¶ **35.** When the dividend is large, how must it be taken? how divided? How is it done when the divisor does not exceed 12? What is the preparation? Where do you begin the division? If you divide units, what will the quotient be? if tens, what? hundreds what? If at any time you have a remainder, what do you do with it In Ex. 5 there are two things to be learned; what is the first thing the second thing? What then is to be done? How do you obviate this difficulty? What does the cipher you write in the quotient show? What is short division? When employed? Repeat the rule.

10. Divide 24108 by 12 *Quotient*, 2009.

¶ 36. 1. A man gave 86 apples to 5 boys; how many apples did each boy receive?

Dividend,
Divisor, 5) 86
Quotient, 17 1 *Remainder*.

SOLUTION. — Here, dividing the number of the apples (86) by the number of boys, (5,) we find that *each* boy's share would be 17 apples; but there is 1 apple left, and this apple, which is called the remainder, is a portion of the dividend yet undivided. Wherefore this 1 apple must be divided equally among the 5 boys. But when a thing is divided into 5 equal parts, one of the parts is called $\frac{1}{5}$, (¶ 30.) So each boy will have $\frac{1}{5}$ of an apple more, or $17\frac{1}{5}$ apples in all. *Ans.* $17\frac{1}{5}$ apples.

NOTE 1. — The 17 (apples) expressing *whole* apples, are called *Integers*, that is, *whole* numbers.

Integers are numbers expressing *whole* things; thus, 86 oranges, 4 dollars, 5 days, 75, 268, &c., are integers, or whole numbers.

NOTE 2. — The $\frac{1}{5}$ (1 fifth) of an apple given to each boy, expressing *part* of a *divided* apple, is called a *Fraction*, or *broken* number.

Fractions are the *parts* into which a unit or whole thing may be divided. Thus, $\frac{1}{2}$ (1 half) of an apple, $\frac{2}{3}$ (2 thirds) of an orange, $\frac{4}{7}$ (4 sevenths) of a week, are fractions.

NOTE 3. — A number composed of a whole number and a fraction, is called a *Mixed Number;* thus, the number $17\frac{1}{5}$ (apples) in the above example, is a *mixed number*, being composed of the integers 17 and the fraction $\frac{1}{5}$.

If we examine the fraction, we shall see, that it consists of the remainder (1) for its *numerator*, and the divisor (5) for its *denominator* Therefore, —

If there be a *remainder*, set it down at the right hand of the quotient for the *numerator* of a fraction, under which write the divisor for its *denominator*.

2. Eight men drew a prize of 453 dollars in a lottery, how many dollars did each receive?

Dividend,
Divisor, 8) 453
Quotient, $56\frac{5}{8}$

Here, after carrying the division as far as possible by *whole* numbers, we have a remainder of 5 dollars, which, written as above directed, gives for the answer 56 dollars and $\frac{5}{8}$ (5 eighths) of another dollar, to each man.

Questions. — **¶ 36.** What are integers? fractions? a mixed number? If there be a remainder after division, it is a portion of what? What do you do with it? If you have a quotient of $23\frac{9}{11}$, what was the remainder? What was the divisor?

¶ **37.** Proof.

1. 16 × 5 = 80, *Product.* } Multiplication and division
2. *Dividend,* 80 ÷ 5 = 16. } are the reverse of each other.

We see, in the 2d of the above examples, that the product 80 of the 1st example, divided by **5**, one of its factors, brings out 16, the other factor, and hence that division may be used to prove multiplication. We see, also, in the 1st example, that the divisor and quotient of the 2d example, multiplied together, reproduce the dividend, and hence that multiplication may be used to prove division. Hence the

RULE.

To prove multiplication by division. — Divide the product by *one* factor, and, if the work be right, the quotient will be the other factor.

To prove division by multiplication. — Multiply the divisor and quotient together, and if the work be right, the *product* will be equal to the dividend.

Note 1. — *To prove division, if there be a remainder.* Multiply the integers of the quotient by the divisor, and to the product add the remainder. If the work be right, their sum will be equal to the dividend.

Example. — Divide 1145 by **7**.

OPERATION.	PROOF.	
7)1145	163	*integers of the quotient.*
	7	*divisor.*
$163\frac{4}{7}$.		
	1141	
	4	*remainder added.*
	1145 =	*the dividend.*

Note 2. — *Proof by excess of nines.* Find the excess of nines in the divisor, write it before the sign of multiplication, also in the quotient, and write it after the sign; multiply together these excesses, and write the excess of nines in their product over the sign; subtract the remainder, if any, from the dividend, and write the excess of nines in what is left under the sign. If the numbers under and over the sign be alike, the work is presumed to be right, in accordance with principles explained in multiplication, ¶ 23, note 3.

Questions. — ¶ **37.** To what, in multiplication, does the dividend in division answer? To what, the divisor and quotient? How, then, is multiplication proved by division? How division by multiplication? How, when there is a remainder?

Let the pupil be required to prove the examples which follow.

EXAMPLES FOR PRACTICE.

1. Divide 1005903360 by 2, 3, 4, 5, 6, 7, 8, 9, 10, 11 and 12.

2. If 2 pints make a quart, how many quarts in 8 pints? —— in 12 pints? —— in 20 pints? —— in 24 pints? —— in 248 pints? —— in 3764 pints? —— in 47632 pints?
Ans. to the last, 23816 quarts.

3. Four quarts make a gallon; how many gallons in 8 quarts? —— in 12 quarts? —— in 20 quarts? —— in 36 quarts? —— in 368 quarts? —— in 4896 quarts? —— in 5436144 quarts? *Ans.* to the last, 1359036 gallons.

4. There are 7 days in a week; how many weeks in 365 days? *Ans.* $52\frac{1}{7}$ weeks.

5. When flour is worth 6 dollars a barrel, how many barrels may be bought for 25 dollars? how many for 50 dollars? —— for 487 dollars? —— for 7631 dollars?

6. Divide 640 dollars among 4 men.
$640 \div 4$, or $\frac{640}{4} = 160$ dollars, *Ans.*

7. $678 \div 6$, or $\frac{678}{6} =$ how many? *Ans.* 113.

8. $\frac{5040}{5} =$ how many? *Ans.* 1008.

9. $\frac{7234}{7} =$ how many? *Ans.* $1033\frac{3}{7}$.

10. $\frac{3464}{9} =$ how many? *Ans.* $384\frac{8}{9}$.

11. $\frac{2764}{11} =$ how many?

12. $\frac{40301}{8} =$ how many?

13. $\frac{2014012}{12} =$ how many?

¶ 38. 1. Divide 4478 dollars equally among 21 men.

Solution. — When, as in this example, the divisor exceeds 12, the decomposing into parts cannot be done in the mind as in short division, but the whole process must be written down at length in the following manner.

OPERATION.

Div'r. Div'd. Quot.
21) 4478 ($213\frac{5}{21}$
42 1*st part.*
——
27
21 2*d part.*
——
68
63 3*d part.*
——
5 *Remainder*

We say, 21 in 44, (hundreds,) 2 (hundred) times, and write 2 on the right hand of the dividend for the first figure of the quotient. That is, we have 2 hundred dollars for each of 21 men, requiring 21×2 (hundred) = 42 hundred in all. This is the first part divided. The 42 hundred must now be subtracted from the hundreds in the dividend, and we find 2 (hundred) remaining, to which, bringing down the 7 tens, the whole is 27 tens. 21 in 27, (tens,) 1 (ten) time. Each man has now 10 dollars more, which require 21 tens, the second part, and taking this from the 27 tens, and bringing down the 8 units, we have 68 dollars yet to be divided. 21 in 68, 3 times, that is, each man will have 3 dollars, which will require 63 dollars, the third part, and there are 5 dollars left. This will not give each man a whole dollar, but $\frac{5}{21}$ of a dollar. So each man has 2 hundred, 1 ten, $3\frac{5}{21}$ dollars; that is, $213\frac{5}{21}$ dollars, *Ans.*

PROOF.

1st part,	4200	*dollars.*
2d part,	210	"
3d part,	63	"
Remainder,	5	"
	4478	"

The parts into which the dividend is decomposed, are 42 hundreds, which contain the divisor 2 (hundred) times; 21 tens, which contain the divisor 1 (ten) time; and 63, which contain the divisor 3 (units) times, or 213 times in all, and the remainder 5. We here see that the parts added make the whole sum.

This method of performing the operation is called *Long Division.* It consists in writing down the whole work of dividing, multiplying, and subtracting.

From the illustrations now given, we deduce the following

RULE.

To perform Long Division.

I. Place the divisor at the left hand of the dividend, and separate them by a curved line, and draw another curved line on the *right* of the dividend, to separate it from the quotient.

II. Take as many figures on the *left* of the dividend as will contain the divisor one or more times; find how *many* times they contain it, and put the *answer* at the right hand of the dividend for the first figure in the quotient.

III. Multiply the divisor by this quotient figure, and set the product under that part of the dividend which you divided.

IV. Subtract this product from the figures over it, and to the *remainder* bring down the next figure in the dividend.

V. Divide the number this makes up as before. Continue to bring down and divide until *all* the figures in the dividend have been brought down and divided.

PROOF.—Long division may be proved by multiplication, by the excess of nines, by adding up the parts into which the

---

Questions. — ¶ **38.** What cannot be done, when the divisor exceeds 12? Into what parts is the dividend, in the first example, decomposed? How, and for what, is 42 obtained? 21? 63? Explain the proof. What is long division, and in what does it consist? Give the rule. Name the different methods of proof. Give the substance of note 1; note 2; note 3.

dividend is decomposed, or by subtracting the remainder from the dividend and dividing what is left by the quotient, which if the work is right, will bring the divisor.

NOTE 1. — Having brought down a figure to the remainder, if the number it makes up will not *contain* the divisor, write a *cipher* in the quotient, and bring down the next figure.

NOTE 2. — When we multiply the divisor by any quotient figure, and the product is *greater* than the number we divided, the quotient figure is *too large*, and must be diminished.

NOTE 3 — If the remainder, at any time, be *greater* than the divisor, or *equal* to it, the quotient figure is *too small*, and must be increased.

EXAMPLES FOR PRACTICE.

1. How many hogsheads of molasses, at 27 dollars a hogshead, may be bought for 6318 dollars?
Ans. 234 hogsheads.

2. If a man's income be 1248 dollars a year, how much is that per week, there being 52 weeks in a year?
Ans. 24 dollars per week.

3. What will be the quotient of 153598, divided by 29?
Ans. $5296\frac{14}{29}$.

4. How many times is 63 contained in 30131?
Ans. $478\frac{17}{63}$ times; that is, 478 times, and $\frac{17}{63}$ of another time.

5. What will be the several quotients of 7652, divided by 16, 23, 34, 86, and 92? *Ans.* to the last, $83\frac{16}{92}$.

6. If a farm, containing 256 acres, be worth 7168 dollars, what is that per acre? *Ans.* 28 dollars.

7. What will be the quotient of 974932, divided by 365?
Ans. $2671\frac{17}{365}$.

8. Divide 3228242 dollars equally among 563 men; how many dollars must each man receive? *Ans.* 5734 dollars.

9. If 57624 be divided into 216, 586, and 976 equal parts what will be the magnitude of one of each of these equal parts?
Ans The magnitude of one of the last of these equal parts will be $59\frac{40}{976}$.

10. How many times does 1030603615 contain 3215?
Ans. 320561 times.

11. The earth, in its annual revolution round the sun, is said to travel 596088000 miles · what is that per hour, there being 8766 hours in a year? *Ans.* 68000 miles.

12. $\frac{1234567890}{1307}$ = how many? *Ans.* $944581\frac{523}{1307}$.

13. $\frac{40703020}{7812}$ = how many? *Ans.* $5210\frac{2500}{7812}$.

14. $\frac{987649031}{9124}$ = how many? *Ans.* $108247\frac{3403}{9124}$.

¶ 40. Contractions in Division.

¶ 39. I. *When the divisor is a composite number.*

1. Bought 18 yards of cloth for 72 dollars; how much was that a yard?

SOLUTION.—This example is the reverse of Ex. 1, ¶ 24. It was there shown that 3 and 6 are factors of 18 ($3 \times 6 = 18$.)

OPERATION.

3)72 *dollars, cost of* 18 *yards.*
6)24 *dollars, cost* 1 *piece* = 6 *yards.*
—
Ans. 4 *dollars, cost of* 1 *yard.*

If the 18 yards be divided into 3 pieces, then the cost of 1 piece would be one third as much as the cost of 3 pieces, that is, $72 \div 3 = 24$ dollars and the cost of 1 yard would be one sixth of the cost of 6 yards, that is, $24 \div 6 = 4$ dollars. That is, we divide the price of 18 yards by 3, and get the price of one third of 18, or 6 yards, and divide the price of 6 yards by 6, and get the price of 1 yard. *Ans.* 4 dollars.

Hence, *To perform division when the divisor is a composite number,*

RULE.

I. Divide the dividend by one of the component parts, and the quotient arising from that division, by the other.

II. *If the component parts be* MORE *than two.*—Divide by each of them in order, and the last quotient will be the quotient required.

EXAMPLES FOR PRACTICE.

2. If a man travel 28 miles a day, how many days will it take him to travel 308 miles? $4 \times 7 = 28$. *Ans.* 11 days.

3. Divide 576 bushels of wheat equally among 48 (8×6) men. *Ans.* 12 bushels each.

4. Divide 1260 by 63 ($= 7 \times 9$) Quotient 20. *Ans.*

5. Divide 2430 by 81 (=) " 30. *Ans.*

6. Divide 448 by 56 (=) " 8. *Ans.*

¶ 40. It not unfrequently happens that there are *remainders after the several divisions*, as in the following example.

1. A man wished to carry 783 bushels of wheat to market, how many loads would he have, allowing 36 bushels to a load?

Questions.—¶ **39.** How may you contract the operation in division when the divisor is a composite number? Which factor should you divide by first? Repeat the rule

Solution. — First, suppose his wheat put into barrels, each barrel containing 4 bushels. It would take as many barrels as 4 is contained times in 783.

4)783
———
195 3 *rem.*

It would take 195 barrels, and leave a remainder of 3 bushels.

Next, suppose he takes 9 barrels at each load; 9 (barrels) × 4 (the number of bushels in each barrel) = 36 bushels at a load, and he would have as many loads as the number of times 9 barrels are contained in 195 barrels

9)195
———
21 6 *rem.*

Hence we see, that he would have 21 loads, and leave a remainder of 6 barrels; also, a former remainder of 3 bushels.

The whole operation stands thus:

4)783 bushels.
———
9)195 barrels, and 3 bushels remainder.
———
21 loads, and 6 barrels remainder.

Our object now is to find the *true remainder.* The last remainder, 6 barrels, multiplied by the first divisor, 4, which is the number of bushels in a barrel, gives a product of 24 bushels. To this add the first remainder, 3 bushels, and we have the *true* remainder, 27 bushels.

Therefore, *When there are remainders in dividing by* TWO *component parts of a number, to get the* TRUE *remainder,*

RULE.

I. Multiply the *last* remainder by the *first* divisor, and to the product *add* the *first* remainder; the sum will be the *true* remainder.

II. *When there are* MORE *than* TWO *divisors.*—Multiply each remainder, except that from the first divisor, by all the divisors preceding the divisor which gave it; to the *sum* of their products *add* the remainder from the first divisor, if any, and the *amount* will be the true remainder.

2. 5783 ÷ 108 = how many? 3 × 4 × 9 = 108; hence, *three* divisors.

Questions. — ¶ **40.** When there are remainders in dividing by two component parts, how do you find the true remainder? When there are more than two, how? If there be a remainder by the first divisor only, what is the true remainder? if by the 2d divisor and none by the 1st, how do you obtain the true remainder? if by the 3d, and none by the 2d and 1st, how? if by the 1st and 3d, and none by the 2d, how? Repeat the rule.

OPERATION

3) 5783

4) 1927 *and* 2, 1*st rem.*

9) 481 *and* 3, 2*d rem.*

53 *and* 4, 3*d rem.*

3*d rem.* 4 × 4 (2*d div.*) × 3 (1*st div.*)	= 48
2*d rem.* 3 × 3 (1*st div.*)	= 9
1*st rem.* 2	*added* 2
	True rem. 59

Ans. $53\frac{59}{108}$.

NOTE. — The remainder by the 1st divisor, if there be no other, is the *true* remainder.

EXAMPLES FOR PRACTICE.

3. Divide 26406 by 42 = 6 × 7; what will be the true remainder? *Ans.* 30.

4. Divide 64823 by 3 component parts, the continued product of which is 96; what will be the true remainder? *Ans.* 23.

5. What is the quotient of 6811 divided by the component parts of 81? *Ans.* $84\frac{7}{81}$.

6 Divide 25431 by the component parts 3 × 4 × 8 = 96, first, in the order here given; secondly, in a reversed order, 8, 4, 3; and lastly, in the order 4, 3, 8, and bring out the true quotient in each case. Quotient, $264\frac{87}{96}$.

¶ **41.** *When the divisor is* 10, 100, 1000, *&c.*

1. A prize of 2478 dollars is drawn by 10 men; what is each man's share?

OPERATION.

10) 2478

$247\frac{8}{10}$.

Shorter way

QUOT. REM.

247 | 8

SOLUTION. — It has been shown (¶ 25) that annexing a cipher to any number is the same as multiplying it by 10; the reverse of this is equally true if we cut off the right hand figure from any number it is the same as *dividing* it by 10; the figures at the *left* will be the quotient. The figure 8, at the right, being an undivided part, is the remainder, and may be written over the divisor (¶ 36) thus, $\frac{8}{10}$. We see that 7, which was tens before, is made units; 4, which was hundreds, is tens, &c. On the same principle, if we cut off two figures it is the same as dividing by 100 · it three figures, the same as dividing by 1000, &c.

Hence, *To divide by* 1 *with any number of ciphers annexed,*

RULE.

Cut off, by a line, as many figures from the right hand of the dividend as there are ciphers in the divisor.

The figures at the left of the line will be the *quotient*, and those at the right the *remainder*.

EXAMPLES FOR PRACTICE.

2. A manufacturer bought 42604 pounds of wool in 100 days; how many pounds did he average each day?

42604 ÷ 100 = $426\frac{4}{100}$, or 426 | 04 = $426\frac{4}{100}$ pounds, *Ans.*

3. In one dollar are 100 cents; how many dollars in 42425 cents? *Ans.* $424\frac{25}{100}$; that is, 424 dollars, 25 cents.

4. 1000 mills make *one* dollar; how many dollars in 4000 mills? —— in 25000 mills? —— in 845000?

Ans. to the last, 845 dollars.

5. In one cent are 10 mills; how many cents in 40 mills? —— in 400 mills? —— in 20 mills? —— in 468 mills? —— in 4603 mills? *Ans.* to the last, $460\frac{3}{10}$ cents.

¶ 42. III. *When there are ciphers on the right hand of the divisor.*

1. A general divided a prize of 749346 dollars equally among an army of 8000 men; what did each receive?

8|000)749|346

93 5 *rem.*

5 (2*d rem.*) × 1000 = 5000, *and* 5000 + 346 (1*st rem.*) = 5346, *true remainder.*

SOLUTION. — The divisor 8000 is a composite number, of which 8 and 1000 are component parts. Dividing what 8000 men receive by 1000, which we do by cutting off the three right hand figures of the dividend, we get 749 dollars, which 8 men will receive, with a remainder of 346 dollars; and dividing 749 dollars, which 8 men receive, by 8, we get what 1 man receives, which is 93 dollars, and a remainder of 5. The 5 must be multiplied by the first divisor, 1000, and the first remainder added to the product; or, which is the same thing, (¶ 25,) the first remainder, 346, may be annexed to the 5, and we have the *Ans.* $93\frac{5346}{8000}$ dolls.

Questions. — ¶ **41.** If we annex one cipher to any number, how does it affect it? if two ciphers, how? three? &c. If we remove the right hand figure from any number, what is the result? How do you divide by 1 with any number of ciphers annexed? What will express the remainder? How do you divide by 10? by 100? by 1000? by 10000 &c.

Hence *When there are ciphers on the right hand of the divisor,*

RULE.

I. Cut them off, and also, as many figures from the right hand of the dividend.

II. Divide the remaining figures in the dividend by the remaining figures in the divisor.

III. Annex the figures cut off from the dividend to the remainder for the *true* remainder.

EXAMPLES FOR PRACTICE.

2. In 1 square mile are 640 square acres; how many square miles in 23040 square acres?

Ans. 36 square miles.

3. Divide 46720367 by 4200000. Quot. $11\frac{520367}{4200000}$.

4. How many acres of land can be bought for 346500 dollars, at 20 dollars per acre? *Ans.* 17325 acres.

5. Divide 76428400 by 900000. Quot. $84\frac{828400}{900000}$.

6. Divide 345006000 by 84000. Quot. $4107\frac{18000}{84000}$.

7. Divide 4680000 by 20, 200, 2000, 20000, 300, 4000, 50, 600, 70000, and 80. *Ans.* to 9th, $66\frac{60000}{70000}$.

¶ 43. Review of Division.

Questions. — What is division? In what does the process consist? Define it. The dividend answers to what in subtraction and multiplication, and why? the divisor? the quotient? In what ways is division expressed? Apply the diagram of stars to division. How does long division differ from short division? Why the difference? Rule for short division — for long division. Give the different methods of proof introduced in both. To what does a remainder give rise, and how written? What are fractions? When the divisor is a composite number, how do you proceed? How are the remainders treated? How divide by 10, 100, &c.? How, when there are ciphers at the right hand of the divisor?

EXERCISES.

1. An army of 1500 men, having plundered a city, took 2625000 dollars; what was each man's share?

Ans. 1750 dollars.

2. A certain number of men were concerned in the payment of 18950 dollars, and each man paid 25 dollars; what was the number of men? *Ans.* 758.

Questions. — ¶ 42. When there are ciphers on the right hand of the divisor, what do you do first? How do you divide? How do you find the true remainder? Repeat the rule.

3. If 7412 eggs be packed in 34 baskets, how many in a basket? *Ans.* 218.

4. What number must I multiply by 135, that the product may be 505710? *Ans.* 3746.

5. Light moves with such amazing rapidity, as to pass from the sun to the earth in about 8 minutes. Admitting the distance, as usually computed, to be 95,000,000 miles, at what rate per minute does it travel? *Ans.* 11875000 miles.

6. If 2760 men can dig a certain canal in one day, how many days would it take 46 men to do the same? How many men would it take to do the work in 15 days? —— in 5 days? —— in 20 days? —— in 40 days? —— in 120 days?

7. If a carriage wheel turns round 32870 times in running from New York to Philadelphia, a distance of 95 miles, how many times does it turn in running 1 mile? *Ans.* 346.

8. Sixty seconds make one minute; how many minutes in 3600 seconds? —— in 86400 seconds? —— in 604800 seconds? —— in 2419200 seconds?

9. Sixty minutes make one hour; how many hours in 1440 minutes? —— in 10080 minutes? —— in 40320 minutes? —— in 525960 minutes?

10. Twenty-four hours make a day; how many days in 168 hours? —— in 672 hours? —— in 8766 hours?

11. How many times can I subtract forty-eight from four hundred and eighty? *Ans.* 10 times.

12. How many times 3478 is equal to 47854?

Ans. $13\frac{2640}{3478}$ times.

13. A bushel of grain is 32 quarts; how many quarts must I dip out of a chest of grain to make one half ($\frac{1}{2}$) of a bushel? —— for one fourth ($\frac{1}{4}$) of a bushel? —— for one eighth ($\frac{1}{8}$) of a bushel? *Ans.* to the last, 4 quarts.

14. Divide 9302688 by 648. Quot. 14356.

15. Divide 1030603615 by 3215. Quot. 320561.

16. Divide 5221580 by 68705. Quot. 76.

17 Divide 2764503721 by 83000.

Quot. $33307\frac{22721}{83000}$.

18. If the dividend be 275868665090130, and the quotient 562916859, what was the divisor? *Ans.* 490070

MISCELLANEOUS EXERCISES,

INVOLVING THE PRINCIPLES OF THE PRECEDING RULES.

¶ 44. The four preceding rules, viz., Addition, Subtraction, Multiplication, and Division, are called the *Fundamental Rules of Arithmetic*, for numbers can be neither increased nor diminished but by one of these rules; hence these four rules are the foundation of all arithmetical operations.

EXERCISES FOR THE SLATE.

1. A man bought a chaise for 218 dollars, and a horse for 142 dollars; what did they both cost?

2. If a horse and chaise cost 360 dollars, and the chaise cost 218 dollars, what is the cost of the horse?

3. If the horse cost 142 dollars, what is the cost of the chaise?

4. If the sum of two numbers be 487, and the greater number be 348, what is the less number?

5. If the less number be 139, what is the greater number?

6. If the minuend be 7842, and the subtrahend 3481, wha is the remainder?

7. If the remainder be 4361, and the minuend be 7842, what is the subtrahend?

8. If the subtrahend be 3481, and the remainder 4361, what is the minuend?

9. The *sum* of two numbers is 48, and *one* of the numbers is 19; what is the *other?*

10. The *greater* of two numbers is 29, and their *difference* 10; what is the *less* number?

11. The *less* of two numbers is 19, and their *difference* is 10; what is the *greater?*

12. The sum of two numbers is 136, their difference is 28, what are the two numbers? *Ans.* { Greater number, 82. Less number, 54.

MENTAL EXERCISES.

1. When the minuend and the subtrahend are given, how do you find the remainder? Ex. 6.

NOTE.—The pupil may be required to give written answers to these mental exercises, or he may answer *orally;* in either case, let

6*

him turn to the exercise for the slate to which reference is made, and let him apply it in illustration of the answer he gives. Thus —

Ans. Subtract the subtrahend from the minuend, and the difference will be the remainder, as Ex. 6, (slate,) where the minuend and subtrahend are given to find the remainder, — we subtract the subtrahend 3481 from the minuend 7842, and the difference, 4361, is the remainder.

2. When the minuend and remainder are given, how do you find the subtrahend? Ex. 7.

3. When the subtrahend and the remainder are given, how do you find the minuend? Ex. 8.

4. When you have the *sum* of two numbers, and *one* of them given, how do you find the other? Ex. 9.

5. When you have the *greater* of two numbers, and their *difference* given, how do you find the *less* number? Ex. 10.

6. When you have the *less* of two numbers, and their *difference* given, how do you find the *greater* number? Ex. 11.

7. When the sum and difference of two numbers are given, how do you find the two numbers? Ex. 12.

EXERCISES FOR THE SLATE.

¶ 45. 1. If the multiplicand (squares in a row) be 754, and the multiplier (rows of squares) be 25, what will be the product (no. of squares)? *See Diagram, page* 69.

2. If the product (no. of squares) be 18850, and the multiplicand (squares in a row) be 754, what must have been the multiplier (rows of squares)?

3. If the product (no. of squares) be 18850, and the multiplier (rows of squares) be 25, what must have been the multiplicand (squares in a row)?

4. If the dividend (no. of squares) be 144, and the divisor (squares in a row) be 8, what is the quotient (no. of rows)?

5. If the dividend (no. of squares) be 144, and the quotient (no. of rows) be 18, what must have been the divisor (squares in a row)?

6. If the divisor (squares in a row) be 8, and the quotient (rows of squares) be 18, what must have been the dividend (no. of squares)?

7. The product of three numbers is 525, and two of the numbers are 5 and 7, what is the other number? *Ans.* 15.

MENTAL EXERCISES.

When the factors are given, how do you find the product? Ex. 1.

When the product and one factor are given, how do you find the other? Ex. 2 and 3.

When the dividend and quotient are given, how do you find the divisor? Ex. 5.

When the divisor and quotient are given, how do you find the dividend? Ex. 6.

When the product of three numbers and two of them are given, how do you find the other? Ex. 7.

EXERCISES FOR THE SLATE.

¶ 46. 1. What will be the cost of 15 pounds of butter, at 13 cents a pound?

2. A man bought 15 pounds of butter for 195 cents; what was that a pound?

3. A man buying butter, at 13 cents a pound, paid out 195 cents; how many pounds did he buy?

4. When rye is 75 cents a bushel, what will be the cost of 984 bushels? how many dollars will it be?

5. If 984 bushels of rye cost 738 dollars, (73800 cents,) what is the price of 1 bushel?

6. A man bought rye to the amount of 738 dollars, (73800 cents,) at 75 cents a bushel; how many bushels did he buy?

7. If 648 pounds of tea cost 284 dollars, (28400 cents,) what is the price of 1 pound? $28400 \div 648 =$ how many?

MENTAL EXERCISES.

1. When the price of *one* pound, *one* bushel, &c., of any commodity is given, how do you find the cost of *any number* of pounds, or bushels, &c., of that commodity? Ex. 1 and 4. If the price of the 1 pound, &c., be in cents, in what will the whole cost be? —— if in dollars, what? —— if in shillings? —— if in pence? &c.

2. When the cost of *any given number* of pounds, or bushels, &c., is given, how do you find the price of *one* pound, or bushel, &c.? Ex. 2, 5, and 7. In what kind of money will the answer be?

3. When the *cost of a number* of pounds, &c., is given, and also the *price of one* pound, &c. how do you find the number of pounds, &c.? Ex. 3 and 6.

EXERCISES FOR THE SLATE.

¶ 47. 1. A boy bought a number of apples; he gave away ten of them to his companions, and afterwards bought thirty-four more, and divided one half of what he then had among four companions, who received 8 apples each; how many apples did the boy first buy?

Let the pupil take the last number of apples, 8, and reverse the process. *Ans.* 40 apples.

2. There is a certain number, to which if 4 be added, and from the sum 7 be subtracted, and the difference be multiplied by 8, and the product divided by 3, the quotient will be 64, what is that number? *Ans.* 27.

3. If a man save six cents a day, how many cents would he save in a year, (365 days?) —— how many in 45 years? how many dollars would it be? how many cows could he buy with the money, at 12 dollars each?

Ans. to the last, 82 cows, and 1 dollar 50 cents remainder.

4. A man bought a farm for 22464 dollars; he sold one half of it for 12480 dollars, at the rate of 20 dollars per acre; how many acres did he buy? and what did it cost him per acre? *Ans.* to the last, 18 dollars.

5. How many pounds of pork, worth 6 cents a pound, can be bought for 144 cents?

6. How many pounds of butter, at 15 cents per pound, must be paid for 25 pounds of tea, at 42 cents per pound?

7. A man married at the age of 23; he lived with his wife 14 years; she then died, leaving him a daughter 12 years of age; 8 years after, the daughter was married to a man 5 years older than herself, who was 40 years of age when the father died; how old was the father at his death?

Ans. 60 years.

8. The earth, in moving round the sun, travels at the rate of 68000 miles an hour; how many miles does it travel in one day, (24 hours?) how many miles in one year, (365 days?) and how many days would it take a man to travel this last distance, at the rate of 40 miles a day? how many years?

Ans. to the last, 40800 years.

Problems in the Measurement of Rectangles and Solids.

NOTE. — A rectangle is a figure having four sides, and each of the four corners a square corner or right angle.

PROBLEM I.

¶ 48. *The length and breadth of a rectangle given, to find the square contents.*

1. How many square rods in a plat of ground 5 rods long and 3 rods wide?

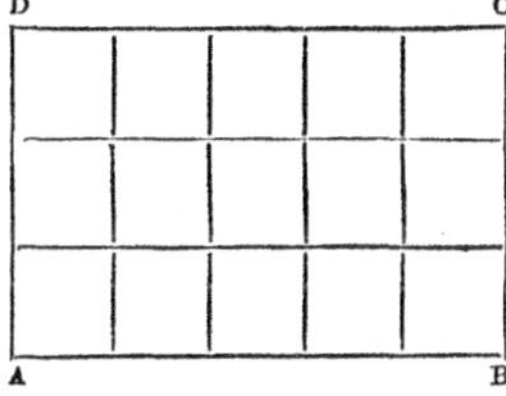

SOLUTION. — A square rod is a square measuring 1 rod on each side like one of those in the annexed diagram. We see from the diagram that there are as many squares in a row as there are rods on one side, and as many rows as there are rods on the other side; that is, 5 rows of 3 squares in a row, or 3 rows of 5 squares in a row. We multiply the number of squares in one row by the number of rows; $5 \times 3 = 15$ square rods, *Ans.*

Hence the

RULE.

Multiply the length by the breadth, and the product will be the square contents.

NOTE. — Three times a line 5 rods long is a line 15 rods long. Hence the pupil must not fail to notice, that we multiply the number of square rods in a piece of ground 1 rod wide and of the given length by the number of rods in the width.

EXAMPLES.

2. How many square rods in a piece of ground 160 rods long (squares in a row) and 8 rods wide (rows of squares)? *Ans.* 1280 square rods.

3. How many square feet in a floor 32 feet long and 23 feet wide? *Ans.* 736.

4. How many yards of carpeting, 1 yard wide, will it take

Questions. — ¶ 48. Describe a rectangle; a square rod. How do you determine the number of squares in a row, and the number of rows? Give the rule. What is the quantity really multiplied? What absurdity in considering it otherwise?

to cover the floors of two rooms, one 8 yards long and 7 yards wide, and the other 6 yards long and 5 yards wide?

Ans. 86 yards.

5. How many square feet of boards will it take for the floor of a room 16 feet long and 15 feet wide, if we allow 12 quare feet for waste? *Ans.* 252.

6. There is a room 6 yards ong and 5 yards wide; how many yards of carpeting, a yard wide, will be sufficient to cover the floor, if the hearth and fireplace occupy 3 square yards? *Ans.* 27

Problem II.

¶ 49. *The square contents and width given, to find the length.*

1. What is the length of a piece of ground 3 rods wide, and containing 15 square rods?

Solution. — In this example we have 15, the number of squares in several rows, (see the diagram, problem I.,) and 3, the number of squares in 1 row, *to find the number of rows.* We divide the squares in the number of rows by the squares in 1 row. Hence,

RULE.

Divide the square contents by the width, and the quotient will be the length.

Note. — Or, really, since the divisor and dividend must be of the same denomination, we divide the whole number of square rods by the square rods in a piece of land 3 rods long by 1 rod wide; thus, 15 ÷ 3 = 5 rods in length, *Ans.*

EXAMPLES.

2. A piece of ground containing 1280 square rods, is 8 rods in width; what is its length? *Ans.* 160 rods.

3. A floor containing 736 square feet, is 23 feet wide; what is its length? *Ans.* 32 feet.

Problem III.

¶ 50. *The square contents and length given, to find the width.*

1. What is the width of a piece of ground, 5 rods long, and containing 15 square rods?

Questions. — ¶ 49. Repeat the 2d problem; the example. What two things are given in the example, and what required? Give the rule. What is really the divisor, and why?

SOLUTION.—We divide the square contents by the length, or really by the square contents of a portion of the ground 5 rods long and 1 rod wide. *Ans.* 3 rods.

Hence,

RULE.

Divide the square contents by the length, and the quotient will be the width.

EXAMPLES.

2. A piece of ground containing 1280 square rods, is 160 rods in length; what is its width? *Ans.* 8 rods.

3. What is the width of a field 186 rods long, and containing 13392 square rods? *Ans.* 72 rods.

PROBLEM IV.

¶ 51. *The length, breadth, and hight, or thickness given, to find the contents of a cubic body.*

1. How many solid feet of wood in a pile 5 feet long, 3 feet wide, and 4 feet high?

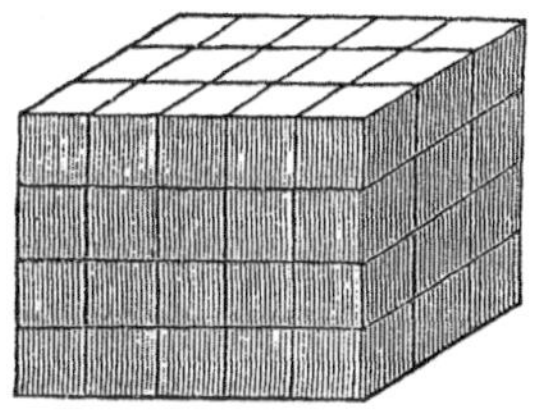

SOLUTION.—A solid foot is a solid 1 foot long, 1 wide, and 1 high. By carefully inspecting the diagram, we may see that a portion of wood 5 feet long, 1 foot wide, and 1 high, will contain 5 solid feet. Multiplying 5 solid feet by 3, we get the contents of a portion 5 feet long, 3 feet wide, and 1 foot high. $5 \times 3 = 15$ solid feet; and multiplying 15 solid feet by 4, we get the contents of the whole pile, $15 \times 4 = 60$ solid feet, *Ans.* These are the quantities multiplied, but for convenience we adopt the following

RULE.

Multiply the length by the breadth, and the resulting product by the hight.

EXAMPLES.

2. A laborer engaged to dig a cellar 27 feet long, 21 feet

Questions.—¶ **50.** Repeat the 3d problem; the example; rule What is really the divisor, and why?

¶ **51.** What is the 4th problem? the first example? Describe a solid foot. What quantity do you multiply in the first multiplication? in the second? What rule do you adopt for convenience?

wide, and 6 feet deep; how many solid feet must he remove? *Ans.* 3402 solid feet.

3. A farmer has a mow of hay 28 feet long, 14 feet wide, and 8 feet high; how many solid feet does it contain?
Ans. 3136 solid feet.

Problem V.

¶ 52. *The solid contents, length, and breadth given, to find the hight.*

1. A pile of wood 5 feet long and 3 feet wide, contains 60 solid feet; what is its hight?

Solution. — Since the divisor must be of the same denomination as the dividend, (solid feet,) we have given the solid contents of a pile 5 feet long, 3 feet wide, and several feet high, which we divide by the solid contents of a portion having the same length and breadth, and 1 foot high, to get the number of feet in the hight of the pile. Thus, $5 \times 3 = 15$, and $60 \div 15 = 4$ feet in hight, *Ans.* Hence,

RULE.

Divide the solid contents by the product of the length multiplied by the breadth.

EXAMPLES.

2. A man dug a cellar 27 feet long, and 21 feet wide, and removed 3402 solid feet of earth; what was its depth?
Ans. 6 feet.

3. A mow of hay, 28 feet long and 14 feet wide, contains 3136 solid feet; what is its hight? *Ans.* 8 feet.

Note. — In a similar manner we may find the breadth or the length, when the solid contents and the other two dimensions are given.

4. A pile of wood, 4 feet wide and 6 feet high, contains 360 solid feet; what is its length? *Ans.* 15 feet.

5. A stick of timber, 78 inches long and 8 inches thick, contains 6864 solid inches; what is its width?
Ans. 11 inches.

Questions. — ¶ 52. What is the 5th problem? the first example? solution? rule? When the solid contents, width, and hight are given, how may the length be found? When the solid contents, length, and hight are given, how may the width be found?

¶ **53.** *General questions to be answered mentally, or by the slate.*

1. If the number of squares be 84, and the squares in a row be 14, how many will be the rows of squares?

2. If the number of squares be 9500, and the rows of squares be 76, how many will be the squares in a row?

3. Were you required to form an oblong field containing 96 square rods, what, and how many ways might you vary the figure, (rows of squares and squares in a row,) each figure to contain just 96 square rods?

4. There is a frame, 40 feet square and 18 feet high, the sides of which are to be covered with boards 13 feet long, 1 foot wide; what number of these boards will it take, allowing only 7 feet waste? *Ans.* 222 boards.

5. A room, in a furniture warehouse, is 36 feet long and 29 feet wide; how many tables, 3 feet square, can be set in it, leaving a space 2 feet wide on one of the sides?

Ans. 108 tables.

¶ 54. Definitions.

Integers are distinguished as prime, composite, even, and odd.

1. A *Prime* number is one that cannot be divided by any integral number except itself and a unit, without a fraction in the quotient; as, 1, 2, 3, 5, 7.

NOTE. — Two numbers are *prime to each other* when a unit is the only whole number that will divide them both without a fractional quotient; as, 8 and 15, 25 and 36.

2. A *Composite number*, see ¶ 24.

3. An *Even number* is one which is exactly divisible by 2.

4. An *Odd number* is one which is *not* exactly divisible by 2.

¶ **55.** 1. One number is a *Measure* of another when it *divides* it *without a remainder*. Thus, 2 is a measure of 18; 5 of 45; 16 of 64.

2. A number is a *Common Measure* of two or more numbers when it divides *each* of them without a remainder. Thus 3 is a common measure of 6 and 18; 7 of 28 and 42; 4 of 12, 20 and 32; 5 of 10, 15, 20, 25.

Questions. — ¶ **54.** How are integers distinguished? What is a prime number? composite number? even number? odd number?

3. One number is a *Multiple* of another when it can be divided by it *without a remainder*. Thus 8 is a multiple of 2; 15 of 5; 33 of 11.

4. A number is a *Common Multiple* of two or more numbers when it can be divided by *each* of them without a remainder. Thus, 15 is a common multiple of 3 and 5; 16 of 2, 4 and 8; 28 of 4 and 7; 54 of 2, 3, 6, 9, 18 and 27.

5. An *Aliquot*, or *even* part, is any number which is contained in *another* number exactly 2, 3, 4, 5, &c., times. Thus, 3 is an aliquot part of 15, so also is 5. Each of the numbers, 2, 3, 4, 6, 8, and 12, is an aliquot part of 24.

6. The *Reciprocal* of a number is a *unit*, or 1, divided by the number. Thus, $\frac{1}{2}$ is the reciprocal of 2; $\frac{1}{3}$ of 3; $\frac{1}{4}$ of 4; $\frac{1}{9}$ of 9, &c.

NOTE. — Any number is *a multiple of all its measures* and *a measure of all its multiples.*

¶ 56. General Principles of Division.

The value of the quotient in division evidently depends on the relative values of the dividend and divisor.

EXAMPLE. — Let the dividend be 24, the divisor 6, and the quotient will be 4. Multiplying the dividend by 2, we in effect multiply the quotient by 2. Thus, $24 \times 2 = 48$, and $48 \div 6 = 8$, which is 2 times 4, the quotient of $24 \div 6$.

Again, dividing the divisor by 2, we in effect multiply the quotient by 2. Thus, $6 \div 2 = 3$, and $24 \div 3 = 8$, which is 2 times 4, the quotient of $24 \div 6$, the same as before. Hence,

PRINCIPLE I. Multiplying the dividend, or dividing the divisor, by any number, is in effect multiplying the quotient by that number.

¶ 57. Example as above, namely, dividend 24, divisor 6, and quotient 4. Dividing the dividend by 2, we in effect divide the quotient by 2. Thus, $24 \div 2 = 12$, and $12 \div 6 = 2$, which is equal to 1 half of the quotient of $24 \div 6$.

Again, multiplying the divisor by 2, we in effect divide the quotient by 2. Thus, $6 \times 2 = 12$, and $24 \div 12 = 2$, which is equal to 1 half of the quotient of $24 \div 6$, the same as before. Hence,

---

Questions. — ¶ 55. What is a measure? common measure? multiple? common multiple? an aliquot part? the reciprocal of a quantity?
¶ 56. On what does the value of the quotient in division depend? What is the 1st principle?

PRINCIPLE II. Dividing the dividend, or multiplying the divisor by any number, is in effect dividing the quotient by that number.

¶ 58. Example, the same as before. Multiplying both dividend and divisor by 2 does not alter the quotient. Thus, $24 \times 2 = 48$; $6 \times 2 = 12$; and $48 \div 12 = 4$, which is equal to the quotient of $24 \div 6$.

Again, dividing both dividend and divisor by 2 does not alter the quotient. Thus, $24 \div 2 = 12$; $6 \div 2 = 3$; and $12 \div 3 = 4$, which is equal to the quotient of $24 \div 6$, the same as before. Hence,

PRINCIPLE III. Multiplying or dividing both dividend and divisor by the same number does not alter the quotient.

¶ 59. EXAMPLE. It is required to multiply 24 by 6, and divide the product by 6. $24 \times 6 = 144$, and the product $144 \div 6 = 24$, which is equal to the number multiplied. Hence,

PRINCIPLE IV. If a number be multiplied, and the product divided by the same number, the quotient will be the number.

This result depends upon the principle that if the product be divided by the multiplier, the quotient will be the multiplicand.

¶ 60. Cancelation.

1. How many oranges, at 4 cents apiece, can be bought for 4 dimes, or 4 ten cent pieces?

SOLUTION. — We multiply 10 by 4 to get the number of cents, $10 \times 4 = 40$; then as many times as 4 is contained in 40 so many oranges can be bought. But multiplying 10 and dividing the product by the same number does not change it, (¶ 59;) hence, we may omit both operations, taking 10 for the result, as follows:

OPERATION.

$$\frac{10 \times \not{4}}{\not{4}} = 10$$

Writing 10 and the multiplier 4 above, and the divisor 4 below a horizontal line, we strike out 4 above and below the line, and we have 10 for the result. *Ans.* 10 oranges.

NOTE. — This process of omitting 4 is called cancelation. When we cancel a number, we usually draw an oblique line across it.

Questions. — **¶ 57.** What is the 2d principle?
¶ 58. What is the 3d principle?
¶ 59. What is the 4th principle?

2. A farmer sold 15 cows for 24 dollars apiece, and took his pay in sheep at 5 dollars apiece; how many sheep did he receive?

SOLUTION. — We see that 24 is to be multiplied by the composite number 15 = 3 × 5, and the product divided by 5. Using the component parts of the multiplier, we multiply 24 by 3. Now the product of 24 × 3 is to be multiplied and the result divided by 5, which operations we may omit, as follows:

OPERATION.

$$\frac{24 \times \overset{3}{\cancel{15}}}{\cancel{5}} = 72$$

Writing the numbers as already described, we strike out 5 below, and 15 = 3 × 5 above the line, and above 15 set the factor 3, by which we multiply 24. Since there is no number by which to divide this product, it is the result required.

Ans. 72 sheep.

3. Multiply 165 by 33, and divide the product by 31; multiply the quotient by 16 and divide the product by 99; multiply the quotient by 62 and divide the product by 55; multiply the quotient by 8 and divide the product by 20.

OPERATION.

$$\frac{\overset{3}{\cancel{165}} \times \cancel{33} \times \overset{4}{\cancel{16}} \times \overset{2}{\cancel{62}} \times 3}{\cancel{31} \times \underset{3}{\cancel{99}} \times \cancel{55} \times \underset{5}{\cancel{20}}} = \frac{24}{5} = 4\frac{4}{5}$$

By closely inspecting these numbers, we see that all the factors above the line are canceled except 4, 2 and 3, which must be multiplied together; and that all the factors below the line are canceled except 5, by which the product of the remaining factors above the line is to be divided.

NOTE 1. — It is plain that 16 above and 20 below the line have the factor 4 common, for 16 = 4 × 4 and 20 = 4 × 5; we therefore cancel the factor 4 from 16 and 20; this we do if we erase the two numbers, and write 4 the other factor of 16 over it, and 5 the other factor of 20 under it. We see also that 3, the reserved factor of 165, cancels 3, the reserved factor of 99.

NOTE 2. — If the pupil will perform the operations at length, of multiplying and dividing, in this example, he will see how much is saved by cancelation.

Cancelation, then, is the method of erasing, or rejecting, a factor or factors, from any number or numbers. It may be applied for shortening the operation where both multiplication and division are required, by rejecting equal factors from the numbers to be multiplied and the divisors.

RULE.

I. Write down the numbers to be multiplied together *above*, and the divisors *below*, a horizontal line.

II. Cancel all the factors common to the numbers to be multiplied and the divisors.

III. Proceed with the remaining numbers as required by the question.

NOTE. — One factor on one side of the line will cancel *only one like factor* on the other side.

EXAMPLES FOR PRACTICE.

4. A man sold 35 barrels of flour at 5 dollars per barrel, and took his pay in salt at 3 dollars per barrel; he sold the salt at 4 dollars per barrel, and took his pay in broadcloth at 7 dollars per yard; he sold the broadcloth at 8 dollars per yard, and took his pay in sheep at 2 dollars a head; he sold the sheep at 3 dollars a head, and took his pay in land at 15 dollars per acre; how many acres of land did he purchase?

If like factors be canceled from the numbers to be multiplied and the divisors, there will remain of the numbers to be multiplied $5 \times 4 \times 4 = 80$, and of the divisors 3; and $\frac{80}{3} = 26\frac{2}{3}$. *Ans.* $26\frac{2}{3}$ acres.

5. What is the quotient of $36 \times 8 \times 4 \times 8 \times 2$ divided by $6 \times 5 \times 3 \times 4 \times 2$?

NOTE. — The remaining factors of the numbers to be multiplied are 2, 8 and 8, and of the divisors, 5.

6. In a certain operation the numbers to be multiplied are 27, 14, 40, 8 and 6, and the divisors are 7, 10, 12 and 15; what is the quotient?

$9 \times 2 \times 2 \times 8 = 288$, and $288 \div 5 = 57\frac{3}{5}$, *Ans.*

7. What is the quotient of $4 \times 7 \times 18 \times 10 \times 8 \times 9$, divided by $24 \times 72 \times 3$?

NOTE. — All the divisors cancel. *Ans.* 70.

8. If the numbers to be multiplied are 14, 5, 3 and 28, and the divisors 15 and 9; what is the quotient?

NOTE. — The remaining factor of the divisors is 9. *Ans.* $43\frac{5}{9}$.

Questions. — ¶ 60. If a number be multiplied and the product divided by the same number, what is the result? When such operations are to be performed, how may they be contracted? What is this process called? How do you indicate that a number is canceled? What is cancelation? When may it be applied? Repeat the rule. Explain the operation in Ex. 5; in Ex. 6, &c.

¶ 61. *To find a common divisor of two or more numbers*

1. Find a common divisor of 6, 9 and 12.

OPERATION $\begin{cases} 6 = 3 \times 2 \\ 9 = 3 \times 3 \\ 12 = 3 \times 4 \end{cases}$ The factor 3, which is common to the several numbers, must be a common divisor of them. Hence the

RULE.

Separate each number into two factors, *one* of which shall be *common* to all the numbers.

The common factor will be their common divisor.

EXAMPLES FOR PRACTICE.

2. Find a common divisor of 4, 16, 24, 36 and 8.
Ans. 4.

3. Find a common divisor, or common measure, (which terms mean the same thing,) of 22, 44, 66, and 88.
Ans. 11.

4. Required the length of a rod which will be a common measure of two pieces of cloth, one of them 25 feet, the other 30 feet long. *Ans.* 5 feet.

¶ 62. *To find the greatest common divisor, or measure, of two or more numbers.*

Find the greatest common divisor of 128 and 160.

FIRST METHOD.

OPERATION.

$128 = 2 \times 2 \times 2 \times 2 \times 2 \times 2 \times 2$
$160 = 2 \times 2 \times 2 \times 2 \times 2 \times 5$

SOLUTION. — The greatest common divisor of two or more numbers is their greatest common measure, (¶ 55,) and is the greatest factor common to them. By separating the numbers into their prime factors, we find the factor 2 occurs 7 times in 128, and 5 times in 160. As no number will contain another, unless it contains all its prime factors, the product of all the prime factors common to both numbers must be the greatest common divisor.

$2 \times 2 \times 2 \times 2 \times 2 = 32$, *Ans.* Or,

SECOND METHOD.

By a sort of trial. The greatest common divisor cannot exceed the less number, for it must contain it. We will try, therefore, if the *less* number, 128, which measures itself, will also divide, or measure, 160.

```
128)160(1
    128
    ———
     32)128(4
        128
```

128 in 160, 1 time, and 32 *remain;* 128, therefore, is not a divisor of 160. We will now try whether this *remainder* be not the divisor sought; for, if 32 be a divisor of 128, the former divisor, it must also be of 160, which consists of 128 + 32; 32 in 128, 4 times, *without any remainder.* Consequently it is contained in 160 = 128 + 32, just 5 times; that is, once more than in 128. And as no number greater than 32, the difference of the two numbers, is contained once more in the greater, it is the greatest common divisor. Hence,

RULE.

I. Separate each number into its prime factors, and the product of all the prime factors common to the several numbers will be the greatest common divisor. Or,

II. Divide the greater number by the less, and that divisor by the remainder, and so on, always dividing the last divisor by the last remainder, till nothing remain. The last *divisor* will be the greatest common divisor required.

NOTE 1 — When we would find the greatest common divisor of *more* than *two* numbers, we may first find the greatest common divisor of *two* numbers, and then of *that* common divisor and one of the *other* numbers, and so on to the last number. Then will the greatest common divisor *last* found be the answer.

NOTE 2. — Two numbers which are *prime to each other*, of course, can have no common divisor greater than 1.

EXAMPLES FOR PRACTICE.

1. Apply the foregoing rule to find the greatest common divisor of 21 and 35.

2. Find the greatest common divisor of 96 and 544.
Ans. 32.

3. Find the greatest common divisor of 468 and 1184.
Ans. 4.

4. What is the greatest common divisor of 32, 80, and 256? *Ans.* 16.

5. What is the greatest common divisor of 75, 200, 625, and 150? *Ans.* 25.

6. A certain tract of land containing 100 acres, is 160 rods long and 100 wide; what is the length of the longest chain that will exactly measure both its length and breadth?
Ans. 20 rods.

7. A has 2640 dollars, B 1680 dollars, and C 756 dollars, which they agree to lay out for land at the greatest price per acre that will allow each to expend the whole of his money; what was the price per acre, and how many acres did each man buy?

Ans. A bought 220 acres, B 140 acres, and C 63 acres, at 12 dollars per acre.

Questions. — ¶ 62. What is the greatest common divisor of two or more numbers? Describe the process of finding it for two numbers? rule? How found when the numbers are more than two? What is the greatest common measure of numbers that are prime to each other?

COMMON FRACTIONS.

¶ 63. When whole numbers, which are called integers (¶ 36,) are subjects of calculations in arithmetic, the operations are called operations in whole numbers. But it is often necessary to make calculations in regard to *parts* of a thing or unit. We may not only have occasion to calculate the price of 3 barrels, 5 barrels, or 8 barrels of flour, but of *one third* of a barrel, *two fifths* of a barrel, or *seven eighths* of a barrel.

When a unit or whole thing is divided or broken into any number of equal parts, the parts are called *fractions*, or *broken numbers*, (from the Latin word, *frango, I break.*) If it be divided into 3 equal parts, the parts are called *thirds;* if into 7 equal parts, *sevenths;* if into 12 equal parts, *twelfths.* The fraction takes its *name*, or *denomination*, from the *number of parts* into which the *unit* or whole thing is divided.

If the unit or whole thing be divided into 16 equal parts, the parts are called *sixteenths*, and 5 of these parts would be 5 sixteenths.

Fractions are of three kinds, *Common*, (sometimes called *Vulgar*,) *Decimal*, and *Duodecimal.*

Common fractions are always expressed by two numbers, one above the other, with a horizontal line between them; thus, $\frac{1}{2}$, $\frac{2}{3}$, $\frac{3}{7}$.

The number *below* the line is called the *Denominator*, because it gives *name* to the parts.

The number *above* the line is called the *Numerator*, because it *numbers* the parts.

The denominator shows into how many parts a thing or unit is divided; and

The numerator shows how many of these parts are contained in the fraction. Thus, in the fraction $\frac{3}{8}$, the denominator, 8, shows that the unit or whole thing is divided into 8 equal parts, and the numerator, 3, shows that 3 of these parts are contained in the fraction. The numerator, 3, *numbers* the parts; the denominator, 8, gives them their *denomination* or

---

Questions. — ¶ **63.** What are integers? What fractions, and whence their necessity? Whence do fractions take their name? How many kinds of fractions? Name them. How are common fractions written? What is the lower number called, and why? What does it show? What is the upper number called, and why? What determines the size of the parts, and why? What are the terms of a fraction What are the terms of the fraction $\frac{7}{10}$? $\frac{11}{18}$? &c.

name, and shows their size or magnitude; for if a thing be divided into 8 equal parts, the parts are but half as large as if divided into but 4 equal parts. It will evidently take 2 eighths to make 1 fourth.

The numerator and denominator, taken together, are called the *terms* of the fraction. Thus, the terms of the fraction $\frac{7}{10}$ are 7 and 10; of $\frac{2}{8}$, 2 and 8.

¶ 64. It is important to bear in mind, that fractions arise from division, and that the *numerator* may be considered a *dividend*, and the *denominator* a *divisor*, and the *value* of the fraction the *quotient;* thus, $\frac{1}{2}$ is the quotient of 1 (the numerator) divided by 2, (the denominator;) $\frac{1}{4}$ is the quotient arising from 1 divided by 4; and $\frac{3}{4}$ is 3 times as much, that is, 3 divided by 4; thus, 1 fourth part of 3 is the same as 3 fourths of 1.

Hence, a common fraction is always expressed by the *sign of division*, the numerator being written in the place of the upper dot, and the denominator in the place of the lower dot.

$\frac{3}{4}$ expresses the quotient, of which $\begin{cases} 3 \text{ is the } \textit{dividend} \text{ or } \textit{numerator}. \\ 4 \text{ is the } \textit{divisor} \text{ or } \textit{denominator}. \end{cases}$

1. If 4 oranges be equally divided among 6 boys, what part of an orange is each boy's share?

A sixth part of 1 orange is $\frac{1}{6}$, and a sixth part of 4 oranges is 4 such pieces, $= \frac{4}{6}$. *Ans.* $\frac{4}{6}$ of an orange.

2. If 3 apples be equally divided among 5 boys, what part of an apple is each boy's share? if 4 apples, what? if 2 apples, what? if 5 apples, what?

3. What is the quotient of 1 divided by 3? —— of 2 by 3? —— of 1 by 4? —— of 2 by 4? —— of 3 by 4? —— of 5 by 7? —— of 6 by 8? —— of 4 by 5? —— of 2 by 14?

4. What part of an orange is a third part of 2 oranges? —— one *fourth* of 2 oranges? —— $\frac{1}{4}$ of 3 oranges? —— $\frac{1}{5}$ of 3 oranges? —— $\frac{1}{5}$ of 4? —— $\frac{1}{6}$ of 2? —— $\frac{1}{7}$ of 5? —— $\frac{1}{7}$ of 3? —— $\frac{1}{8}$ of 2?

¶ 65. A fraction being part of a whole thing, is properly less than a unit, and the numerator will be less than the denominator, since the denominator shows how many parts

---

Questions. — ¶ 64. From what do fractions always arise? What may the numerator be considered? the denominator? What is the value of the fraction? Of what is $\frac{1}{2}$ the quotient? $\frac{3}{4}$? $\frac{11}{16}$? $\frac{1}{5}$ of 3 is what part of 1? $\frac{1}{12}$ of 7 is what part of 1? By what is a common fraction always expressed?

make a whole thing, and there must not be so many of the parts taken as will make a whole thing.

But we call an expression written in the fractional form a fraction, though its numerator equals or exceeds the denominator, and its value, consequently, equals or exceeds a unit; but since there is not a strict propriety in the name, it is called an improper fraction. Hence,

A Proper Fraction is one that is less than a unit, its numerator being less than the denominator.

An Improper Fraction is one that equals or exceeds a unit, its numerator equaling, or exceeding the denominator. Thus, $\frac{13}{8}$, $\frac{21}{12}$, are improper fractions.

A Simple Fraction is a single fraction, either proper or improper. Thus, $\frac{3}{4}$, $\frac{9}{9}$, $\frac{16}{15}$, are simple fractions.

A Compound Fraction is a fraction of a fraction, or several fractions connected by the word *of.* Thus, $\frac{1}{3}$ of $\frac{5}{9}$, $\frac{3}{5}$ of $\frac{11}{2}$, $\frac{2}{5}$ of $\frac{1}{2}$ of $\frac{16}{9}$, are compound fractions.

A Complex Fraction is one which has a fraction, either simple or compound, or a mixed number, for its numerator, or for its denominator, or for both. Thus, $\frac{\frac{3}{4}}{2\frac{5}{7}}$, $\frac{4\frac{3}{5}}{\frac{7}{9}}$, $\frac{\frac{2}{3} \text{ of } \frac{4}{5}}{2\frac{5}{7}}$, are complex fractions.

A Mixed Number, as already shown, is one composed of a whole number and a fraction. Thus, $14\frac{1}{2}$, $13\frac{7}{8}$, &c., are mixed numbers.

A father bought 4 oranges, and cut each orange into 6 equal parts; he gave to Samuel 3 pieces, to James 5 pieces, to Mary 7 pieces, and to Nancy 9 pieces; what was each one's fraction?

Was James' fraction *proper* or *improper?* Why?

Was Nancy's fraction proper or improper? Why?

If an orange be cut into 5 equal parts, by what fraction is 1 part expressed? —— 2 parts? —— 3 parts? —— 4 parts? —— 5 parts? How many parts will make unity or a whole orange?

If a pie be cut into 8 equal pieces, and two of these pieces be given to Harry, what will be his fraction of the pie? if 5

Questions. — ¶ **65.** What is a proper fraction, and why so called? its value? What is an improper fraction, and why so called? When is its value a unit? When greater than a unit? Why? What is a simple fraction? a simple proper fraction? a simple improper fraction? a compound fraction? a complex fraction? a mixed number? What kind of a fraction is $\frac{2}{5}$ of $\frac{1}{2}$ of $\frac{9}{16}$? *More questions of this character.*

pieces be given to John, what will be his fraction? what fraction or part of the pie will be left?

¶ 66. Reduction of Fractions.

Reduction of fractions is changing them from one form to another without altering their value.

To reduce an improper fraction to a whole or mixed number.

1. In 4 halves ($\frac{4}{2}$) of an apple how many whole apples?

SOLUTION. — Since 2 halves ($\frac{2}{2}$) of an apple are equal to 1 whole apple, 4 halves ($\frac{4}{2}$) are equal to as many apples as the number of times 2 halves are contained in 4 halves, which is 2 times. *Ans.* 2 apples.

3. In $\frac{6}{2}$ of an apple how many whole apples? —— in $\frac{8}{2}$? —— in $\frac{10}{2}$? —— in $\frac{20}{2}$? —— in $\frac{48}{2}$? —— in $\frac{120}{2}$? —— in $\frac{984}{2}$?

5. How many yards in $\frac{3}{3}$ of a yard? —— in $\frac{6}{3}$ of a yard? —— in $\frac{8}{3}$? —— in $\frac{9}{3}$? —— in $\frac{10}{3}$? —— in $\frac{11}{3}$? —— in $\frac{15}{3}$? —— in $\frac{17}{3}$? —— in $\frac{20}{3}$? —— in $\frac{48}{3}$?

7. How many bushels in 8 pecks? that is, in $\frac{8}{4}$ of a bushel? —— in $\frac{10}{4}$? —— in $\frac{11}{4}$? —— in $\frac{13}{4}$? —— in $\frac{24}{4}$? —— in $\frac{100}{4}$? —— in $\frac{31}{4}$?

9. If I give 27 children $\frac{1}{4}$ of an orange each, how many oranges will it take?

To reduce a whole or mixed number to an improper fraction.

2. In 2 whole apples how many halves?

SOLUTION. — In 2 apples are two times as many halves as there are in 1 apple. Since there are 2 halves ($\frac{2}{2}$) in 1 apple, there are 2 times 2 halves in 2 apples, = 4 halves, that is, $\frac{4}{2}$, *Ans.*

4. In 3 apples how many halves? in 4 apples? in 6 apples? in 10 apples? in 24? in 60? in 170? in 492?

6. Reduce 2 yards to *thirds*. *Ans.* $\frac{6}{3}$. Reduce $2\frac{2}{3}$ yards to thirds. *Ans.* $\frac{8}{3}$. Reduce 3 yards to thirds. —— $3\frac{1}{3}$ yards. —— $3\frac{2}{3}$ yards. —— 5 yards. —— $5\frac{2}{3}$ yards —— $6\frac{2}{3}$ yards.

8. Reduce 2 bushels to *fourths*. —— $2\frac{2}{4}$ bushels. —— 6 bushels. —— $6\frac{1}{4}$ bushels. —— $7\frac{3}{4}$ bushels. —— $25\frac{2}{4}$ bushels.

10. In $6\frac{3}{4}$ oranges how many fourths of an orange?

OPERATION.

$$4\overline{)27}$$

Ans. $6\frac{3}{4}$ oranges.

It will take $\frac{27}{4}$; and it is evident, that dividing the numerator, 27, (= the number of parts contained in the fraction,) by the denominator, 4, (= the number of parts in 1 orange,) will give the number of *whole* oranges, and the remainder, written over the denominator, will express the fractional part. Hence,

To reduce an improper fraction to a whole or mixed number,

RULE.

Divide the numerator by the denominator; the quotient will be the whole or mixed number.

OPERATION.

$6\frac{3}{4}$ *oranges.*
4

24 *fourths in* 6 *oranges.*
3 " *cont'd in the fraction.*

$27 = \frac{27}{4}$, *Ans.*

Since there are 4 fourths in 1 orange, in 6 oranges there are 6 times 4 fourths = 24 fourths, and 24 fourths + 3 fourths = 27 fourths. Hence,

To reduce a mixed number to an improper fraction,

RULE.

Multiply the denominator of the fraction by the whole number; to the product add the numerator, and write the result over the denominator.

NOTE 1. — A whole number may be reduced to the form of an improper fraction, by writing 1 under it for a denominator.

NOTE 2. — A whole number may be reduced to a fraction having a specified denominator, by multiplying the whole number by the given denominator, and taking the product for a numerator.

EXAMPLES FOR PRACTICE.

11. In $\frac{83}{6}$ of a dollar, how many dollars?

12. In $13\frac{5}{6}$ dollars, how many sixths of a dollar?

13. In $\frac{1407}{60}$ of an hour, how many hours?

14. What is the improper fraction equivalent to $23\frac{27}{60}$ hours?

15. In $\frac{8763}{12}$ of a shilling, how many shillings?

16. Reduce $730\frac{3}{12}$ shillings to an improper fraction.

Questions. — ¶ 66. What is reduction of fractions? To what is the value of a fraction equal? What is the rule for reducing an improper fraction to a whole or mixed number? a mixed number to an improper fraction? How may a whole number be reduced to the form of an improper fraction? How to a fraction having a specified denominator?

17. In $\frac{3761}{24}$ of a day, how many days?

18. In $156\frac{17}{24}$ days, how many 24ths of a day?
Ans. $\frac{3761}{24} = 3761$ hours.

19. In $\frac{1371}{4}$ of a gallon, how many gallons?

20. In $342\frac{3}{4}$ gallons, how many 4ths of a gallon?
Ans. $\frac{1371}{4}$ of a gallon $=$ 1371 quarts.

21. Reduce $\frac{36}{20}$, $\frac{706}{40}$, $\frac{875}{100}$, $\frac{4786}{1000}$, $\frac{3465}{450}$, to whole or mixed numbers.

22. Reduce $1\frac{16}{20}$, $17\frac{26}{40}$, $8\frac{75}{100}$, $4\frac{786}{1000}$, and $7\frac{315}{450}$ to improper fractions.

¶ 67. *To reduce a fraction to its lowest or most simple terms.*

If $\frac{1}{2}$ of an apple be divided into 2 equal parts, it becomes $\frac{2}{4}$ The effect on the fraction is evidently the same as if we had multiplied both of its terms by 2. In either case, *the parts are made* 2 *times as* MANY *as they were before; but they are only* HALF AS LARGE; for it will take 2 times as many *fourths* to make a whole one as it will take *halves;* and hence it is that $\frac{2}{4}$ is the same in value or quantity as $\frac{1}{2}$.

$\frac{2}{4}$ is 2 parts; and if each of these parts be again divided into 2 equal parts, that is, if both terms of the fraction be multiplied by 2, it becomes $\frac{4}{8}$.

Now if we reverse the above operation, and divide both terms of the fraction $\frac{4}{8}$ by 2, we obtain its equal, $\frac{2}{4}$; dividing again by 2, we obtain $\frac{1}{2}$, which is the *most simple* form of the fraction, because the terms are the *least* possible by which the fraction can be expressed. Hence, $\frac{1}{2} = \frac{2}{4} = \frac{4}{8}$, and the reverse of this is evidently true, that $\frac{4}{8} = \frac{2}{4} = \frac{1}{2}$.

It follows, therefore, *by multiplying or dividing both terms of the fraction by the same number, we change its terms without altering its value.* (¶ 58.)

The process of changing $\frac{4}{8}$ into its equal $\frac{1}{2}$, is called *reducing the fraction to its lowest terms.*

A fraction is said to be in its lowest terms when no number greater than 1 will divide its numerator and denominator without a remainder.

1. Reduce $\frac{128}{160}$ to its lowest terms.

OPERATION.

$$4)\frac{128}{160} = \frac{32}{40} = \frac{4}{5}\ \textit{Ans.} \quad (8)$$

We find, by trial, that 4 will exactly measure both 128 and 160, and, dividing, we change the fraction to its equal $\frac{32}{40}$. Again, we find that 8 is a divisor common to both terms, and, dividing, we reduce the fraction to its equal $\frac{4}{5}$, which is now in its lowest terms, for no greater number than 1 will again measure them.

NOTE 1. — Any fraction may evidently be reduced to its lowest terms by a single division, if we use the *greatest* common divisor of the two terms. Thus, we may divide by 32, which we found (¶ 62) to be the greatest common divisor of 128 and 160. $32)\frac{128}{160}=\frac{4}{5}$ *Ans.*

Hence, *To reduce a fraction to its lowest terms,*

RULE.

Cancel all the factors common to both terms of the fraction.

NOTE 2. — By referring to ¶ 60, the pupil will perceive that to cancel all the factors common to both terms consists in dividing both terms by their greatest common divisor (¶ 62), or by any common divisor (¶ 61), and the quotients thence arising by any other, and so on, until the terms are prime to each other (¶ 54, Note).

EXAMPLES FOR PRACTICE.

2. Reduce $\frac{156}{468}$ to its lowest terms. *Ans.* $\frac{1}{3}$.
3. Reduce $\frac{400}{500}$, $\frac{45}{600}$, $\frac{165}{275}$, and $\frac{21}{35}$ to their lowest terms. *Ans.* $\frac{4}{5}$, $\frac{3}{40}$, $\frac{3}{5}$, $\frac{3}{5}$.

NOTE 3. — Let the following examples be wrought by both methods; by several divisors, and also by finding the greatest common divisor.

4. Reduce $\frac{450}{900}$, $\frac{99}{297}$, $\frac{140}{160}$, and $\frac{1644}{2192}$ to their lowest terms. *Ans.* $\frac{1}{2}$, $\frac{1}{3}$, $\frac{7}{8}$, and $\frac{3}{4}$.
5. Reduce $\frac{384}{1152}$ to its lowest terms. *Ans.* $\frac{1}{3}$.
6. Reduce $\frac{114}{285}$ to its lowest terms. *Ans.* $\frac{2}{5}$.
7. Reduce $\frac{468}{1184}$ to its lowest terms. *Ans.* $\frac{117}{296}$.
8. Reduce $\frac{1429}{2858}$ to its lowest terms. *Ans.* $\frac{1}{2}$.

NOTE 4. — The reduction of compound fractions to simple ones is presented in ¶ 79; the reduction of fractions to a common denominator, in ¶¶ 70, 71; and the reduction of complex fractions to simple ones, in ¶ 85 (2), for the reason that the pupil is not sufficiently advanced to understand them at this place.

Questions. — ¶ **67.** Give the illustration with the half apple. Reverse the operation. What follows? What is the process mentioned, and what is it called? When is a fraction in its lowest terms? Explain Ex. 1. How can a fraction be reduced by a single division? Rule. Give the note by which you determine by what number you divide.

Addition and Subtraction of Fractions.

COMMON DENOMINATOR.

¶ 68. 1. A boy gave to one of his companions $\frac{2}{8}$ of an orange, to another $\frac{4}{8}$, to another $\frac{1}{8}$; what part of an orange did he give to all? $\frac{2}{8}+\frac{4}{8}+\frac{1}{8}=$ how much?

SOLUTION. — The adding together of $\frac{2}{8}$, $\frac{4}{8}$ and $\frac{1}{8}$ of an orange is the same as the adding of 2 oranges, 4 oranges, and 1 orange, which would make 7 oranges. The 8 is called the common denominator, as it is common to the several fractions; and we write over it the sum of the numerators, to express the answer. *Ans.* $\frac{7}{8}$.

2. A boy had $\frac{7}{10}$ of a dollar, of which he expended $\frac{3}{10}$; what had he left?

SOLUTION. — $\frac{1}{10}$ of a dollar is one dime, or ten cent piece. The operation, then, is to subtract 3 ten cent pieces from 7 ten cent pieces, which will leave 4 ten cent pieces, or, *Ans.* $\frac{4}{10}$.

3. $\frac{1}{3}+\frac{2}{3}+\frac{1}{3}=$ how much? $\frac{3}{4}-\frac{1}{4}=$ how much?

4. $\frac{1}{20}+\frac{7}{20}+\frac{9}{20}+\frac{14}{20}+\frac{2}{20}=$ how much? $\frac{14}{18}-\frac{3}{18}=$ how much?

5. A boy, having $\frac{3}{4}$ of an apple, gave $\frac{1}{4}$ of it to his sister; what part of the apple had he left? $\frac{3}{4}-\frac{1}{4}=$ how much?

¶ 69. 1. A boy, having an orange, gave $\frac{3}{4}$ of it to his sister, and $\frac{1}{8}$ of it to his brother; what part of the orange did he give away?

SOLUTION. — The fractions $\frac{3}{4}$ and $\frac{1}{8}$ of an orange can no more be added than 3 oranges and 1 apple, which would make neither 4 oranges nor 4 apples, as they are of different kinds, (¶ 12.) But if 1 orange made 2 apples, the 3 oranges would make 6 apples, and the 1 apple being added we should have 7 apples. Now $\frac{1}{4}$ does make just $\frac{2}{8}$, and consequently $\frac{3}{4}$ make $\frac{6}{8}$, to which if $\frac{1}{8}$ be added we shall have the *Ans.* $\frac{7}{8}$.

The denominator, 4, of the fraction $\frac{3}{4}$, is a factor of 8, the denominator of the fraction $\frac{1}{8}$. And if each term of the fraction $\frac{3}{4}$ be multiplied by 2, the remaining factor of 8 it will be reduced to 8^{ths}, ($\frac{6}{8}$,) without altering its value. (¶ 67.) Hence, *if the denominator of one fraction be a factor of the denominator of another fraction, and both its terms be multi-*

Questions. — ¶ 68. Like what, is the process of adding eighths What is the 8 called, and why? What is the tenth of a dollar?

plied by the remaining factor, it will be reduced to the same denominator with the latter fraction, without altering its value. (¶ 58.) For example:

2. How many 12ths in $\frac{3}{4}$?

SOLUTION. — The factors of 12 are 3 and 4, the latter of which is the denominator of $\frac{3}{4}$, and multiplying both terms of $\frac{3}{4}$ by 3, the other factor, we have $\frac{9}{12}$, a fraction of the same value as $\frac{3}{4}$, but having a different denominator. *Ans.* $\frac{9}{12}$.

3. A man has $\frac{5}{12}$ of a barrel of sugar in one cask, $\frac{4}{6}$ in another, and $\frac{3}{4}$ in another; how much in all?

SOLUTION. — The denominator 6, of the second fraction, is a factor of 12, the denominator of the first; and if both terms of $\frac{4}{6}$ be multiplied by the other factor, 2, it will become $\frac{8}{12}$. Also 4, the denominator of the third fraction, is a factor of 12, and if both its terms be multiplied by 3, the other factor, it will be $\frac{9}{12}$. And $\frac{5}{12}+\frac{8}{12}+\frac{9}{12}=\frac{22}{12}=1\frac{10}{12}$ barrels. *Ans.*

4. What is the amount of $\frac{1}{4}$, $\frac{2}{6}$, and $\frac{5}{9}$?

SOLUTION. — As the denominators are not factors of each other, we must take some number of which each is a factor. 36 is such a number. The first denominator, 4, being a factor of 36, both terms of $\frac{1}{4}$ may be multiplied by 9, the other factor, and we shall have $\frac{9}{36}$. In like manner, both terms of $\frac{2}{6}$ being multiplied by 6, we have $\frac{12}{36}$ and both terms of $\frac{5}{9}$ being multiplied by 4, we have $\frac{20}{36}$; then, $\frac{9}{36}+\frac{12}{36}+\frac{20}{36}=\frac{41}{36}=1\frac{5}{36}$. *Ans.*

The process in the above examples is called *reducing fractions to a common denominator*, and is necessary when we wish to add or subtract those of different denominators. The common denominator, it will be perceived, ***must contain, as a factor, each of the other denominators.***

It is not always manifest what number will contain all the denominators. There are two methods of finding such a number.

FIRST METHOD.

¶ 70. If several numbers are multiplied together, each will evidently be a factor of the product. We have, then, the following

---

Questions. — ¶ 69. How can eighths and fourths be added? When, and how, can one fraction be reduced to the denominator of another? Explain the third example; the fourth. What is the process called? When is this necessary? What must the common denominator contain? What is *not* manifest? How many methods of finding it?

RULE.

Multiply the numerator and denominator of each fraction by the product of the other denominators.

NOTE 1. —The several new denominators will be products of the same numbers, and, therefore, will be alike ; and the numerator and denominator of each fraction being multiplied by the same number, its value is not altered. See ¶ 58.

NOTE. 2 The common denominator of two or more fractions is the common multiple of all their denominators ; see ¶ 55.

EXAMPLES.

1. Reduce $\frac{2}{3}$, $\frac{3}{4}$ and $\frac{4}{5}$ to equivalent fractions having a common denominator.

Each term of $\frac{2}{3}$ being multiplied by 4×5, or 20, we have $\frac{40}{60}$
" " $\frac{3}{4}$ " " 3×5, or 15, " " $\frac{45}{60}$.
" " $\frac{4}{5}$ " " 3×4, or 12, " " $\frac{48}{60}$.

The terms of each fraction are changed, while its value is not altered.

2. Reduce $\frac{1}{2}$, $\frac{2}{3}$, $\frac{7}{8}$, and $\frac{4}{5}$ to equivalent fractions, having a common denominator. *Ans.* $\frac{120}{240}$, $\frac{160}{240}$, $\frac{210}{240}$, $\frac{192}{240}$.

3. Reduce to equivalent fractions, of a common denominator, and add together, $\frac{1}{3}$, $\frac{3}{5}$, and $\frac{1}{4}$.
Ans. $\frac{20}{60}+\frac{36}{60}+\frac{15}{60}=\frac{71}{60}=1\frac{11}{60}$, Amount.

4. Add together $\frac{3}{4}$ and $\frac{6}{7}$. Amount, $1\frac{17}{28}$.

5. What is the amount of $\frac{1}{2}+\frac{1}{3}+\frac{1}{7}+\frac{1}{5}$?
Ans. $\frac{247}{210}=1\frac{37}{210}$.

6. What are the fractions of a common denominator equivalent to $\frac{3}{4}$ and $\frac{5}{6}$? *Ans.* $\frac{18}{24}$ and $\frac{20}{24}$, or $\frac{9}{12}$ and $\frac{10}{12}$.

SECOND METHOD.

¶ 71. While we can always find a common denominator by the above rule, it will not always give us the least common denominator. In the last example, 12 as well as 24 is a common denominator of $\frac{3}{4}$ and $\frac{5}{6}$. Let us see how the 12 is obtained.

One number will contain another having several factors, when it contains all these factors. For example, let 18 be resolved into the factors $2 \times 3 \times 3$, which, multiplied together, will produce it. It contains 6, the factors of which, 2 and 3, are the first and second factors of 18. It also contains

---

Questions. — ¶ 70. What is the rule in the first method ? Whence its propriety ? What is a common multiple ? Explain the first example

9, the factors of which, 3 and 3, are the second and third factors of 18. But it will not contain $8 = 2 \times 2 \times 2$, for 2 is only once a factor in 18.

Now 12, the factors of which are $2 \times 2 \times 3$, will contain $4 = 2 \times 2$, since these factors are the first and second factors of 12. It will in like manner contain 6. And it is the least number that will contain 6 and 4, for 2 must be twice a factor, or it will not contain 4, and 3 must be a factor, or it will not contain 6. Hence, no one of these factors can be spared. But $24 = 2 \times 2 \times 2 \times 3$, has, it is seen, 2 three times as a factor; so one 2 can be omitted, and we have the factors of 12 as before. We have 2 as a factor once more than necessary, because it is a factor in both 4 and 6. Hence, *when several of the denominators have the same factor we need retain it but once in the common denominator.*

¶ 72. The process of omitting the needless factors is called getting the least common denominator of several fractions, and is as follows:

```
2 | 4 . 6
  |------
  | 2 . 3
2 × 2 × 3 = 12.
```

4 and 6 are each divided by 2; and the divisor and remainders being taken for the factors of the common denominator, we have rejected 2 once.

1. Find the least common denominator of $\frac{1}{2}$, $\frac{3}{4}$, $\frac{3}{6}$, $\frac{5}{8}$, $\frac{7}{10}$.

```
2 | 2 . 4 . 6 . 8 . 10
  |-------------------
2 | 1 . 2 . 3 . 4 . 5
  |-------------------
  | 1 . 1 . 3 . 2 . 5
2 × 2 × 3 × 2 × 5 = 120
```

SOLUTION. — We write the denominators in a line, and divide as here seen. By the first division, 2 existing as a factor in each of the five numbers, is rejected four times, being retained once; as the divisor is substituted for the five factors 2, which we should have had by multiplying all the numbers together. But 2 being a factor in two of the remainders, it is rejected once more by a second division.

Ans. 120.

2. Find the least common denominator of $\frac{2}{3}$, $\frac{7}{12}$, and $\frac{13}{24}$.

Questions. — ¶ **71.** Why the necessity of a second method? When will one number contain another? What numbers will 18 contain, and why? What will it not contain? why? Why will 12 contain the denominators of both $\frac{3}{4}$ and $\frac{5}{6}$? Why is it the least number that will contain them? Why is a factor in 24 once more than necessary? What may then be done?

FIRST OPERATION.

```
12 | 8 . 12 . 24
   |-----------
 2 | 8 . 1 . 2
   |-----------
   | 4 . 1 . 1
```

12 × 2 × 4 = 96, *Ans.*

SECOND OPERATION.

```
4 | 8 . 12  24
  |-----------
3 | 2 . 3 . 6
  |-----------
2 | 2 . 1 . 2
  |-----------
  | 1 . 1 . 1
```

4 × 3 × 2 = 24, *Ans.*

THIRD OPERATION.

```
2 | 8 . 12 . 24
  |-----------
2 | 4 . 6 . 12
  |-----------
3 | 2 . 3 . 6
  |-----------
2 | 2   1 . 2
  |-----------
  | 1 . 1 . 1
```

2 × 2 × 3 × 2 = 24, *Ans*

It may be seen that the product of the factors rejected by the first operation is 24, while it is 96 by the second and third. The answer by the first is consequently four times greater, and is not the least common denominator.

Care must be taken to avoid this error in practice. The divisor should not be too large. It may always safely, though not necessarily, be the smallest number that can divide any two or more of the denominators without a remainder.

NEW NUMERATORS.

¶ 73. 1. Reduce the fractions $\frac{1}{2}$, $\frac{2}{3}$, $\frac{3}{4}$, and $\frac{2}{6}$ to equivalent fractions having the least common denominator.

```
2 | 2 . 3 . 4 . 6
  |--------------
3 | 1 . 3 . 2 . 3
  |--------------
  | 1 . 1 . 2 . 1
```

2 × 3 × 2 = 12

SOLUTION. — The new denominator being found, as above, to be 12, the denominator, 2, of the first fraction, has been really multiplied by 6, and to preserve the equality of the fraction, the numerator must be multiplied by the same number, and $\frac{1}{2}$ becomes $\frac{6}{12}$. So $\frac{2}{3}=\frac{8}{12}$, $\frac{3}{4}=\frac{9}{12}$, and $\frac{2}{6}=\frac{4}{12}$, and hence the fractions are $\frac{6}{12}$, $\frac{8}{12}$, $\frac{9}{12}$, $\frac{4}{12}$.

NOTE 1. — The factor by which the numerator of any fraction is to be multiplied, may be found by dividing the common denominator by the denominator of this fraction.

Hence, — *For reducing fractions to their lowest terms.*

RULE.

Write down all the denominators in a line, and divide by the smallest number greater than 1 that will divide two or more of them without a remainder. Having written the quo

Questions. — ¶ **72.** What is the process called? In the first example, what factors are omitted, and what substituted, by the first division? What by the second? Explain the second example.

tients and undivided numbers beneath, divide as before; and so on till there are no two numbers which can be divided by a number greater than 1.

The continued product of the quotients and divisors will be the denominator required. Then multiply each numerator by the number by which its denominator has really been multiplied.

NOTE 2. — The least common denominator of two or more fractions is the least common multiple of all their denominators. See ¶ 55.

2. Reduce $\frac{1}{2}$, $\frac{3}{8}$, and $\frac{5}{6}$ to fractions having the least common denominator, and add them together.

NOTE 3. — In writing fractions for addition and subtraction which have a common denominator, the numerators may be written in a line, connected by the appropriate signs, one line extended under them all, and the denominator written under this line but once. Thus, in the last example,

$$\frac{1}{2} + \frac{3}{8} + \frac{5}{6} = \frac{12+9+20}{24} = \frac{41}{24} = 1\frac{17}{24} \text{ amount.}$$

3. Reduce $\frac{1}{6}$ and $\frac{1}{9}$ to fractions of the least common denominator, and subtract one from the other.

Ans. $\frac{3}{18} - \frac{2}{18} = \frac{1}{18}$, difference.

4. There are 3 pieces of cloth, one containing $7\frac{3}{4}$ yards another $13\frac{5}{6}$ yards, and the other $15\frac{7}{8}$ yards; how many yards in the 3 pieces?

Before adding, reduce the fractional parts to their least common denominator; this being done, we shall have, —

$$\left.\begin{aligned} 7\tfrac{3}{4} &= 7\tfrac{18}{24} \\ 13\tfrac{5}{6} &= 13\tfrac{20}{24} \\ 15\tfrac{7}{8} &= 15\tfrac{21}{24} \end{aligned}\right\}$$

Ans. $37\frac{11}{24}$ *yds.*

Adding together all the 24ths, viz., 18 + 20 + 21, we obtain 59, that is, $\frac{59}{24} = 2\frac{11}{24}$. We write down the fraction $\frac{11}{24}$ under the other fractions, and reserve the 2 integers to be carried to the amount of the other integers, making in the whole $37\frac{11}{24}$ *Ans.*

5. There was a piece of cloth containing $34\frac{3}{8}$ yards, from which were taken $12\frac{2}{3}$ yards; how much was there left?

$$\begin{aligned} 34\tfrac{3}{8} &= 34\tfrac{9}{24} \\ 12\tfrac{2}{3} &= 12\tfrac{16}{24} \end{aligned}$$

Ans. $21\frac{17}{24}$ *yds.*

We cannot take 16 twenty-fourths, ($\frac{16}{24}$,) from 9 twenty-fourths, ($\frac{9}{24}$); we must, therefore, borrow 1 integer, = 24 twenty-fourths ($\frac{24}{24}$,) which, with $\frac{9}{24}$, makes $\frac{33}{24}$; we can now

Questions. — ¶ 73. In getting 12 as the common denominator of the fractions in the first example, by what number has the denominator of $\frac{3}{4}$ been multiplied? By what, then, must the numerator be multiplied? The same questions in regard to $\frac{5}{6}$; in regard to $\frac{2}{6}$. How is this multiplier found? Give the rule. What is the least common multiple? What is done with the sum of the fractions in the fourth example? Explain the borrowing in the fifth example.

subtract $\frac{16}{24}$ from $\frac{33}{24}$, and there will remain $\frac{17}{24}$; and taking 12 integers from 33 integers, we have 21 integers remaining. *Ans.* $21\frac{17}{24}$.

¶ 74. We have, then, *for the addition and subtraction of fractions, this general*

RULE.

Add and subtract their numerators when they have a common denominator; otherwise, they must first be reduced to a common denominator.

EXAMPLES FOR PRACTICE.

1. What is the amount of $\frac{6}{7}$, $4\frac{2}{3}$, and 12? *Ans.* $17\frac{11}{21}$.
2. A man bought a ticket, and sold $\frac{3}{8}$ of it; what part of the ticket had he left? *Ans.* $\frac{5}{8}$.
3. Add together $\frac{1}{2}$, $\frac{5}{8}$, $\frac{1}{4}$, $\frac{7}{10}$, $\frac{1}{5}$, and $\frac{14}{20}$. *Amount,* $2\frac{39}{40}$.
4. What is the difference between $14\frac{8}{11}$ and $16\frac{7}{33}$? *Ans.* $1\frac{16}{33}$
5. From $1\frac{1}{2}$ take $\frac{3}{4}$. *Remainder,* $\frac{3}{4}$
6. From 3 take $\frac{1}{3}$. *Rem.* $2\frac{2}{3}$.
7. From $147\frac{1}{3}$ take $48\frac{4}{9}$. *Rem.* $98\frac{8}{9}$.
8. Add together $112\frac{1}{2}$, $311\frac{2}{3}$, and $1000\frac{3}{4}$. *Ans.* $1424\frac{11}{12}$.
9. Add together 14, 11, $4\frac{2}{3}$, $\frac{1}{18}$, and $\frac{1}{2}$. *Ans.* $30\frac{2}{9}$.
10. From $\frac{3}{4}$ take $\frac{1}{2}$. From $\frac{7}{8}$ take $\frac{3}{4}$.
11. What is the difference between $\frac{1}{2}$ and $\frac{1}{3}$? $\frac{2}{3}$ and $\frac{1}{2}$? and $\frac{2}{3}$? $\frac{4}{5}$ and $\frac{3}{4}$? $\frac{5}{6}$ and $\frac{4}{5}$? $\frac{5}{6}$ and $\frac{3}{4}$?
12. How much is $1-\frac{1}{4}$? $1-\frac{1}{2}$? $1-\frac{3}{8}$? $1-\frac{5}{8}$? $2-\frac{2}{3}$? $2-\frac{4}{7}$? $2\frac{1}{4}-\frac{2}{3}$? $3\frac{4}{5}-\frac{1}{10}$? $1000-\frac{1}{10}$?

Multiplication of Fractions.

¶ 75. I. *To multiply a fraction by a whole number.*

1. If 1 yard of cloth cost $\frac{1}{3}$ of a dollar, what will 2 yards cost? $\frac{1}{3}\times 2=$ how much?
2. If a cow consume $\frac{1}{4}$ of a bushel of meal in 1 day, how much will she consume in 3 days? $\frac{1}{4}\times 3=$ how much?
3. A boy bought 5 cakes, at $\frac{2}{7}$ of a dollar each; what did he give for the whole? $\frac{2}{7}\times 5=$ how much?
4. How much is 2 times $\frac{1}{3}$? —— 3 times $\frac{1}{4}$? —— 2 times $\frac{2}{5}$?

Questions. — ¶ 74. Give the rule. How may $\frac{1}{2}$ be reduced to the denominator of $\frac{3}{4}$? $\frac{3}{4}$ to the denominator of $\frac{7}{8}$? (¶ 69.

5. Multiply $\frac{2}{7}$ by 3. —— $\frac{3}{8}$ by 2. $\frac{1}{6}$ by 7.

6. A woman gives to her son $\frac{3}{8}$ of an apple, and to her daughter twice as much; what part of an apple does the daughter receive?

SOLUTION. — She gives the son 3 pieces of an apple that had been cut into 8 pieces, and she may give to the daughter twice the number of the same size, that is, 6 pieces, $\frac{3}{8} \times 2 = \frac{6}{8}$. We multiply the numerator without changing the denominator.

Or, she may give the daughter 3 pieces of an apple that had been cut into half as many, that is, 4 pieces, each piece being *twice as large.* We divide the denominator by 2, without changing the numerator, showing that, as 2 small pieces make 1 large piece, the 8 small pieces will make 4 large ones. *Ans.* $\frac{6}{8}$, or $\frac{3}{4}$.

Hence, dividing the denominator, which is the divisor, has the same effect on the value of the fraction as multiplying the numerator, which is the dividend. (¶ 56.)

Hence, *there are* TWO *ways to multiply a fraction by a whole number:* —

I. *Divide the denominator* by the whole number, (when it can be done without a remainder,) and over the quotient write the numerator. — Otherwise,

II. *Multiply the numerator* by the whole number, and under the product write the denominator.

If then the product be an improper fraction, it may be reduced to a whole or mixed number.

EXAMPLES FOR PRACTICE.

1. If 1 man consume $\frac{5}{36}$ of a barrel of flour in a month, how much will 18 men consume in the same time? —— 6 men? —— 9 men? *Ans.* to the last, $1\frac{1}{4}$ barrels.

2. What is the product of $\frac{71}{120}$ multiplied by 40? $\frac{71}{120} \times 40 =$ how much? *Ans.* $23\frac{2}{3}$.

3. Multiply $\frac{13}{144}$ by 12. —— by 18. —— by 21. —— by 36. —— by 48. —— by 60.

NOTE 1. When the multiplier is a composite number, we may first multiply by one component part, and that product by the other. (¶ 24.) Thus, in the last example, the multiplier, 60 is equal to 12×5; therefore, $\frac{13}{144} \times 12 = \frac{13}{12}$, and $\frac{13}{12} \times 5 = \frac{65}{12} = 5\frac{5}{12}$, *Ans.*

Questions. — ¶ 75. Repeat the sixth example. Why is a fraction multiplied by multiplying the numerator? Why by dividing the denominator? Give the rule. How may we proceed when the multiplier is a composite number? How is a mixed number multiplied?

4. Multiply $5\frac{3}{4}$ by 7.

NOTE 2. The mixed number may be reduced to an improper fraction, and multiplied, as in the preceding examples; but the operation will usually be shorter to multiply the fraction and whole number *separately*, and add the results together. Thus, 7 times 5 are 35; and 7 times $\frac{3}{4}$ are $\frac{21}{4}=5\frac{1}{4}$, which, added to 35, make $40\frac{1}{4}$, *Ans.*

Or, we may multiply the *fraction* first, and, writing down the fraction, reserve the integers, to be carried to the product of the whole number.

5. What will $9\frac{13}{20}$ tons of hay come to, at 17 dollars per ton? *Ans.* $164\frac{1}{20}$ dollars.

6. If a man travel $2\frac{6}{40}$ miles in 1 hour, how far will he travel in 5 hours? —— in 8 hours? —— in 12 hours? —— in 3 days, supposing he travel 12 hours each day?
Ans. to the last, $77\frac{2}{5}$ miles.

¶ 76. II. *To multiply a whole number by a fraction.*

1. If 36 dollars be paid for a piece of cloth, what costs $\frac{3}{4}$ of it?

SOLUTION. — If the price of 1 piece of cloth had been given to find the price of several pieces, we should multiply the price of 1 piece by the number of pieces, and we must consequently multiply the price of 1 piece by the fraction of a piece where the price of a fraction is required.

The price of 1 piece, 36 dollars, must be multiplied by $\frac{3}{4}$. One fourth of the cloth would cost $\frac{1}{4}$ of 36, or 9 dollars, and $\frac{3}{4}$ would cost 3 times as much, $9\times3=27$. *Ans.* 27 dollars.

The product is $\frac{3}{4}$ of the multiplicand, a part denoted by the multiplying fraction.

Multiplication, therefore, when applied to fractions, does not always imply increase, as in whole numbers; for, when the multiplier is less than *unity*, it will always require the product to be less than the *multiplicand*, to which it would be equal if the multiplier were 1.

There are two operations, a division and a multiplication. But it is matter of indifference, as it respects the *result*, which of these operations precedes the other, for $36\times3\div4=27$, the same as $36\div4\times3=27$.

Hence, *To multiply by a fraction*, we have this

RULE.

Divide the multiplicand by the denominator of the multi-

Questions. — **¶ 76.** Why must 36 be multiplied by $\frac{3}{4}$? How does the product compare with the multiplicand, and why? Give the rule.

plying fraction, and multiply the quotient by the numerator; or, when there would be a remainder by division, first multiply by the numerator, and divide the product by the denominator.

2. What is the product of 90 multiplied by $\frac{1}{2}$? *Ans.* 45.

3. Multiply 369 by $\frac{2}{3}$.

4. Multiply 45 by $\frac{7}{10}$. *Product*, $31\frac{1}{2}$.

5. Multiply 210 by $\frac{4}{9}$.

6. Multiply 1326 by $\frac{2}{11}$. *Prod.* $241\frac{1}{11}$.

NOTE. — As either factor may be the multiplier, (¶ 21,) we may multiply by the whole number, making the fraction the multiplicand. Hence, the examples in this and ¶ 75, may be performed by the same rule

¶ 77. 1. At 40 dollars for 1 acre of land, what will $\frac{4}{5}$ of an acre cost? $40 \times \frac{4}{5} =$ how much?

In this example, the price of 1 acre, 40 dollars, is multiplied by the fraction of an acre, $\frac{4}{5}$. *Ans.* 32 dollars.

Hence, *When the price of unity is given, to find the cost of any quantity, less or more than unity,*

RULE.

Multiply the price by the quantity.

EXAMPLES FOR PRACTICE.

2. If a ship be worth 1367 dollars, what is $\frac{2}{9}$ of it worth? *Ans.* $303\frac{7}{9}$ dollars.

3. What cost $\frac{11}{13}$ of a ton of butter, at 225 dollars per ton? *Ans.* $190\frac{5}{13}$ dollars.

¶ 78. III. *To multiply one fraction by another.*

1. At $\frac{4}{5}$ of a dollar for one bushel of corn, what will $\frac{2}{3}$ of a bushel cost? $\frac{4}{5} \times \frac{2}{3} =$ how much?

SOLUTION. — The price of one bushel, $\frac{4}{5}$, is to be multiplied by the fraction of a bushel, $\frac{2}{3}$, (¶ 77.)

We first divide $\frac{4}{5}$ by 3, to get the price of $\frac{1}{3}$ of a bushel. This we can do by multiplying the denominator by 3, thus making the parts of a dollar only one third as large, (15ths,) while the same number, 4, is taken. $\frac{4}{5} \div 3 = \frac{4}{15}$ of a dollar, the price of one

---

Questions. — ¶ 77. Explain the first example. What two things are given, and what required? Rule.

third; and $\frac{4}{15} \times 2 = \frac{8}{15}$ of a dollar, price of $\frac{2}{3}$ of a bushel *Ans.* $\frac{8}{15}$ of a dollar.

The denominator 5 of the multiplicand is multiplied by 3, the denominator of the multiplier, and 4, the numerator of the multiplicand, by 2, the numerator of the multiplier.

Hence, *To multiply one fraction by another,*

RULE.

Multiply the denominators together for the *denominator* of the product, and the numerators for the *numerator* of the product.

By this process the multiplicand is divided by the denominator of the multiplying fraction, and the quotient multiplied by the numerator, as in ¶ 76.

EXAMPLES FOR PRACTICE.

2. Multiply $\frac{7}{8}$ by $\frac{6}{7}$. Multiply $\frac{9}{10}$ by $\frac{2}{7}$. *Product,* $\frac{9}{35}$.

3. At $\frac{6}{25}$ of a dollar a yard, what will $\frac{7}{8}$ of a yard of cloth cost?

4. At $6\frac{3}{8}$ dollars per barrel for flour, what will $\frac{7}{16}$ of a barrel cost?

NOTE.—Mixed numbers must be reduced to improper fractions.

$6\frac{3}{8} = \frac{51}{8}$; then $\frac{51}{8} \times \frac{7}{16} = \frac{357}{128} = 2\frac{101}{128}$ dollars, *Ans.*

5. At $\frac{5}{6}$ of a dollar per yard, what cost $7\frac{3}{4}$ yards? *Ans.* $6\frac{11}{24}$ dollars.

6. At $2\frac{1}{4}$ dollars per yard, what cost $6\frac{5}{8}$ yards? *Ans.* $14\frac{29}{32}$ dollars.

¶ 79. 1. What will $\frac{2}{3}$ of a yard of cloth cost at $\frac{4}{5}$ of a dollar per yard?

SOLUTION.—We multiply the price of 1 yard, $\frac{4}{5}$, by $\frac{2}{3}$, the fraction of a yard, (¶ 78.) Getting the price of $\frac{2}{3}$ of a yard is getting $\frac{2}{3}$ of $\frac{4}{5}$ of a dollar. $\frac{2}{3}$ of $\frac{4}{5}$ is an expression called a compound fraction, (¶ 65.) The reducing of a compound fraction to a simple one is, then, the same as multiplying fractions together.

2. What is $\frac{3}{5}$ of $\frac{7}{8}$? $\frac{3}{7}$ of $\frac{2}{13}$? $\frac{1}{2}$ of $\frac{1}{3}$?

3. How much is $\frac{2}{3}$ of $\frac{4}{5}$ of $\frac{6}{9}$?

$\frac{2}{3}$ of $\frac{4}{5}$ we have found to be $\frac{8}{15}$, and $\frac{8}{15}$ of $\frac{6}{9}$ by the above rule is $\frac{16}{45}$, *Ans.* Hence,

---

Questions.—¶ 78. What is the first operation, Ex. 1? Whence its propriety? Second operation? Rule. What is done by the first operation required in the rule? by the second operation?

The word *of* between fractions implies their continued multiplication. If there are more than two fractions, we multiply together the several numerators and the several denominators.

4. How much is $\frac{5}{13}$ of $\frac{2}{9}$ of $\frac{3}{4}$ of $\frac{6}{8}$? *Ans.* $\frac{180}{3744}=\frac{5}{104}$.
5. How much is $\frac{4}{5}$ of $\frac{2}{3}$ of $\frac{7}{8}$ of $\frac{3}{4}$? *Ans.* $\frac{7}{20}$.

¶ 80. 1. How much is $\frac{9}{10}$ of $\frac{5}{6}$ of $\frac{2}{4}$ of $\frac{2}{3}$?

Since the numerators are to be multiplied together, and their product to be divided by the product of the denominators,

The operation may be shortened by Cancelation, (¶ 60.)

OPERATION.

$$\overset{\cancel{3}}{\frac{9}{\underset{2}{\cancel{10}}}} \text{ of } \frac{\cancel{5}}{\underset{2}{\cancel{6}}} \text{ of } \frac{\cancel{2}}{\underset{\cancel{2}}{\cancel{4}}} \text{ of } \frac{2}{\cancel{3}} = \frac{1}{4}.$$

SOLUTION.—Performing this operation, as described in ¶ 60, we have canceled all the factors of the numerators, and have the factors 2, 2, remaining of the denominators. But the numerator $2=2\times1$, the numerator $5=5\times1$, $3=3\times1$, &c., and we have in reality the factors 1, 1, 1, and 1, left in the numerators. $1\times1\times1\times1=1$ for the new numerator, and $2\times2\times1\times1=4$ for the new denominator. Hence, when all the factors excepting the 1's in the numerators or denominators cancel, the new numerator or denominator, as the case may be, will be 1.

EXAMPLES FOR PRACTICE IN CANCELATION.

2. $\frac{3}{4}$ of $\frac{4}{5}$ of $\frac{6}{7}$ of $\frac{5}{6}$ of $\frac{9}{10}$ of $\frac{7}{8}$ of $\frac{8}{9}$ = how much? *Ans.* $\frac{3}{10}$.
3. What is the continual product of 7, $\frac{1}{2}$, $\frac{5}{7}$ of $\frac{3}{8}$ and $3\frac{1}{9}$?

NOTE.—The integer 7 may be reduced to the form of an improper fraction, by writing a unit under it for a denominator, thus, $\frac{7}{1}$.

Ans. $2\frac{11}{12}$.

4. What is the continued product of 3, $\frac{2}{5}$, $\frac{5}{9}$ of $\frac{3}{4}$, $2\frac{5}{7}$, and $\frac{11}{12}$ of $\frac{6}{7}$ of $\frac{4}{5}$? *Ans.* $\frac{209}{245}$.
5. Reduce $\frac{3}{4}$ of $\frac{4}{5}$ of $\frac{6}{7}$ of $\frac{5}{6}$ of $22\frac{1}{9}$ to a simple fraction. *Ans.* $9\frac{10}{21}$.
6. A horse consumed $\frac{6}{7}$ of $\frac{1}{3}$ of 8 tons of hay in one winter; how many tons did he consume? *Ans.* $2\frac{2}{7}$ tons.
7. Reduce $\frac{7}{8}$ of $\frac{3}{4}$ of $\frac{6}{7}$ of $\frac{5}{6}$ of $\frac{8}{9}$ of 1 to a simple fraction. *Ans.* $\frac{5}{12}$.

Questions.—**¶ 79.** How does it appear that we have a compound fraction in the first example? What does the word *of* between fractions imply? What is done when there are more than two fractions?

¶ 80. Why can cancelation be applied to the multiplication of fractions? Explain the process. What is done with integers when occurring with fractions?

¶ 80. (2.) PROMISCUOUS EXAMPLES IN THE MULTIPLICATION OF FRACTIONS.

1. At $\frac{3}{4}$ dollars per yard, what cost 4 yards of cloth? —— 5 yards? —— 6 yards? —— 8 yards? —— 20 yards?
Ans. to the last, 15 dollars.

2. Multiply 148 by $\frac{1}{2}$. —— by $\frac{7}{8}$. —— by $\frac{3}{20}$. —— by $\frac{3}{10}$
Last product, $44\frac{4}{10}$.

3. If $2\frac{9}{10}$ tons of hay keep 1 horse through the winter, how much will it take to keep 3 horses the same time? —— 7 horses? —— 13 horses? *Ans.* to last, $37\frac{7}{10}$ tons.

4. What will $8\frac{7}{12}$ barrels of cider come to, at 3 dollars per barrel? *Ans.* $25\frac{3}{4}$ dollars.

5. At $14\frac{3}{4}$ dollars per cwt., what will be the cost of 147 cwt.? *Ans.* $2168\frac{1}{4}$ dollars.

6. A owned $\frac{3}{5}$ of a ticket; B owned $\frac{6}{15}$ of the same; the ticket was so lucky as to draw a prize of 1000 dollars; what was each one's share of the money?
Ans. A's share, 600 dollars; B's share, 400 dollars.

7. Multiply $\frac{1}{2}$ of $\frac{2}{3}$ by $\frac{3}{4}$ of $\frac{4}{5}$. *Product*, $\frac{1}{5}$.
8. Multiply $7\frac{1}{2}$ by $2\frac{1}{15}$. " $15\frac{1}{2}$.
9. Multiply $\frac{7}{8}$ by $2\frac{2}{3}$. " $2\frac{1}{3}$.
10. Multiply $\frac{3}{4}$ of 6 by $\frac{2}{9}$. " 1.
11. Multiply $\frac{3}{4}$ of 2 by $\frac{1}{2}$ of 4. " 3.
12. Multiply continually together $\frac{1}{9}$ of 8, $\frac{2}{3}$ of 7, $\frac{3}{8}$ of 9, and $\frac{1}{7}$ of 10. *Product*, 20.
13. Multiply 1000000 by $\frac{5}{9}$. *Product*, $555555\frac{5}{9}$.

Division of Fractions.

¶ 81. I. *To divide a fraction by a whole number.*

1. If 2 yards of cloth cost $\frac{2}{3}$ of a dollar, what does 1 yard cost? how much is $\frac{2}{3}$ divided by 2?

2. If a cow consume $\frac{3}{4}$ of a bushel of meal in 3 days, how much is that per day? $\frac{3}{4} \div 3 =$ how much?

3. If a boy divide $\frac{4}{8}$ of an orange among 2 boys, how much will he give each one? $\frac{4}{8} \div 2 =$ how much?

4. A boy bought 5 cakes for $\frac{10}{12}$ of a dollar; what did 1 cake cost? $\frac{10}{12} \div 5 =$ how much?

5. If 2 bushels of apples cost $\frac{2}{6}$ of a dollar, what is that per bushel?

1 bushel is the half of 2 bushels; the half of $\frac{2}{6}$ is $\frac{1}{6}$.
Ans. $\frac{1}{6}$ dollar

6. If 3 horses consume $\frac{12}{13}$ of a ton of hay in a month, what will 1 horse consume in the same time?

Solution.—$\frac{12}{13}$ are 12 parts; if 3 horses consume 12 such parts in a month, as many times as 3 are contained in 12, so many parts 1 horse will consume. *Ans.* $\frac{4}{13}$ of a ton.

Hence, *we divide a fraction by dividing the numerator without changing the denominator*, taking a less number of parts of the same size.

7. A woman would divide $\frac{3}{4}$ of a pie equally between her two children; how much does each receive?

Solution.—She cannot divide the 3 pieces into 2 equal parts and leave them all whole. But as the denominator 4 shows into how many parts the pie is cut, multiplying it by 2 is equivalent to cutting the pie into twice as many, or 8 pieces of half the size. That is, we may cut each piece into 2 equal parts, and give 1 of them to each child, who will then have the same number of pieces, 3, only half as large. *Ans.* $\frac{3}{8}$.

Hence, *a fraction is divided by multiplying its denominator without changing its numerator*, as the parts are made smaller, while the same number is taken.

Multiplying the denominator, then, which is the divisor, has the same effect on the fraction as dividing the numerator, which is the dividend, (¶ 57.)

Note 1.—By comparing this, and ¶ 75, we shall see that where either term of a fraction is to be multiplied or divided, the contrary operation may be performed on the other term.

Hence, we have two *ways to divide a fraction by a whole number:*—

I. *Divide the numerator* by the whole number, (if it will contain it without a remainder,) and under the quotient write the denominator.—Otherwise,

II. *Multiply the denominator* by the whole number, and over the product write the numerator.

Questions.—**¶ 81.** How are $\frac{12}{13}$ divided by 4? What difficulty in dividing 3 pieces of pie among 2 children? How then may $\frac{3}{4}$ be divided? Why does multiplying the denominator divide the fraction? In what two ways is a fraction divided? Apply ¶ 57 to the operation. What appears from comparing this with ¶ 75? Repeat the rule. How do you divide a fraction by a composite number? How divide a mixed number? How, when the mixed number is large?

EXAMPLES FOR PRACTICE.

8. If 7 pounds of coffee cost $\frac{21}{25}$ of a dollar, what is that per pound? $\frac{21}{25} \div 7 =$ how much? *Ans.* $\frac{3}{25}$ of a dollar.

9. If $\frac{18}{20}$ of an acre produce 24 bushels, what part of an acre will produce 1 bushel? $\frac{18}{20} \div 24 =$ how much?

10. If 12 skeins of silk cost $\frac{10}{11}$ of a dollar, what is that a skein? $\frac{10}{11} \div 12 =$ how much?

11. Divide $\frac{8}{9}$ by 16.

NOTE 2. — When the divisor is a composite number, we can first divide by *one* component part, and the quotient thence arising by the *other*, (¶ 39.) Thus, in the last example, $16 = 8 \times 2$, and $\frac{8}{9} \div 8 = \frac{1}{9}$, and $\frac{1}{9} \div 2 = \frac{1}{18}$. *Ans.* $\frac{1}{18}$.

12. Divide $\frac{4}{10}$ by 12. Divide $\frac{7}{40}$ by 21. Divide $\frac{36}{49}$ by 24.

13. If 6 bushels of wheat cost $4\frac{7}{8}$ dollars, what is it per bushel?

NOTE 3. — The mixed number may be reduced to an improper fraction, and divided as before.

Ans. $\frac{39}{48} = \frac{13}{16}$ of a dollar, expressing the fraction in its *lowest* terms.

14. Divide $4\frac{11}{13}$ dollars by 9. *Quot.* $\frac{7}{13}$ of a dollar.

15. Divide $12\frac{6}{7}$ by 5. *Quot.* $\frac{18}{7} = 2\frac{4}{7}$.

16. Divide $14\frac{3}{4}$ by 8. *Quot.* $1\frac{27}{32}$.

17. Divide $184\frac{1}{2}$ by 7. *Ans.* $26\frac{5}{14}$.

NOTE 4. — When the mixed number is *large*, it will be most convenient, first, to divide the *whole* number, and then reduce the remainder to an improper fraction; and, after dividing, annex the quotient of the fraction to the quotient of the whole number; thus, in the last example, dividing $184\frac{1}{2}$ by 7, as in whole numbers, we obtain 26 integers with $2\frac{1}{2} = \frac{5}{2}$ remainder, and, dividing this by 7, we have $\frac{5}{14}$, and $26 + \frac{5}{14} = 26\frac{5}{14}$. *Ans.*

18. Divide $2786\frac{1}{4}$ by 6. *Ans.* $464\frac{3}{8}$.

19. How many times is 24 contained in $7646\frac{11}{24}$? *Ans.* $318\frac{347}{576}$.

20. How many times is 3 contained in $462\frac{1}{3}$? *Ans.* $154\frac{1}{9}$.

¶ 82. II. *To divide a whole number by a fraction.*

1. A man would divide 9 dollars among some poor persons giving them $\frac{3}{4}$ of a dollar each; how many will receive the money?

SOLUTION. — We wish to see how many times $\frac{3}{4}$ of a dollar is contained in 9 dollars. But as the divisor is 4ths, (25 cent. pieces,) we must reduce the dividend to 4ths, as both must be of the same denomination, (¶ 33;) thus, we multiply 9 by 4, to reduce it to 4ths, since there are 4 fourths in one dollar. Then, as many times as 3 fourths are contained in 36 fourths, so many persons will receive the money.

OPERATION.

$$\begin{array}{r} 9 \\ \underline{4} \\ \text{4ths of a dollar, } 3)\underline{36}\text{ 4ths of a dollar.} \\ 12 \textit{ persons.} \end{array}$$

We find the number to be 12 persons, a number greater than the dividend or number of dollars. Division, then, when applied to fractions, does not always imply decrease. The quotient is greater than the dividend when the divisor is less than 1, to which it is just equal when the divisor is 1.

Hence, *To divide a whole number by a fraction,*

RULE.

Multiply the dividend by the denominator of the dividing fraction, (thereby reducing the dividend to parts of the same magnitude as the divisor,) and divide the product by the numerator.

EXAMPLES FOR PRACTICE.

2. How many times is $\frac{1}{2}$ contained in 7? $7 \div \frac{1}{2} =$ how many?

3. How many times can I draw $\frac{1}{4}$ of a gallon of wine out of a cask containing 26 gallons?

4. Divide 3 by $\frac{3}{4}$. —— 6 by $\frac{2}{3}$. —— 10 by $\frac{2}{5}$.

5. If a man drink $\frac{9}{16}$ of a quart of beer a day, how long will 3 gallons last him? *Ans.* $21\frac{1}{3}$ days.

6. If $2\frac{3}{4}$ bushels of oats sow an acre, how many acres will 22 bushels sow? $22 \div 2\frac{3}{4} =$ how many times?

NOTE.—Reduce the mixed number to an improper fraction, $2\frac{3}{4} = \frac{11}{4}$. *Ans.* 8 acres

7. How many times is $\frac{36}{5}$ contained in 6? *Ans.* $\frac{5}{6}$ of 1 time.

8. How many times is $8\frac{5}{9}$ contained in 53? *Ans.* $6\frac{15}{77}$ times.

Questions.—¶ **82.** How is the principle that the divisor and dividend must be of the same denomination applied to the first example When is the quotient greater than the dividend, and when equal to it Give the rule for dividing a whole number by a fraction.

¶ 83. 1. At $\frac{2}{3}$ of a dollar per yard, how much cloth can be bought for 12 dollars?

SOLUTION. — As many times as $\frac{2}{3}$ of a dollar is contained in (or can be subtracted from) 12 dollars, so many yards can be bought.

Ans. 18 yards.

Hence, *When the price of unity and the price of any quantity are given, to find the quantity,*

RULE.

Divide the price of the quantity by the price of unity.

EXAMPLES FOR PRACTICE.

2. At $4\frac{2}{5}$ dollars a yard, how many yards of cloth may be bought for 37 dollars? *Ans.* $8\frac{9}{22}$ yards.

3. At $\frac{96}{103}$ of a dollar a pound, how many pounds of tea may be bought for 84 dollars? *Ans.* $90\frac{1}{8}$ pounds.

4. At $\frac{5}{6}$ of a dollar for building 1 rod of stone wall, how many rods may be built for 87 dollars? $87 \div \frac{5}{6} =$ how many times? *Ans.* $104\frac{2}{5}$ rods.

¶ 84. III. *To divide one fraction by another.*

1. At $\frac{2}{9}$ of a dollar per bushel, how many bushels of oats can be bought for $\frac{5}{6}$ of a dollar?

SOLUTION. — We are to divide $\frac{5}{6}$ by $\frac{2}{9}$. (¶ 83.) To divide by a fraction we multiply the dividend by the denominator of the dividing fraction, and divide the product by the numerator. (¶ 82.)

$\frac{5}{6} \times 9 = \frac{45}{6}$, and $\frac{45}{6} \div 2 = \frac{45}{12} = 3\frac{3}{4}$ bushels, *Ans.*

Hence,

RULE.

Invert the divisor, and multiply together the two upper terms for a numerator, and the two lower terms for a denominator.

NOTE 1. — In the above rule it will be seen that the numerator of the dividend is multiplied by the denominator of the divisor, and thus the dividend is multiplied by this denominator, and the denominator of the dividend is multiplied by the numerator of the divisor, and thus the dividend is divided by this numerator, as in ¶ 82.

Questions. — ¶ **83.** What two things are given, and what required. Ex. 1? Rule.

¶ **84.** How do we divide by a fraction? How multiply a fraction How divide a fraction? Rule for dividing one fraction by another What thereby is done?

EXAMPLES FOR PRACTICE.

2. Divide $\frac{1}{2}$ by $\frac{1}{2}$. *Quot.* 1. Divide $\frac{1}{2}$ by $\frac{1}{4}$. *Quot.* 2.
3. Divide $\frac{3}{4}$ by $\frac{1}{4}$. *Quot.* 3. Divide $\frac{7}{8}$ by $\frac{9}{10}$. *Quot.* $\frac{35}{36}$
4. If $4\frac{3}{5}$ pounds of butter serve a family 1 week, how many weeks will $36\frac{7}{8}$ pounds serve them?

NOTE. 2 The mixed numbers, it will be recollected, may be reduced to improper fractions. *Ans.* $8\frac{3}{184}$ weeks.

5. Divide $2\frac{1}{4}$ by $1\frac{1}{2}$. *Quot.* $1\frac{1}{2}$. Divide $10\frac{3}{8}$ by $2\frac{1}{8}$. *Quot.* $4\frac{15}{17}$.
6. How many times is $\frac{1}{10}$ contained in $\frac{2}{5}$? *Ans.* 4 times.
7. How many times is $\frac{3}{7}$ contained in $4\frac{7}{8}$? *Ans.* $11\frac{3}{8}$ times.
8. Divide $\frac{2}{3}$ of $\frac{3}{4}$ by $\frac{7}{8}$ of $\frac{1}{7}$. *Quot.* 4.

¶ 85. 1. If $\frac{7}{8}$ of a yard of cloth cost $\frac{3}{5}$ of a dollar, what is that per yard?

SOLUTION.—Had the price of several yards been given, we would divide it by the number of yards, to find the price of 1 yard, and, in like manner, we must divide the price of the fraction of a yard ($\frac{3}{5}$ of a dollar) by the fraction of a yard, ($\frac{7}{8}$,) to find the price of 1 yard. *Ans.* $\frac{24}{35}$ of a dollar per yard.

Hence, *When the price of any quantity less or more than unity is given, to find the price of unity,*

RULE.

Divide the price by the quantity.

EXAMPLES FOR PRACTICE.

2. At $\frac{7}{8}$ of a dollar for $3\frac{1}{2}$ bushels of apples, what does 1 bushel cost? *Ans.* $\frac{1}{4}$ of a dollar.

3. At $\frac{7}{8}$ of a dollar for $4\frac{2}{3}$ bushels of oats, what does 1 bushel cost? *Ans.* $\frac{3}{16}$ of a dollar.

¶ 85. (2.) *Reduction of complex to simple fractions.*

1. What simple fraction is equivalent to the complex fraction $\frac{\frac{2}{3}}{\frac{4}{5}}$?

SOLUTION.—Since the numerator of a fraction is a dividend of

Questions.—¶ 85. What two things are given, and what required, Ex. 1? Give the rule.

which the denominator is the divisor, we may divide $\frac{2}{3}$ by $\frac{4}{5}$, by the rule, ¶ 84. $\frac{2}{3} \div \frac{4}{5} = \frac{10}{12}$. *Ans.*

2. What simple fraction is equal to $\frac{4\frac{3}{4}}{7}$?

OPERATION.
$4\frac{3}{4} = \frac{19}{4}$, *and*
$\frac{19}{4} \div 7 = \frac{19}{28}$, *Ans.*

SOLUTION. — We reduce $4\frac{3}{4}$ to the improper fraction $\frac{19}{4}$, which we divide by 7, according to the rule, ¶ 81.

The above illustrations are sufficient to establish the following

RULE.

Reduce any mixed number which may occur in the complex fraction to the fractional form, or any compound fraction to a simple one, after which divide the numerator by the denominator, according to the ordinary rules for the division of fractions.

EXAMPLES FOR PRACTICE.

3. Reduce the complex fraction $\frac{7\frac{1}{5}}{\frac{3}{9}}$ to a simple one.

$\frac{324}{15} = \frac{108}{5} = 21\frac{3}{5}$

4. What is the value of $\frac{6\frac{3}{9}}{\frac{1}{3}}$? *Ans.* 19.

5. What simple fraction is equal to $\frac{3\frac{4}{7}}{9}$? *Ans.* $\frac{25}{63}$

6. What simple fraction is equal to $\frac{\frac{7}{12}}{4\frac{3}{5}}$? *Ans.* $\frac{35}{276}$

7. What simple fraction is equal to $\frac{7\frac{11}{16}}{9\frac{2}{3}}$? *Ans.* $\frac{369}{464}$

8. What simple fraction is equal to $\frac{10}{\frac{3}{8}}$? *Ans.* $\frac{80}{3} = 26\frac{2}{3}$.

9. What simple fraction is equal to $\frac{5}{7\frac{2}{3}}$? *Ans.* $\frac{15}{23}$.

10. What simple fraction is equal to $\frac{\frac{6}{24}}{16}$? *Ans.* $\frac{1}{64}$.

11. What simple fraction is equal to $\frac{\frac{3}{8} \text{ of } \frac{4}{5} \text{ of } 9}{3\frac{7}{10} \times 2\frac{1}{3}}$?

Ans. $\frac{144}{259}$.

Questions. — ¶ **85.** (2.) How is the complex fraction, Ex. 1, reduced to a simple one? why? Give the rule for reducing complex to simple fractions.

¶ 85. (3.) PROMISCUOUS EXAMPLES IN THE DIVISION OF FRACTIONS.

1. If 7 lb. of sugar cost $\frac{63}{100}$ of a dollar, what is it per pound? $\frac{63}{100} \div 7 =$ how much? $\frac{1}{7}$ of $\frac{63}{100}$ is how much?

2. At $\frac{1}{8}$ of a dollar for $\frac{3}{8}$ of a barrel of cider, what is that per barrel? *Ans* $\frac{1}{3}$ of a dollar.

3. If 4 pounds of tobacco cost $\frac{7}{8}$ of a dollar, what does 1 pound cost? *Ans.* $\frac{7}{32}$ doll.

4. If $\frac{7}{8}$ of a yard cost 4 dollars, what is the price per yard? *Ans.* $4\frac{4}{7}$ dollars.

5. If $14\frac{3}{8}$ yards cost 75 dollars, what is the price per yard? *Ans.* $5\frac{5}{23}$ dollars.

6. At $31\frac{1}{2}$ dollars for $10\frac{1}{2}$ barrels of cider, what is that per barrel? *Ans.* 3 dollars.

7. How many times is $\frac{3}{8}$ contained in 746? *Ans.* $1989\frac{1}{3}$

8. Divide $\frac{1}{2}$ of $\frac{2}{3}$ by $\frac{3}{4}$. *Quot.* $\frac{4}{9}$. Divide $\frac{7}{8}$ by $\frac{4}{7}$ of $\frac{2}{5}$. *Quot.* $3\frac{53}{64}$.

9. Divide $\frac{1}{2}$ of $\frac{4}{5}$ by $\frac{5}{6}$ of $\frac{2}{3}$. *Quot.* $\frac{18}{25}$.

10. Divide $\frac{1}{5}$ of 4 by $\frac{4}{15}$. *Quot.* 3.

11. Divide $4\frac{5}{9}$ by $\frac{5}{9}$ of 4. *Quot.* $2\frac{1}{20}$

12. Divide $\frac{5}{9}$ of 4 by $4\frac{5}{9}$. *Quot.* $\frac{20}{41}$.

13. Divide $\frac{8\frac{3}{7}}{9\frac{2}{5}}$ by $\frac{\frac{2}{3}}{7}$. *Quot.* $9\frac{39}{94}$.

¶ 86. Review of Common Fractions.

Questions. — What are *fractions?* Whence is it that the parts into which any thing or any number may be divided, take their name? What determines the *magnitude* of the parts? Why? How does increasing the denominator affect the value of the fraction? Increasing the numerator affects it how? How is an improper fraction reduced to a whole or mixed number? How is a mixed number reduced to an improper fraction? a whole number? How is a fraction reduced to its most *simple* or *lowest* terms? How is a common divisor found? (¶ 61.) the greatest common divisor? (¶ 62.) Whence the necessity of reducing fractions to a common denominator? When may one fraction be reduced to the denominator of another? What must the common denominator be? (¶ 69.) Give the first method of finding it, and the principles on which it is founded; the second method, and the principles. What is understood by a *multiple?* by a *common multiple?* by the *least* common multiple? What is the process of finding it? (¶ 72.) How are fractions added and subtracted? How many ways are there to multiply a fraction by a whole number? How does it appear, that *dividing the denominator multiplies the fraction?* How is a *mixed* number multiplied? What is implied in multiplying by a fraction? Of what operations does it consist? When the multiplier is *less* than a unit, what is the product compared with the multiplicand? What two things are given, and what required in ¶ 77? What in ¶ 83? What in ¶ 85 Explain the principle of multiplying one fraction by another. Of dividing one fraction by another. How do you multiply a mixed number by a mixed number? How does it appear, that in multiplying both terms of the fraction by the same number, the value of the fraction is not altered? How many, and what are the ways of dividing a fraction by a whole number? How does it appear that a *fraction is divided* by *multiplying its denominator?* How does *dividing* by a fraction differ from *multiplying* by a fraction? When the *divisor* is *less* than a unit, what is the quotient compared with the dividend? How do you divide a whole number by a fraction?

EXERCISES.

1. What is the amount of $\frac{5}{6}$ and $\frac{3}{8}$? —— of $\frac{1}{2}$ and $\frac{2}{3}$? —— of $12\frac{1}{2}$, $3\frac{2}{3}$, and $4\frac{3}{4}$? *Ans.* to the last, $20\frac{11}{12}$.

2. How much is $\frac{1}{4}$ less $\frac{1}{8}$? $\frac{3}{10}-\frac{1}{5}$? $\frac{3}{14}-\frac{8}{40}$? $14\frac{1}{2}-4\frac{1}{7}$? $6-4\frac{3}{8}$? $\frac{109}{110}-\frac{1}{2}$ of $\frac{2}{3}$ of $\frac{3}{4}$? *Ans.* to the last, $\frac{163}{220}$.

3. What fraction is that, to which if you add $\frac{2}{5}$ the sum will be $\frac{5}{6}$? *Ans.* $\frac{13}{30}$.

4. What number is that, from which if you take $\frac{3}{5}$ the remainder will be $\frac{1}{8}$? *Ans.* $\frac{29}{40}$.

5. What number is that, which being divided by $\frac{3}{4}$ the quotient will be 21? *Ans.* $15\frac{3}{4}$.

6. What number is that, which multiplied by $\frac{2}{3}$ produces $\frac{1}{4}$? *Ans.* $\frac{3}{8}$.

7. What number is that, from which if you take $\frac{2}{5}$ of itself the remainder will be 12? *Ans.* 20.

8. What number is that, to which if you add $\frac{2}{5}$ of $\frac{5}{3}$ of itself the whole will be 20? *Ans.* 12.

9. What number is that, of which 9 is the $\frac{2}{3}$ part? *Ans.* $13\frac{1}{2}$.

10. At $\frac{5}{8}$ of a dollar per yard, what costs $\frac{3}{4}$ of a yard of cloth? *Ans.* $\frac{15}{32}$ of a dollar.

11. At $5\frac{6}{7}$ dollars per barrel, what costs $18\frac{1}{2}$ barrels of flour? *Ans.* $108\frac{5}{14}$ dollars.

12. What costs 84 pounds of cheese, at $\frac{3}{22}$ of a dollar per pound? *Ans.* $11\frac{5}{11}$ dollars.

13. What cost 45 yards of gingham, at $\frac{5}{8}$ of a dollar per yard? *Ans.* $28\frac{1}{8}$ dollars.

14. What must be paid for $\frac{7}{16}$ of a yard of velvet, at 5 dollars per yard? *Ans.* $2\frac{3}{16}$ dollars.

15. If $\frac{7}{8}$ of a pound of tea cost $\frac{11}{15}$ of a dollar, what is that per pound? *Ans.* $\frac{88}{105}$ of a dollar.

16. If $7\frac{1}{5}$ barrels of pork cost $73\frac{4}{5}$ dollars, what is that per barrel? *Ans.* $10\frac{1}{4}$ dollars.

17. If 4 acres of land cost $82\frac{9}{16}$ dollars, what is that per acre? *Ans.* $20\frac{41}{64}$ dollars.

18. At $\frac{5}{12}$ of a dollar for $3\frac{1}{2}$ bushels of lime, what costs 1 bushel? *Ans.* $\frac{5}{42}$ of a dollar.

19. Paid $4\frac{3}{8}$ dollars for coffee, at $\frac{3}{20}$ of a dollar per pound, how many pounds did I buy? *Ans.* $29\frac{1}{6}$ pounds.

20. At $1\frac{2}{5}$ dollars per bushel, how much wheat may be bought for $82\frac{3}{4}$ dollars? *Ans.* $59\frac{3}{28}$ bushels.

21. If $8\frac{3}{4}$ yards of silk make a dress, and 9 dresses be made from a piece containing 80 yards, what will be the remnant left? *Ans.* $1\frac{1}{4}$ yards.

NOTE.—Let the pupil reverse and prove this, and the following example.

22. How many vests, containing $\frac{7}{8}$ of a yard each, can be made from 22 yards of vesting? what remnant will be left? *Ans.* 25 vests. Remnant, $\frac{1}{8}$ yard.

23. What number is that, which being multiplied by 15 the product will be $\frac{3}{4}$? *Ans.* $\frac{1}{20}$.

24. What is the product of $\dfrac{\frac{2}{8} \text{ of } \frac{7}{10}}{7}$ into $\dfrac{6}{3\frac{4}{11}}$?

Ans. $\frac{33}{740}$.

25. Which of the eleven numbers, 8, 9, 11, 12, 14, 15, 16, 18, 20, 22, 24, have all their factors the same as factors in 72? (¶ 61.)

NOTE.—The 72 must be resolved into the greatest number of factors possible, which are 2, 2, 2, 3, 3. In like manner, each of the other numbers must be resolved. *Ans.* 8, 9, 12, 18, 24.

26. What is the quotient of $\dfrac{\frac{3}{4} \text{ of } 9\frac{2}{3}}{\frac{8}{11} \text{ of } 16}$ divided by $\dfrac{\frac{2}{6} \text{ of } \frac{1}{3} \text{ of } 2\frac{1}{2}}{\frac{1}{4} \text{ of } 19\frac{1}{2}}$?

Ans. $10\frac{9569}{10240}$.

DECIMAL FRACTIONS.

¶ **87.** We have seen (¶ 69) that fractions having different denominators, as thirds, sevenths, elevenths, &c., cannot be added and subtracted until they are changed to *equal* fractions, having a *common denominator*—a process which is often somewhat tedious. To obviate this difficulty, Decimal fractions have been devised, founded on the Arabic system of notation.

If a unit or whole thing be divided into 10 equal parts, each of those parts will be 1 tenth, thus, $\frac{1}{10}$ of $1 = \frac{1}{10}$; and if each tenth be divided into 10 equal parts, the 10 tenths will make 100 parts, and each part will be 1 hundredth of a whole thing, thus, $\frac{1}{10}$ of $\frac{1}{10} = \frac{1}{100}$. In like manner, if each hundredth be divided into ten equal parts, the parts will be thousandths, $\frac{1}{10}$ of $\frac{1}{100} = \frac{1}{1000}$, and so on. Such are called Decimal Fractions, from the Latin *decem*, meaning ten.

Common fractions, then, are the common divisions of a unit or whole thing into halves, thirds, fourths, or *any* number of parts into which we *choose* to divide it.

Decimal fractions are the divisions of a unit or whole thing first into 10 equal parts, then each of these into 10 other equal parts, or hundredths, and each hundredth into 10 other equal parts, or thousandths, and so on.

The parts of a unit, thereby, increase and decrease in a ten fold ratio in the same manner as whole numbers.

The following examples will show the convenience of decimal fractions.

1. Add together $\frac{2}{10}$ and $\frac{15}{100}$.

Solution. — Since 1 tenth makes 10 hundredths, we may reduce the tenths to hundredths by annexing a cipher which, in effect, multiplies them by 10.

Thus, $\frac{2}{10} = 20$ hundredths, ($\frac{20}{100}$,)
and adding 15 hundredths, ($\frac{15}{100}$,)

we have 35 hundredths, ($\frac{35}{100}$.)

2. From $\frac{36}{100}$ take $\frac{187}{1000}$.

$\frac{36}{100} = 360$ thousandths.
187 thousandths.

173 thousandths.

Solution. — We reduce the 36 hundredths to thousandths by annexing a cipher to multiply it by 10. Then subtracting and borrowing as in whole numbers, we have left 173 thousandths, ($\frac{173}{1000}$.)

Questions. — **¶ 87.** What are fractions? What occasions the chief difficulty in operations with common fractions? To what has this difficulty led? What are decimal fractions? Why are they so called? What are the divisions and subdivisions of a unit in decimal fractions? and what are the parts of the 1st, 2d, &c., divisions called? With what system of notation do these divisions of a unit correspond? What is the law of increase and decrease in the Arabic system of notation? What, then, do you say of the increase of the whole numbers, and of the parts? How do these divisions of a unit in decimal fractions differ from the divisions of a unit in common fractions Give examples showing the superiority of decimal fractions.

NOTE. — The pupil will notice, that in thus reducing the fraction $\frac{36}{100}$, the $\frac{6}{100}$ makes 60 thousandths, and the $\frac{30}{100}$ makes 300 thousandths.

Notation of Decimal Fractions.

¶ 88. 1. Let it be required to find the amount of $325\frac{5}{10} + 16\frac{78}{100} + 4\frac{379}{1000} + \frac{25}{1000}$, and express the fractions decimally.

SOLUTION. — Since 1 hundred integers make 10 tens, 1 ten 10 units, 1 unit 10 tenths, 1 tenth 10 hundredths, &c., *decreasing uniformly* from left to right, we may write down the *numerators* of the fractions, placing tenths after units, hundredths after tenths, and so on, in this way indicating their values without expressing their denominators. We place a point (') called the *Decimal* point, or *Separatrix*, on the left of tenths to separate the fraction from units, or whole numbers.

hundreds.	tens.	units.	tenths.	hundredths.	thousandths.
3	2	5'	5		
	1	6'	7	8	
		4'	3	7	9
		'	0	2	5
3	4	6'	6	8	4

or,

hundreds.	tens.	units.	tenths.	hundredths.	thousandths.
3	2	5'	5	0	0
	1	6'	7	8	0
		4'	3	7	0
		'	0	2	5
3	4	6'	6	8	4

As 10 in each right hand make 1 in the next left hand column, the adding and carrying will be the same throughout as in whole numbers.

The reducing to a common denominator, it will be seen, is simply filling up the vacant places with ciphers, which are omitted in the first operation without affecting the result, each figure being written in its proper place.

The denominator to a decimal fraction, although not expressed, is always understood, and is 1 with as many ciphers annexed as there are places at the right hand of the point. Thus, '684 in the last example, is a decimal of 3 places; consequently 1, with 3 ciphers annexed, (1000,) is its proper denominator. Any decimal may be expressed in the form of a common fraction by writing under it its proper denominator. Thus, '684, expressed in the form of a common fraction, is $\frac{684}{1000}$.

Questions. — ¶ 88. Have decimal fractions numerators and denominators, and both expressed? How can you write the numerators so as to indicate the value of the fraction, without expressing the denominator? How can the denominator be known, if it is not expressed? What is the separatrix and its use? How can a decimal be expressed in form of a common fraction? How many, and what advantages have decimal over common fractions?

NOTE. — The decimal point can never be safely omitted in operations with decimals.

Decimal fractions have two advantages over common fractions.

First, — They are more readily reduced to a common denominator.

Second, — They may be added and subtracted like whole numbers, without the formal process of reducing them to a common denominator.

The names of the places to ten-millionths, and, generally how to read or write decimal fractions, may be seen from the following

TABLE.

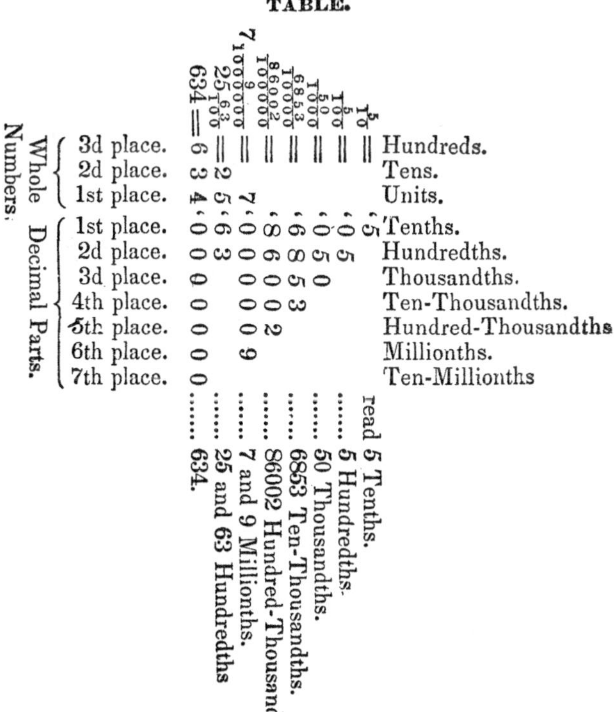

		634 =	$25\frac{63}{100}$ =	$7\frac{9}{1000000}$ =	$\frac{86002}{100000}$ =	$\frac{6853}{10000}$ =	$\frac{50}{1000}$ =	$\frac{5}{100}$ =	$\frac{5}{10}$ =	
Whole Numbers.	3d place.	6								Hundreds.
	2d place.	3	2							Tens.
	1st place.	4	5	7						Units.
Decimal Parts.	1st place.	,0	,6	,0	,8	,6	,0	,0	,5	Tenths.
	2d place.	0	3	0	6	8	5	5		Hundredths.
	3d place.	0		0	0	5	0			Thousandths.
	4th place.	0		0	0	3				Ten-Thousandths.
	5th place.	0		0	2					Hundred-Thousandths
	6th place.	0		9						Millionths.
	7th place.	0								Ten-Millionths
		...... 634.	...... 25 and 63 Hundredths	...... 7 and 9 Millionths.	...... 86002 Hundred-Thousandths.	...... 6853 Ten-Thousandths.	...... 50 Thousandths.	...... 5 Hundredths.	read 5 Tenths.	

¶ 89. *To read Decimals.*—As every fraction has a numerator and a denominator, to read the decimal fraction, of which the denominator is not expressed, *requires two enumerations*,—one from left to right to ascertain the denominator, that is, the name or denomination of the parts, and another from right to left, to ascertain the numerator, that is, the number of parts.

Take, for instance, the fraction '00387. We begin at the first place on the right hand of the decimal point and say, as in the table, tenths, hundredths, &c., to the last figure, which we ascertain to be hundred-thousandths, and that is the name or denomination of the fraction. Then, to know how many there are of this denomination, that is, to determine the numerator, we begin as in whole numbers, and say units, (7,) tens, (8,) hundreds, (3,) which being the highest significant figure, we proceed no further; and find that we have 387 hundred-thousandths, ($\frac{387}{100000}$,) the numerator being 387.

In this way a mixed number may be read as a fraction. Take 25'634. Beginning at the first place at the right of the point we have tenths, hundredths, thousandths, (the lowest denomination;) then beginning at the right with 4, we say, units, tens, &c., as in whole numbers, and find that we have 25634 thousandths, which, expressed as a common fraction, is $\frac{25634}{1000}$.

¶ 90. *To write Decimal Fractions.*

I. Write the given decimal in such a manner that each figure contained in it may occupy the place corresponding to its value.

II. Fill the vacant places, if any, with ciphers, and put the decimal point in its proper place.

Forty-six and seven tenths $= 46\frac{7}{10} =$ 46'7.

Write the following numbers in the same manner:

Eighteen and thirty-four hundredths.
Fifty-two and six hundredths.
Nineteen and four hundred eighty-seven thousandths.
Twenty and forty-two thousandths.
One and five thousandths.
135 and 3784 ten-thousandths.
9000 and 342 ten-thousandths.

Questions.—¶ **89.** To read decimals requires what? how made and for what purposes? How may a mixed number be read as a fraction?

¶ **90** What is the rule for writing decimal fractions?

10000 and 15 ten-thousandths.
974 and 102 millionths.
320 and 3 tenths, 4 hundredths, and 2 thousandths
500 and 5 hundred-thousandths.
47 millionths.
Four hundred and twenty-three thousandths.

Reduction of Decimal Fractions

¶ 91. The value of every figure is determined by its place from *units.* Consequently, ciphers annexed to decimals do *not* alter their value, since every significant figure continues to possess the same place from unity. Thus, '5, '50, '500, are all of the same value, each being equal to $\frac{5}{10}$, or $\frac{1}{2}$.

But every cipher prefixed to a decimal *diminishes* it tenfold, by removing the significant figures one place further from unity, and consequently making each part only one tenth as large. Thus, '5, '05, '005, are of different value, '5 being equal to $\frac{5}{10}$, or $\frac{1}{2}$; '05 being equal to $\frac{5}{100}$, or $\frac{1}{20}$; and '005 being equal to $\frac{5}{1000}$, or $\frac{1}{200}$.

A whole number is reduced to a decimal by annexing ciphers; to tenths by annexing 1 cipher, since this is multiplying by 10; to hundredths by annexing two ciphers, &c. Thus, if 1 cipher be annexed to 25 it will be 25'0, (250 tenths;) if 2 ciphers, it will be 25'00, (2500 hundredths.)

Several numbers may be reduced to decimals, having the same or a common denominator, by annexing ciphers till all have the same number of decimal places. Thus, 15'7, '75, 12'183, 9'0236 and 17' are reduced to ten thousandths, the lowest denominator contained in them, as follows:

15'7 = 15'7000, annexing three ciphers
'75 = '7500, " two "

Questions.—¶ **91.** How is the value of every decimal figure determined? How do ciphers at the *left* of a decimal affect its value? at the right, how? In the fraction '02613, what is the value of the 4? of the 2? How does the 0 affect the value of the fraction? In the fraction '15012 is the value of each significant figure affected by the 0? if not, point out the difference, and wherefore? If from the fraction '8634 we withdraw the 6, leaving the fraction to consist of the other three figures only, how much should we deduct from its value? Demonstrate by some process that you are right. How may integers be reduced to decimals? Reduce 46 to thousandths; how many thousandths does the number make? How may several numbers be reduced to decimals having a common denominator? To what denominator should they all be reduced?

12'183 = 12'1830, annexing one cipher.
9'0236 = 9'0236, already ten-thousandths.
17' = 17'0000, annexing four ciphers.

NOTE. — All the numbers should be reduced to the denominator of the one having the greatest number of decimal places.

EXAMPLES.

1. Reduce 7'25, 14'082, 2'3, '00083, and 25 to a common denominator.

2. Reduce 2'1, 3'02, '425, 32'98762, and '3000001, to a common denominator.

¶ **92.** *To change or reduce Common to Decimal Fractions.*

1. A man has $\frac{4}{5}$ of a barrel of flour; what is that, expressed in decimal parts?

As many times as the denominator of a fraction is contained in the numerator, so many whole ones are contained in the fraction. We can obtain no whole ones in $\frac{4}{5}$, because the denominator is not contained in the numerator. We may, however, reduce the numerator to *tenths*, by annexing a cipher to it, (which, in effect, is multiplying it by 10,) making 40 tenths, or 4'0. Then, as many times as the denominator, 5, is contained in 40, so many *tenths* are contained in the fraction. 5 into 40 eight times, and no remainder.

Ans. '8 of a bushel.

2. Express $\frac{3}{4}$ of a dollar in decimal parts.

OPERATION.

```
          Num.
Denom. 4)3'0('75 of a dollar.
         28
         --
          20
          20
```

The numerator, 3, reduced to tenths, is $\frac{30}{10}$, 3'0, which, divided by the denominator, 4, the quotient is 7 tenths, and a remainder of 2. This remainder must now be reduced to *hundredths* by annexing another cipher, making 20 hundredths. Then, as many times as the denominator, 4, is contained in 20, so many *hundredths* also may be obtained. 4 into 20 5 times, and no remainder. $\frac{3}{4}$ of a dollar, therefore, reduced to decimal parts is 7 tenths and 5 hundredths; that is, '75 of a dollar.

Questions. — ¶ **92.** To what is the value of every fraction equal (¶ 64.) How is a common fraction reduced to a decimal? Of how many places must the quotient consist? When there are not so many places how is the deficiency to be supplied? Repeat the rule. What is the course of reasoning advanced to establish this rule?

3. Reduce $\frac{4}{66}$ to a decimal fraction.

The numerator must be reduced to *hundredths*, by annexing two ciphers, before the division can begin.

66) 4'00 ('0606 +, *the Answer.*
    396
    ———
      400
      396
      ——
        4

As there can be no *tenths*, a cipher must be placed in the quotient, in tenths' place.

NOTE. — $\frac{4}{66}$ cannot be reduced *exactly;* for, however long the division be continued, there will still be a remainder.* It is sufficiently exact for most purposes, if the decimal be extended to three or four places.

---

* Decimal figures, which *continually repeat,* like '06, in this example, are called *Repetends,* or *Circulating* Decimals. If only *one figure* repeats, as '3333 or '7777, &c., it is called a *single repetend.* If *two* or *more figures* circulate alternately, as '060606, '234234234, &c., it is called a *compound repetend.* If other figures arise *before* those which circulate, as '743333, '143010101, &c., the decimal is called a *mixed repetend.*

A single repetend is denoted by writing only the *circulating figure* with a point over it: thus, '$\dot{3}$, signifies that the 3 is to be continually repeated, forming an *infinite* or *never-ending series* of 3s.

A compound repetend is denoted by a point over the *first* and *last* repeating *figures:* thus, '$\dot{2}3\dot{4}$ signifies that 234 is to be continually repeated.

It may not be amiss, here, to show how the *value* of any *repetend* may be found, or, in other words, how it may be *reduced to its equivalent vulgar fraction.*

If we attempt to reduce $\frac{1}{9}$ to a *decimal,* we obtain a continual repetition of the figure 1: thus, '11111, that is, the *repetend* '$\dot{1}$. The value of the repetend '$\dot{1}$, then, is $\frac{1}{9}$; the value of '222, &c., the repetend '$\dot{2}$, will evidently be *twice* as much, that is, $\frac{2}{9}$. In the same manner, $\dot{3} = \frac{3}{9}$, and '$\dot{4} = \frac{4}{9}$, and '$\dot{5} = \frac{5}{9}$, and so on to $\dot{9}$, which $= \frac{9}{9} = 1$.

1. What is the value of '$\dot{8}$? *Ans.* $\frac{8}{9}$.

2. What is the value of '$\dot{6}$? *Ans.* $\frac{6}{9} = \frac{2}{3}$. What is the value of '$\dot{3}$? —— of '$\dot{7}$? —— of '$\dot{4}$? —— of '$\dot{5}$? —— of '$\dot{9}$? —— of '$\dot{1}$?

If $\frac{1}{99}$ be reduced to a decimal, it produces '010101, or the repetend '$\dot{0}\dot{1}$. The repetend '$\dot{0}\dot{2}$, being 2 times as much, must be $\frac{2}{99}$ and '$\dot{0}\dot{3} = \frac{3}{99}$, and '$\dot{4}\dot{8}$, being 48 times as much, must be $\frac{48}{99}$, and '$\dot{7}\dot{4} = \frac{74}{99}$, &c.

If $\frac{1}{999}$ be reduced to a decimal, it produces '$\dot{0}0\dot{1}$; consequently, '$\dot{0}0\dot{2} = \frac{2}{999}$, and '$\dot{0}3\dot{7} = \frac{37}{999}$, and '$\dot{4}2\dot{5} = \frac{425}{999}$, &c. As this principle will apply to any number of places, we have this *general* RULE *for reducing a circulating decimal to a common fraction.*

Make the *given repetend* the *numerator,* and the *denominator* will be as many 9s as there are *repeating figures.*

From the foregoing examples we may deduce the following general

RULE.

To reduce a common to a decimal fraction.—Annex one or more ciphers, as may be necessary, to the numerator, and divide it by the denominator. If then there be a remainder, annex another cipher, and divide as before, and so continue to do, so long as there shall continue to be a remainder, or until the fraction shall be reduced to any necessary degree of exactness.

The quotient will be the decimal required, which must consist of as many decimal places as there are ciphers annexed to the numerator; and, if there are not so many figures in the quotient, the deficiency must be supplied by prefixing ciphers.

EXAMPLES FOR PRACTICE.

4. Reduce $\frac{1}{2}$, $\frac{1}{4}$, $\frac{12}{480}$, and $\frac{9}{1129}$ to decimals.
Ans. ‘5; ‘25; ‘025; ‘00797 +.
5. Reduce $\frac{27}{39}$, $\frac{3}{1000}$, $\frac{5}{1785}$, and $\frac{11}{60006}$ to decimals.
Ans. ‘692 +; ‘003; ‘0028 +; ‘000183 +.
6. Reduce $\frac{478}{942}$, $\frac{10}{367}$, $\frac{16}{8600}$ to decimals.
7. Reduce $\frac{4}{9}$, $\frac{5}{99}$, $\frac{8}{999}$, $\frac{1}{3}$, $\frac{2}{3}$, $\frac{1}{11}$, $\frac{4}{11}$, $\frac{1}{909}$ to decimals.
8. Reduce $\frac{1}{8}$, $\frac{3}{8}$, $\frac{5}{8}$, $\frac{1}{5}$, $\frac{2}{5}$, $\frac{3}{5}$ $\frac{4}{5}$, $\frac{1}{20}$, $\frac{1}{25}$, $\frac{3}{75}$ to decimals.

Federal Money.

¶ 93. *Federal Money is the currency of the United States*

The unit of English money is the pound sterling, which is divided into 20 equal parts, (twentieths,) called shillings,

---

3. What is the vulgar fraction equivalent to ‘$\dot{7}0\dot{4}$? *Ans.* $\frac{704}{999}$.

4. What is the value of ‘$\dot{0}0\dot{3}$? —— ‘$\dot{0}1\dot{4}$? —— ‘$\dot{3}2\dot{4}$? —— ‘$\dot{0}102\dot{1}$? —— ‘$\dot{2}46\dot{3}$? —— ‘$\dot{0}0210\dot{3}$? *Ans. to last,* $\frac{701}{333333}$.

5. What is the value of ‘$4\dot{3}$?
In this fraction, the repetend begins in the second place, or place of hundredths. The first figure, 4, is $\frac{4}{10}$, and the *repetend*, 3, is $\frac{3}{9}$ of $\frac{1}{10}$, that is, $\frac{3}{90}$; these two parts must be added together. $\frac{4}{10}+\frac{3}{90}=\frac{39}{90}=\frac{13}{30}$, *Ans.* Hence, to find the value of a *mixed repetend,*—Find the value of the two parts, *separately*, and add them together.

6. What is the value of ‘$15\dot{3}$? $\frac{15}{100}+\frac{3}{900}=\frac{138}{900}=\frac{23}{150}$, *Ans.*

7. What is the value of ‘$004\dot{7}$? *Ans.* $\frac{43}{9000}$.

8. What is the value of ‘138? —— ‘$1\dot{6}$? —— ‘$4\dot{1}2\dot{3}$?
It is plain, that circulates may be added, subtracted, multiplied, and divided by first reducing them to their equivalent vulgar fractions.

each shilling is divided into 12 parts called pence; a penny being $\frac{1}{240}$ of a pound. Each penny again is divided into 4 parts called farthings, a farthing being $\frac{1}{960}$ of a pound. These divisions, therefore, are like those of common fractions, and the same difficulties occur in operations with English money as with common fractions.

The unit of Federal money is the Dollar, divided into 10 parts called *dimes*, from a French word meaning *tenth* (of a dollar); each dime into 10 parts called *cents*, from the French for *hundredth* (of a dollar); and each cent into 10 parts called *mills*, from the French for *thousandth* (of a dollar). These divisions of the money unit are like those of decimal fractions.

Our money, then, has this advantage over the English, viz., that operations in it are as in whole numbers, and we shall therefore consider it in connection with Decimals.

The denominations of Federal money are eagles, dollars, dimes, cents, and mills.

TABLE.

10 *mills*	make	1 *cent.*
10 *cents* (= 100 *mills*)		1 *dime.*
10 *dimes* (= 100 *cents* = 1000 *mills*)		1 *dollar.*
10 *dollars*		1 *eagle.*

NOTE. — Coin is a piece of metal stamped with certain impressions to give it a legal value, and also to serve as a guarantee for its weight and purity.

The mill is so small that it is not usually regarded in business. The eagle is merely the name of a gold coin worth 10 dollars. Dimes are read as 10s of cents. Federal money, then, is calculated in dollars and cents, and accounts are kept in these denominations.

A character, $, which may be regarded a contraction of U. S., placed before a number, signifies that it is Federal, or U. S. money.

Questions. — ¶ **93.** What is Federal money? What is the unit of English money? What are its denominations? and what are they like? What is the unit of Federal money? how divided? and whence the names of the divisions? What advantages has Federal over English money? Repeat the table. What is the eagle? How are dimes read? In what then is Federal money calculated, and accounts kept? What is coin? What is the character for U. S. money, and where placed? Where is the decimal point placed? Where and how many are the places for cents? for mills? Why more places for cents than for mills? If the sum be but 8 cents how may it be written? if three mills only, how? How are 5 mills usually written?

As the dollar is the *unit* of Federal money, the decimal point is placed at the right hand of dollars; and since dimes (tenths) and cents (hundredths) are read together as cents the first two places at the right hand of the point express cents, and the third, mills, (thousandths.) Thus, 25 dollars 78 cents, and 6 mills, are written, $25'786.

If there be no dimes, (tenths, or 10s of cents,) that is, if the cents are less than 10, a cipher is put in the place of tenths; thus, 8 cents are written $'08; and if there are only mills, ciphers must be put in the place of tenths and hundredths; thus, 5 mills are written $'005. But 5 mills are usually expressed as *half a cent;* thus, 12 cents 5 mills, are written 12½ cents, or $'12½.

Reduction of Federal Money.

¶ **94.** It is evident that *dollars are reduced to cents* in the same manner as whole numbers are reduced to *hundredths,* by annexing two ciphers;

To mills or thousandths, by annexing three ciphers.

On the contrary,

Mills are reduced to cents by cutting off the right hand figure;

To dollars, by cutting off three figures from the right, which is dividing by 1000, (¶ 41.)

Cents are reduced to dollars by cutting off two figures from the right, which is dividing by 100.

EXAMPLES.

1. Reduce $34 to cents. *Ans.* 3400 cents.
2. Reduce 48143 mills to dollars. *Ans.* $48'143.
3. Reduce $40'06½ to mills. *Ans.* 40065 mills.
4. Reduce 48742 cents to dollars. *Ans.* $487'42.
5. Reduce $16 to mills. *Ans.* 16000 mills.
6. Reduce 125 mills to cents. $'12½.
7. Reduce $'75 to mills. *Ans.* 750 mills.
8. Reduce 2064½ cents to dollars. *Ans.* $20'64½.
9. Reduce $'007 to mills. *Ans.* 7 mills.
10. Reduce 9 cents to dollars. *Ans.* $'09.

Questions. — ¶ **94.** How are dollars reduced to cents? to mills? cents to mills? mills to cents? to dollars? cents to dollars?

Addition and Subtraction of Decimal Fractions.

¶ 95. As the value of the decimal parts of a unit vary in a tenfold proportion like whole numbers, the addition and subtraction of decimal fractions, and of Federal money, may be performed as in whole numbers.

1. What is the amount of 14·68, 9·045, 38·5, and 9·0025?

2. From 765·06 take 27·6895.

SOLUTION. — As numbers of the same denomination only can be added together, (¶ 12,) or subtracted from each other, the several numbers in each of these examples must be reduced to the lowest denomination contained in any one of the numbers, (¶ 91,) which is ten-thousandths, when the operation may be performed as in simple numbers.

FIRST OPERATION.	FIRST OPERATION
14·6800	765·0600
9·0450	27·6895
38·5000	737·3705, *Ans.*
9·0025	
71·2275, *Ans.*	

Or if only like denominations are written under each other, these alone will be added together, or subtracted from each other, and the operations may be performed without the formality of reducing to a common denominator, since the ciphers, by which the reduction is effected, make no difference in the result: thus,

SECOND OPERATION.	SECOND OPERATION
14·68	765·06
9·045	27·6895
38·5	737·3705
9·0025	
71·2275	

NOTE. — As the decimal point is at the right of units, which are written under each other, the point in the result is directly below the points in the several numbers. Hence,

To add or subtract decimal fractions,

RULE.

Write the numbers under each other, tenths under tenths, hundredths under hundredths, &c., according to the value of their places; add or subtract as in simple numbers, and point off in the result as many places for decimals as are equal to

---

Questions. — ¶ **95.** Why can addition and subtraction of decimals be performed as in whole numbers? What numbers only can be added and subtracted? How is this effected by the first operations? How by the second? How do you write down decimals for addition? How for subtraction? Where place the point in the results? Repeat the rule How do you prove addition of decimals? How subtraction?

the greatest number of decimal places in any of the given numbers.

Proof. — The same as in the addition and subtraction of simple numbers.

EXAMPLES FOR PRACTICE.

3. A man sold wheat at several times as follows, viz., 13‘25 bushels, 8‘4 bushels, 23‘051 bushels, 6 bushels, and ‘75 of a bushel; how much did he sell in the whole?

Ans. 51‘451 bushels.

4. What is the amount of 429, $21\frac{37}{100}$, $355\frac{3}{1000}$, $1\frac{7}{100}$, and $1\frac{7}{10}$? *Ans.* $808\frac{143}{1000}$, or 808‘143.

5. What is the amount of 2 tenths, 80 hundredths, 89 thousandths, 6 thousandths, 9 tenths, and 5 thousandths?

Ans. 2.

6. What is the amount of three hundred twenty-nine and seven tenths, thirty-seven and one hundred sixty-two thousandths, and sixteen hundredths?

7. From thirty-five thousand take thirty-five thousandths.

Ans. 34999·965.

8. From 5‘83 take 4‘2793. *Ans.* 1‘5507.

9. From 480 take 245‘0075. *Ans.* 234·9925.

10. What is the difference between 1793‘13 and 817‘05693? *Ans.* 976‘07307.

11. From $4\frac{8}{100}$ take $2\frac{1}{10}$. *Remainder*, $1\frac{98}{100}$, or 1‘98.

12. What is the amount of $29\frac{3}{10}$, $374\frac{9}{10000000}$, $97\frac{253}{1000}$, $315\frac{4}{1000}$, 27, and $100\frac{4}{10}$? *Ans.* 942·957009.

Examples in Federal Money can evidently be performed in the same way.

1. Bought 1 barrel of flour for 6 dollars 75 cents, 10 pounds of coffee for 2 dollars 30 cents, 7 pounds of sugar for 92 cents, 1 pound of raisins for 12½ cents, and 2 oranges for 6 cents: what was the whole amount? *Ans.* $10‘155.

2. A man is indebted to A, $237‘62; to B, $350; to C, $86‘12½; to D, $9‘62½; and to E, $0‘834; what is the amount of his debts? *Ans.* $684·204.

3. A man has three notes specifying the following sums viz., three hundred dollars, fifty dollars sixty cents, and nine dollars eight cents; what is the amount of the three notes?

Ans. $359·68.

4. What is the amount of $56‘18, $7‘37½, $280, $0‘287 $17, and $90‘413? *Ans.* $451‘255.

5. Bought a pair of oxen for $76'50, a horse for $85, and a cow for $17'25; what was the whole amount?

Ans. $178'75.

6. Bought a gallon of molasses for 28 cents, a quarter of tea for 37½ cents, a pound of saltpetre for 24 cents, 2 yards of broadcloth for 11 dollars, 7 yards of flannel for 1 dollar 62½ cents, a skein of silk for 6 cents, and a stick of twist for 4 cents; how much for the whole? *Ans.* $13'62.

7. A man bought a cow for eighteen dollars, and sold her again for twenty-one dollars thirty-seven and a half cents; how much did he gain? *Ans.* $3'375.

8. A man bought a horse for 82 dollars, and sold him again for seventy-nine dollars seventy-five cents; did he gain or lose? and how much? *Ans.* He lost $2'25.

9. A merchant bought a piece of cloth for $176, which proving to have been damaged, he is willing to lose on it $16'50; what must he have for it? *Ans.* $159'50.

10. A man sold a farm for $5400, which was $725'37½ more than he gave for it; what did he give for the farm?

11. A man, having 500 dollars, lost 83 cents; how much had he left? *Ans.* $499'17.

12. A man's income is $1200 a year, and he spends $800'35; how much does he lay up?

13. Subtract half a cent from seven dollars.

Rem. $6'99½.

14. How much must you add to $16'82 to make $25?

15. How much must you subtract from $250, to leave $87'14?

16. A man bought a barrel of flour for $6'25, 7 pounds of coffee for $1'41, he paid a ten dollar bill; how much must he receive back in change? *Ans.* $2'34.

Multiplication of Decimal Fractions.

¶ 96. 1. Multiply '7 by '3.

'7 $= \frac{7}{10}$ and '3 $= \frac{3}{10}$, then $\frac{7}{10} \times \frac{3}{10} = \frac{21}{100}$ = '21, *Ans.* We here see that *tenths multiplied into tenths produce hundredths*, just as tens, (70,) into tens, (30,) make hundreds, (2100.) We may write down the numerators decimally, thus:

OPERATION.
'7
'3
—
'21, *Ans.*

The 21 must be hundredths as before. The number of figures in the product, it will be seen, is equal to the number in the multiplicand and multiplier; hence, we have as many places for decimals in the product as there are in both the factors.

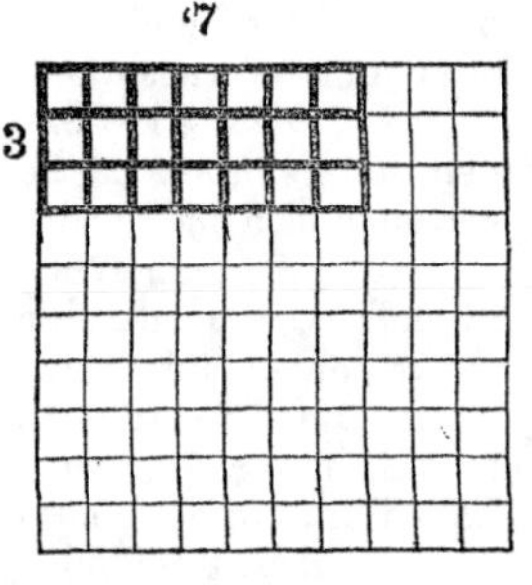

NOTE. — The correctness of the above rule may be illustrated by the annexed diagram. The length of the plot of ground which it represents may be regarded 10 feet and the breadth 10 feet, each division of a line, consequently, being one tenth of the whole line. Multiplying 10 by 10, we have 100 square feet in the plot, each of the small squares being 1 square foot, or one hundredth of the whole plot. Now take the part encircled by the black lines, 7 feet ('7 of the whole line) long, and 3 feet ('3 of the whole line) wide. The contents are 21 square feet, or '21 of the whole plot; hence the product of '7 into '3 is '21 as above.

2. Multiply '125 by '03.

OPERATION.
'125
'03
———
'00375 *Prod.*

Here, as the number of significant figures in the product is not equal to the number of decimals in both factors, the deficiency must be supplied by prefixing ciphers, that is, placing them at the left hand. The correctness of the rule may appear from the following process: '125 is $\frac{125}{1000}$, and '03 is $\frac{3}{100}$: now, $\frac{125}{1000} \times \frac{3}{100} = \frac{375}{100000} =$ '00375, the same as before.

Hence, *To multiply decimal fractions,*

RULE.

Multiply as in whole numbers, and from the right hand of the product point off as many figures for decimals as there are decimal places in the multiplicand and multiplier counted together, and if there are not so many figures in the product, supply the deficiency by prefixing ciphers.

Questions. — ¶ **96.** Tenths × tenths produce what? Illustrate this by a diagram. To what must the number of places in the product be equal? When the number of decimals in the product is less than the number in both factors what do you do? How can you tell of what name or denomination will be the product of one given decimal multiplied into another given decimal, without going through the process of multiplication? Of what denomination, then, will be the product of '46 × '25? of '0005 × '07?

EXAMPLES FOR PRACTICE.

3. Multiply five hundredths by seven thousandths. *Product*, '00035.

4. What is '3 of 116? *Ans.* 34'8.

5. What is '85 of 3672? *Ans.* 3121'2.

6. What is '37 of '0563? *Ans.* '020831.

7. Multiply 572 by '58. *Product*, 331'76.

8. Multiply eighty-six by four hundredths. *Product*, 3'44

9. Multiply '0062 by '0008.

10. Multiply forty-seven tenths by one thousand eighty-six hundredths. *Prod.* 51'042.

EXAMPLES IN FEDERAL MONEY.

¶ 97. 1. If a melon be worth \$'09, what is '7 of it worth? (¶77.) *Ans.* \$'063.

2. What will 250 bushels of rye cost, at \$'88½ per bushel?

3. What is the value of 87 barrels of flour, at \$6'37½ a barrel? *Ans.* \$554'62½.

4. What will be the cost of a hogshead of molasses, containing 63 gallons, at 28½ cents a gallon? *Ans.* \$17'955.

5. If a man spend 12½ cents a day, what will that amount to in a year of 365 days? What will it amount to in 5 years? *Ans.* It will amount to \$228'12½ in 5 years.

6. If it cost \$36'75 to clothe a soldier 1 year, how much will it cost to clothe an army of 17800 men? *Ans.* \$654150.

7. Multiply \$367 by 46.

8. Multiply \$0'273 by 8600. *Ans.* \$2347'80.

9. At \$5'47 per yard, what cost 8'3 yards of cloth? *Ans.* \$45'401

10. At \$'07 per pound, what cost 26'5 pounds of rice? *Ans.* \$1'855.

11. What will be the cost of thirteen hundredths of a ton of hay, at \$11 a ton? *Ans.* \$1'43.

12. What will be the cost of three hundred seventy-five thousandths of a cord of wood, at \$2 a cord? \$'75.

13. If a man's wages be seventy-five hundredths of a dollar a day how much will he earn in 4 weeks, Sundays excepted? *Ans.* \$18

Division of Decimal Fractions.

¶ 98. 1. Divide '21 by '3. '21 $= \frac{21}{100}$, and '3 $= \frac{3}{10}$. Now $\frac{21}{100} \div \frac{3}{10}$; or $\frac{10}{3}$ of $\frac{21}{100} = \frac{210}{300} = \frac{7}{10}$. It appears, then, that hundredths divided by tenths give tenths, just as hundreds (2100) divided by tens (30) give tens, (70.) The numerators may be set down decimally, and the division performed as follows:—

OPERATION.

```
'3)'21
   ---
    '7
```

The 7 must be tenths as before. The dividend, which answers to the product in multiplication, contains two decimal places; and the divisor and quotient, which answer to the factors in multiplication, (¶ 31,) together contain two decimal places. Hence, we see that the number of decimal places in the quotient is equal to the difference between the number in the dividend and divisor.

2. At 4'75 of a dollar per barrel, how many barrels of flour can be bought for $31?

OPERATION.

```
4'75)31'00(6'526+
     2850
     ----
      2500
      2375
      ----
       1250
        950
       ----
        3000
        2850
        ----
         150
```

The 4'75 are 475 hundredths, and, since the dividend and divisor must be of the same denomination, we annex 2 ciphers to 31 and it becomes 3100 hundredths, (¶ 91.) Then there can be as many whole barrels bought as the number of times 475 hundredths can be subtracted from 3100 hundredths. The 6 barrels thus found will cost 2850 hundredths of a dollar, and as 250 hundreths or cents remain, it will buy part of another barrel, which we find by annexing ciphers, and continuing the operation.

We now see that there are 5 decimal places in the dividend, counting all the ciphers that are annexed, and as there are but two in the divisor, we point off 3 in the quotient. There is still a remainder of 150, which, written over the divisor, (¶ 36,) gives $\frac{150}{475}$ of a thousandth of a barrel, a quantity so small that it may be neglected. But we place + at the right of the last quotient figure, to show that there is more flour than indicated by the quotient.

Ans. 6'526+ barrels.

NOTE.—It is sufficiently exact for most practical purposes to carry the division to *three* decimal places.

3. Divide '00375 by '125.

OPERATION.

```
125)'00375('03
      375
      ---
      000
```

The divisor, 125, in 375, goes 3 times, and no remainder. We have only to place the decimal point in the quotient, and the work is done. There are five decimal places in the dividend; consequently there must be five in the divisor and quotient counted together; and as there are *three* in the divisor, there must be *two* in the quotient; and, since we have but one figure in the quotient, the *deficiency* must be supplied by prefixing a cipher.

The operation by vulgar fractions will bring us to the same result. Thus, '125 is $\frac{125}{1000}$, and '00375 is $\frac{375}{100000}$: now, $\frac{375}{100000} \div \frac{125}{1000} = \frac{375000}{12500000} = \frac{3}{100} =$ '03, the same as before.

4. Divide '75 by '005.

OPERATION.

```
        '75
         10
        ----
'005 ) '750
        ----
        150, Quot.
```

SOLUTION. — We cannot divide hundredths by thousandths, until the former are reduced to thousandths by multiplying by 10, or annexing one cipher, when the divisor and dividend will be of the same denomination; and '005 is contained in (can be subtracted from) '750, 150 times, the quotient being a whole number.

These illustrations will establish the following

RULE.

I. Reduce, if necessary, the dividend to the lowest denomination in the divisor, divide as in whole numbers, annexing ciphers to a remainder which may occur, and continuing the operation

II. If the decimal places in the dividend with the ciphers annexed exceed those in the divisor, point off the excess from the right of the quotient as decimals; but if the excess is more than the number of places in the quotient, supply the deficiency by prefixing ciphers.

EXAMPLES FOR PRACTICE.

5. Divide 3156'293 by 25'17. *Quot.* 125'3 +.
6. Divide 173948 by '375. *Quot.* 463861 +.

NOTE. — The pupil will point off the decimal places in the quotient of this and the following example, as directed by the rule.

7. Divide 5737 by 13'3. *Quot.* 431353.
8. What is the quotient of 2464'8 divided by '008? *Ans.* 308100.

Questions. — ¶ **98.** Hundredths, divided by tenths, give what? How is it in integers? Exhibit on the blackboard the process of dividing 7 by 1'25. Why do you annex ciphers to the 7? What is the quotient? Why pointed thus? Give a demonstration by common fractions, as after Ex. 3, and show that this placing of the point is right. The sign of addition, annexed to the quotient, is an indication of what? When there are remainders, to how many places should the division be carried? Why not to more places? Repeat the rule for division

9. Divide 2 by 53'1. *Quot.* '037 +.
10. Divide '012 by '005. *Quot.* 2'4.
11. Divide three thousandths by four hundredths. *Quot.* '075.
12. Divide eighty-six tenths by ninety-four thousandths.
13. How many times is '17 contained in 8?

EXAMPLES IN FEDERAL MONEY.

¶ 99. 1. Divide $59'387 equally among 8 men; how much will each man receive?

OPERATION.
8)59'387

Ans. $7'423$\frac{3}{8}$, that is, 7 dollars, 42 cents, 3 mills, and $\frac{3}{8}$ of another mill. The $\frac{3}{8}$ is the remainder, after the last division, written over the divisor, and expresses such fractional part of another mill.

For most purposes of business, it will be sufficiently exact to carry the quotient only to mills, as the *parts* of a mill are of so little value as to be disregarded.

2. At $'75 per bushel, how many bushels of rye can be bought for $141? *Ans.* 188 bushels.

3. At 12$\frac{1}{2}$ cents per lb., how many pounds of butter may be bought for $37? *Ans.* 296 lbs.

4. At 6$\frac{1}{4}$ cents apiece, how many oranges may be bought for $8? *Ans.* 128 oranges.

5. If '6 of a barrel of flour cost $5, what is that per barrel? *Ans.* $8'333 +.

NOTE.—If the sum to be divided contain only dollars, or dollars and cents, it may be reduced to mills, by annexing ciphers before dividing; or, we may first divide, annexing ciphers to the remainder, if there shall be any, till it shall be reduced to mills, and the result will be the same.

6. If I pay $468'75 for 750 pounds of wool, what is the value of 1 pound? *Ans.* $0'625; or thus, $'62$\frac{1}{2}$.

7. If a piece of cloth, measuring 125 yards, cost $181'25, what is that a yard? *Ans.* $1'45.

8. If 536 quintals of fish cost $1913'52, how much is that quintal? *Ans.* $3'57

9. Bought a farm, containing 84 acres, for $3213; what d it cost me per acre? *Ans.* $38'25.

10. At $954 for 3816 yards of flannel, what is that a yard? *Ans.* $0'25.

11. Bought 72 pounds of raisins for $8; what was that a pound? *Ans.* $0'111$\frac{1}{9}$; or, $0'111 +.

12. Divide $12 into 200 equal parts; how much is *one* of the parts? $\frac{12}{200}$ = how much? *Ans.* $'06.

13. Divide $30 by 750. $\frac{30}{750}$ = how much?

14. Divide $60 by 1200. $\frac{60}{1200}$ = how much?

15. Divide $215 into 86 equal parts; how much will one of the parts be? $\frac{215}{86}$ = how much?

¶ 100. Review of Decimal Fractions.

Questions. — What are decimal fractions? How do they differ from common fractions? How can the proper denominator to a decimal fraction be known, if it be not expressed? What advantages have decimal over common fractions? How is the value of every figure determined? Describe the manner of numerating and reading decimal fractions? of writing them? How are decimals, having different denominators, reduced to a common denominator? How may any whole number be reduced to decimal parts? How can any mixed number be read together, and the whole expressed in the form of a common fraction? What is federal money? What is the money unit, and what are its divisions and subdivisions? How is a common fraction reduced to a decimal? To what do the denominations of federal money correspond? What is the rule for addition and subtraction of decimals? — multiplication? — division?

EXERCISES.

1. A merchant had several remnants of cloth, measuring as follows, viz.:

7 $\frac{3}{8}$ yds. 6 $\frac{5}{8}$ " 1 $\frac{4}{5}$ " 9 $\frac{2}{5}$ " 8 $\frac{1}{4}$ " 3 $\frac{1}{40}$ "	How many yards in the whole, and what would the whole come to, at $3'67 per yard? NOTE. — Reduce the common fractions to decimals. Do the same wherever they occur in the examples which follow. *Ans.* 36'475 yards. $133'863 +, cost.

2. From a piece of cloth, containing 36$\frac{5}{8}$ yards, a merchant sold, at one time, 7$\frac{3}{10}$ yards, and, at another time, 12$\frac{5}{8}$ yards; how much of the cloth had he left? *Ans.* 16'7 yds.

3. A farmer bought 7 yards of broadcloth for 33\frac{1}{3}$, two barrels of flour for 14\frac{6}{15}$, three casks of lime for 7\frac{5}{9}$, and 7 pounds of rice for $$\frac{5}{6}$; what was the cost of the whole?

NOTE. — The following examples are to be performed according to the rule in ¶ 77, or in ¶ 83, or in ¶ 85.

4. At 12$\frac{1}{2}$ cents per lb., what will 37$\frac{3}{4}$ lbs. of butter cost? *Ans.* $4'718$\frac{3}{4}$.

5. At $17'37 per ton for hay, what will $11\frac{5}{8}$ tons cost? *Ans.* $201'92$\frac{5}{8}$.

6. *The above example reversed.* At $201'92$\frac{5}{8}$ for $11\frac{5}{8}$ tons of hay, what is that per ton? *Ans.* $17'37.

7. If '45 of a ton of hay cost $9, what is that per ton? *Ans.* $20.

8. At '4 of a dollar a gallon, what will '25 of a gallon of molasses cost? *Ans.* $'1.

9. What will 2300 lbs. of hay come to, at 7 mills per lb.? *Ans.* $16'10.

10. What will $765\frac{1}{2}$ lbs. of coffee come to, at 18 cents per lb.? *Ans.* $137'79.

11. Bought 23 firkins of butter, each containing 42 pounds, for $16\frac{1}{2}$ cents a pound; what would that be a firkin? and how much for the whole? *Ans.* $159'39 for the whole.

12. A man killed a beef, which he sold as follows, viz., the hind quarters, weighing 129 pounds each, for 5 cents a pound; the fore quarters, one weighing 123 pounds, and the other 125 pounds, for $4\frac{1}{2}$ cents a pound; the hide and tallow, weighing 163 pounds, for 7 cents a pound; to what did the whole amount? *Ans.* $35'47.

13. A farmer bought 25 pounds of clover seed at 11 cents a pound, 3 pecks of herds grass seed for $2'25, a barrel of flour for $6'50, 13 pounds of sugar at $12\frac{1}{2}$ cents a pound; for which he paid 3 cheeses, each weighing 27 pounds, at $8\frac{1}{2}$ cents a pound, and 5 barrels of cider at $1'25 a barrel. The balance between the articles bought and sold is 1 cent; is it *for* or *against* the farmer?

14. A man dies, leaving an estate of $71600; there are demands against the estate, amounting to $39876'74; the residue is to be divided between 7 sons; what will each one receive? *Ans.* $4531'894$\frac{2}{7}$.

15. How much coffee, at 25 cents a pound, may be had for 100 bushels of rye, at 87 cents a bushel? *Ans.* 348 pounds.

16. At $12\frac{1}{2}$ cents a pound, what must be paid for 3 boxes of sugar, each containing 126 pounds? *Ans.* $47'25.

17. If 650 men receive $86'75 each, what will they all receive? *Ans.* $56387'50.

18. A merchant sold 275 pounds of iron at $6\frac{1}{4}$ cents a pound, and took his pay in oats, at $0'50 a bushel; how many bushels did he receive? *Ans.* 34'375 bushels.

19. How many yards of cloth, at $4'66 a yard, must be given for 18 barrels of flour, at $9'32 a barrel? *Ans.* 36 yards.

20. What is the price of three pieces of cloth, the first containing 16 yards, at $3·75 a yard; the second, 21 yards, at $4·50 a yard; and the third, 35 yards, at $5·12½ a yard? *Ans.* $333·87½.

BILLS.

¶ 101. A Bill, in business transactions, is a written list of the articles bought or sold, and their prices, together with the entire cost or amount footed.

No. 1.—*Bill of Sale. Payment received.*

Boston, May 25th, 1847.

James Brown, Esq.

Bought of Hastings & Belding,

6 yards black broadcloth,	@	$3·00	18·00
2½ " cambric,	"	·14	·35
2 dozen buttons,	"	·15	·30
4 skeins sewing silk,	"	·04	·16
25 lbs. brown sugar,	"	·09	2·25
Received payment,			$21·06.

Hastings & Belding.

No. 2.—*Bill of Sale. Charged in account.*

New Orleans, Aug. 1st, 1847.

Gen. Z. Taylor,

To Daniels & Thomas, Dr.

To 278 bbls. beef,	@	$9·75	
" 191 " pork,	"	12·00	
" 250 " flour,	"	5·70½	
" 500 sacks Indian meal,	"	·62½	
Charged in acc't.			Amount, $6741·25

Daniels & Thomas.

No. 3.—*Barter Bill.*

Buffalo, Sept. 15th, 1847.

Mr. D. F. Standart,

To O. B. Hopkins & Co., Dr.

To 15 lbs. brown sugar	@	$ ·10	
" 2 " Y. H. tea,	"	·87½	
" 24 " mackerel,	"	·04½	
" 3 gal. molasses,	"	·42	
" 16 yds. sheeting,	"	·09	
			$

Cr.

By	4 doz. eggs,	@	$ '08	
"	8 lbs. butter,	"	'14	
"	40 " chéese,	"	'07½	
"	note at 30 days, to balance,			2'59
				$7'03

O. B. Hopkins & Co.
by L. D. Swift.

No. 4.—Bill of goods sold at wholesale.

New-York, April 5th, 1847.

Davis & Horton,

Bought of Barnes, Porter & Co.

3 hhds. molasses, 118 gal. each,	@	$ '31	
2 " brown sugar, 975 and 850 lbs.	"	'09½	
3 casks rice, 205 lbs. each,	"	'04½	
5 sacks coffee, 75 " "	"	'11	
1 chest H. tea, 86 " "	"	'92	
			$431'16

Rec'd payment, by note, at 60 days.

For Barnes Porter & Co.
James D. Willard.

It is sometimes practised, in collecting and settling accounts to make a copy of each individual account, and present it to the person for his inspection.

No. 5.—Copy of an individual account.

Frank H. Wright,

In acc't with Edward F. Cooper,

1847.			Dr.	
Jan. 7.	To 125 bushels corn,	@	$ '50	
" "	" 20 " apples,	"	'31	
March 13.	" 12 " rye,	"	'62	
" 20,	" 15¾ lbs. cast steel,	"	'24	
				$

Questions.—¶ 101. What is a bill? If the amount of the bill be paid at the time, how is it shown? Which bill is an example of this? If charged in account, how is it shown? example? How does a barter bill differ from a bill of sale? In what order are the articles bought and sold arranged? What is practised in collecting and settling accounts? How does such a copy differ from a barter bill? To which of the bills must the bill to be made out conform? and what will it be called?

1847.		Cr.		
Feb. 15.	By 3 cows,		"	$17'00
" 22.	" 5 sheep,		"	2'50
				$

Amount due me, $16,42
Edward F. Cooper.

Baltimore, May 9th, 1847.

The pupil is required to make out a bill from the statement contained in the following example.

Wm. Prentiss sold to David S. Platt 780 lbs. of pork, at 6 cents per lb.; 250 lbs. of cheese, at 8 cents per lb.; and 154 lbs. of butter, at 15 cents per lb.; in pay he received 60 lbs. of sugar, at 10 cents per lb.; 15 gallons of molasses, at 42 cents per gallon; ½ barrel of mackerel, $3'75; 4 bushels of salt, at $1'25 per bushel; and the balance in money: how much money did he receive? *Ans.* $68'85.

COMPOUND NUMBERS.

¶ 102. When several abstract numbers, or several denominate numbers of *the same unit value*, are employed in an arithmetical calculation, they are called *simple numbers*, and operations with such numbers are called *operations in simple numbers*. Thus, if it were required to add together 7 gallons, 9 gallons, and 5 gallons, the numbers are simple numbers, being denominate numbers of the same unit value, (1 gal.,) and the operation is an addition of simple numbers. We have had, also, subtraction, multiplication, and division of simple numbers.

But when several numbers of *different unit values* are employed to express one quantity, the whole together is called a *compound number*. Thus, 12 rods, 9 yards, 2 feet, 6 inches employed to express the length of a field, is a compound number. So also, 9 gallons, 2 quarts, 1 pint, employed to express a quantity of water, is a compound number.

Note.—The word *denomination* is used in compound numbers to

Questions.—¶ 102. What are simple numbers? Examples. What are operations in such numbers called? What is a compound number? Give examples other than those in the book. What is meant by the word denomination?

denote the name of the unit considered. Thus, *bushel* and *peck* are names or denominations of measure; *hour*, *minute* and *second* are denominations of time.

¶ 103. The fundamental operations of addition, subtraction, multiplication and division, cannot be performed on compound numbers till we are acquainted with the method of changing numbers of one denomination to another without altering their value, which is called *Reduction*. Thus, we wish to add 2 bushels 3 pecks, and 3 bushels 1 peck, together. They will not make 9 bushels nor 9 pecks, (adding together the several numbers,) since some of the numbers express bushels, and some express pecks. But 2 bushels equal 8 pecks, (2 times 4 pecks, the number of pecks in a bushel,) and 3 pecks added make 11 pecks; 3 bushels equal 12 pecks, and 1 peck added make 13 pecks. Then, 11 pecks + 13 pecks = 24 pecks. Hence, before proceeding further, we must attend to the

Reduction of Compound Numbers.

Sterling or English Money.

¶ 104. Money is expressed in different denominations, and 4 dollars, 3 dimes, 7 cents, 5 mills = $4·375, employed to express one sum in Federal money is a compound number. But as the denominations in Federal money vary uniformly in a tenfold proportion, (¶ 93,) being conformed to the Arabic notation of whole numbers, the operations in it are as in whole numbers.

The denominations in English (called, also, sterling) money, pounds, shillings, pence and farthings, do not vary uniformly, but according to the following

TABLE.

Note 1. — All the tables in Reduction of Compound Numbers must be carefully committed to memory by the pupil.

4 farthings (qrs.) make	1 penny, marked	*d.*
12 pence (plural of penny)	1 shilling, "	*s.*
20 shillings	1 pound, "	*£.*

Note 2. — Farthings are often written as the fraction of a penny, thus, 1 farthing = ¼d., 2 farthings = ½d., 3 farthings = ¾d.

Questions. — ¶ 103. What is reduction? Whence its necessity? Explain by the example of adding bushels and pecks. To what, then, must we attend before proceeding further?

NOTE 3. — The value of these denominations in Federal money is nearly as follows:

1*qr.*	=	$\frac{121}{240}$ *of* 1 *cent.*
1*d.*	=	$2\frac{1}{60}$ *cents.*
1*s.*	=	$24\frac{1}{5}$ *cents.*
1£.	=	$4·84
4*s.* 1*d.* $2\frac{42}{121}$*qrs.*	=	$1·00

There is in England a gold coin, called a sovereign, the value of which is £1.

1. How many farthings in 5 pence?

1ST OPERATION.

$$\begin{array}{r} 4 \\ 5 \\ \hline 20qrs. \end{array}$$

2D OPERATION.

$$\begin{array}{r} 5 \\ 4 \\ \hline 20qrs. \end{array}$$

SOLUTION. — We may multiply the number of farthings (4) in 1 penny by the number of pence, (5.) (¶ 46.) Or, as either factor may be made the multiplicand, (¶ 21,) we may multiply the number of pence (5) by the number of farthings in 1 penny. *Ans.* 20qrs.

2. In 20 farthings, how many pence?

OPERATION.

$$\begin{array}{r} 4)20 \\ \hline 5d. \end{array}$$

SOLUTION. — We have given the number of farthings in 1 penny to find the number of pence in a given number of farthings, (20,) and we divide the number of farthings in the number of pence by the number in 1 penny, (¶ 46.) *Ans.* 5*d.*

3. How many farthings in 3 pence? —— 6 pence? —— 9 pence? —— 7 pence? —— 2 pence? —— 10 pence? —— 11 pence? —— 12 pence = 1 shilling?

5. How many pence in 3 shillings? —— 5 shillings? —— 5s. 8d.? —— 7s.? —— 8s. 4d.? —— 12s.? —— 15s. 6d.?

7. How many shillings in 3£.? —— 5£.? —— 4£. 2s.? —— 6£. 11s.?

4. How many pence in 12 farthings? —— 24 farthings? —— 36 farthings? —— 28 farthings? —— 8 farthings? —— 40 farthings? —— 44 farthings? —— 48 farthings?

6. How many shillings in 36 pence? —— 60 pence? —— 68d.? —— 84d.? —— 100d.? —— 144d.? —— 186d.?

8. How many pounds in 60s.? —— 100s.? —— 82s.? —— 131s.?

Questions. — ¶ **104.** What is said of operations in Federal money? What are the denominations of English money? the signs? How do they vary differently from those of Federal money? Give the table. How are farthings written? What is the value of a pound sterling in Federal money? Explain the first operation of Ex. 1; the second operation Explain Ex. 2. Of how many kinds is reduction? what are they What is reduction descending? — reduction ascending?

The changing of higher denominations to lower, as pounds to shillings, is called *Reduction Descending*, and is performed by multiplication.

The changing of lower denominations to higher, as shillings to pounds, is called *Reduction Ascending*, and is performed by division.

REDUCTION DESCENDING.

¶ 105. 1. In £17 13s. $6\frac{3}{4}$d., how many farthings?

OPERATION.

17£ 13*s.* 6*d.* 3*qrs.*
20

353*s. in* 17£ 13*s.*
12

4242*d. in* 17£ 13*s.* 6*d.*
4

16971*qrs. Ans.*
In 17£ 13*s.* 6*d.* 3*qrs.*

Solution. — We multiply 17£ by 20, the shillings in 1£, and add in the 13s. to get the number of shillings in 17£ 13s., which is 353. This number we multiply by 12, adding in the 6d. given, to get the number of pence, 4242, which we multiply by 4, adding in the 3qrs. given, to get the number of qrs. or farthings, which is 16971qrs.

Hence, *for Reduction Descending*,

RULE.

Multiply each higher denomination by the number which it takes of the next less

REDUCTION ASCENDING.

2. In 16971 farthings, how many pounds?

OPERATION.

Farthings in 1 penny, {4) 16971
Pence in 1 shilling, {12) 4242*d.* 3*qrs*
Shillings in £1, {2 0) 35|3*s.* 6*d.*
17£ 13*s.*
Ans. £17 13*s.* 6*d.* 3*qrs.*

Solution. — We divide the whole number of farthings by 4 the number in 1d., to get the number of pence; for as many times as 4 can be subtracted from 16971, so many pence there will be, which is 4242d. and 3qrs. remaining. On the same principle, dividing the 4242 by 12, the quotient, 353, is shillings, and the remainder, 6, is pence, and dividing 353s. by 20, the quotient, 17, is pounds, and the remainder, 13, is shillings.

Hence, *for Reduction Ascending*,

RULE.

Divide each lower denomination by the number which it takes of it to make one of

---

Questions. — ¶ **105.** Explain the first example. Give the rule for reduction descending. Ex. 2. Give the rule for reduction ascending to make 1 of this higher, increasing the product by the given number, if any, of this lower denomination. Proceed in this way till the work is done.

the next higher. Proceed in this way till the work is done.

EXAMPLES FOR PRACTICE.

3. Reduce 32£. 15s. 8d. to qrs.

4. Reduce 31472 farthings to pounds.

5. Reduce 7£. 14s. 6d. 1 qr. to qrs.

6. Reduce 7417 qrs. to pounds.

7. In 91£. 11s. $3\frac{1}{2}$d., how many farthings?

8. In 87902 farthings, now many pounds?

9. In 40£. 12s. 8d., how many pence?

10. In 9752 pence, how many pounds?

11. In 1£. 18s. $4\frac{1}{2}$d., how many half pence?

12. In 921 half pence, how many pounds?

Weight.

I. Avoirdupois Weight.

¶ 106. Avoirdupois Weight is employed in all the ordinary purposes of weighing. The denominations are tons, pounds, ounces, and drams.

TABLE.

16 drams (drs.)	make	1 ounce,	marked oz.
16 ounces	"	1 pound,	" lb.
2000 pounds	"	1 ton,	" T.

Or, *as was formerly reckoned,*

28 lbs.	1 quarter,	qr.
4 qrs. (=112 lbs.)	1 hundred weight	cwt.
20 cwt. (=2240 lbs.)	1 ton,	T.

By the last table, 2240 lbs. make 1 ton, which is sometimes called the "long ton;" while the ton of 2000 lbs. is called the "short ton." The long ton is still used in the U. S. cus-

---

Questions. — ¶ **106.** What is the use of avoirdupois weight? the denominations? the signs? Repeat the table; the table by the old method. Explain the difference between the long and short ton. When is the long ton used? the short ton?

tom-house operations, in invoices of English goods, and of coal from the Pennsylvania mines. But in selling coal in cities, and in other transactions, unless otherwise stipulated, 2000 lbs. are called a ton.

EXAMPLES FOR PRACTICE.

1. In 14 tons 607 lbs. 6 oz. 12 drs., how many drams?

SOLUTION.—As there are 2000 lbs. in a ton, we multiply 14 by 2000, to get 14 tons to lbs., and add the 607 lbs. to the product. The lbs. we multiply by 16 to get them to oz., adding in 6 oz., and the oz. by 16, adding in 12, and the whole are in drams.

2. In 7323500 drams, how many tons?

SOLUTION.—Dividing the drams by 16, the number in an oz., the quotient is oz. and the remainder drs. Dividing the oz. by 16, the quotient is lbs. and the remainder oz., and dividing the lbs. by 2000, the quotient is tons and the remainder lbs.

3. In 7 tons 665 lbs. of sugar, how many lbs.?

4. In 14665 lbs. of sugar, how many tons?

5. In 12 T. 15 cwt. 1 qr. 19 lbs. 6 oz. 12 drs. of glass, received from an English house, how many drams?

6. In 7323500 drams, how many tons?

7. Received from Birmingham, England, 5 T. 9 cwt. 12 lbs. of iron screws, in packages of 26 lbs. each; how many were the packages?

8. In 470 packages of screws, each containing 26 lbs., how many tons?

II. TROY WEIGHT.

¶ 107. Troy Weight is used where great accuracy is required, as in weighing gold, silver, and jewels. The denominations are pounds, ounces, pennyweights, and grains.

TABLE.

24 grains (grs.) make	1 pennyweight,	marked	pwt.
20 pwts.	1 ounce,	"	oz.
12 oz.	1 pound,	"	lb.

NOTE.—A lb. Troy = 5760 grs., and 1 lb. avoirdupois = 7000 grs. Troy. Hence a quantity expressed in one weight, may be changed to the denominations of the other.

Questions.—¶ 107. For what is Troy weight used? What are the denominations? — the signs? Repeat the table. What difference between the pound Troy and the pound avoirdupois?

EXAMPLES FOR PRACTICE.

1. In 210 lbs. 8 oz. 12 pwts., how many pwts.?

SOLUTION. — Multiply the lbs. by 12, adding the 8 oz. to the product, and the sum is oz., which, multiplying by 20, adding in the 12 pwts., the sum is pwts.

2. In 50572 pwts., how many lbs.?

SOLUTION.—Dividing the pwts. by 20, the quotient is oz., and the remainder pwts., and dividing the oz. by 12, the quotient is lbs., and the remainder oz.

3. In 7 lbs. 11 oz. 3 pwts. 9 grs. of silver, how many grains?

4. In 45681 grains of silver, how many lbs.?

5. Reduce 11 oz. 13 pwts. 13 grs. of gold to grains.

6. Reduce 5605 grs. of gold to ounces.

7. Reduce 28 lbs. avoirdupois to the denominations of Troy weight?

8. Reduce 34 lbs. 6 pwts. 16 grs. Troy to lbs. avoirdupois. (Consult Note.)

III. APOTHECARIES' WEIGHT.

¶ 108. Apothecaries' Weight is used by apothecaries and physicians, in mixing and preparing medicines. But medicines are bought and sold by avoirdupois weight.

The denominations are pounds, ounces, drams, scruples, and grains.

TABLE.

20 grains (grs.)	make	1 scruple,	marked	℈.
3 ℈		1 dram,	"	ʒ.
8 ʒ		1 ounce,	"	℥.
12 ℥		1 pound,	"	℔.

NOTE.—The pound and ounce, Apothecaries' and Troy weight, are the same, but the ounce is differently divided.

EXAMPLES FOR PRACTICE.

1. In 9 lbs. 8 ℥. 1 ʒ. 2 ℈. 19 grs. how many grains?

2. Reduce 55799 grs. to lb.

Questions.—¶ 108. To what is apothecaries' weight limited the denominations? Give the table. Make the sign for each denomination. What is said of the pound and ounce?

Measures of Extension.

Extension has three dimensions, length, breadth, and thickness.

I. Linear Measure.

¶ 109. Linear Measure (the measure of lines) is used when only one dimension is considered, which may be either the length, breadth, or thickness.

The usual denominations are miles, furlongs, rods, yards, feet, inches, and barley-corns.

TABLE.

3 barley-corns (bar.) make	1 inch,	marked	in.
12 inches	1 foot,	"	ft.
3 ft.	1 yard,	"	yd.
$5\frac{1}{2}$ yards, or $16\frac{1}{2}$ ft.,	1 rod,	"	rd.
40 rods	1 furlong,	"	fur.
8 furlongs, or 320 rods,	1 mile,	"	mi.

$69\frac{1}{6}$ common miles,	1 degree, deg., or °, { on the equatorial circumference of the earth.
3 geographical miles,	1 league, L., { used in measuring distances at sea.
60 geographical miles,	1 degree of latitude.
6 feet,	1 fathom, in measuring depths at sea.

Note. — The geographical mile, used in measuring latitude, is not quite uniform, but is always a little less than $1\frac{1}{6}$ common miles, as the degree varies from $68\frac{3}{4}$ to $69\frac{1}{3}$ miles.

The degree of longitude grows shorter towards the poles, where it is nothing. Instead of the barley-corn, inches are now divided into eighths and tenths.

EXAMPLES FOR PRACTICE.

1. How many inches in the equatorial circumference of the earth, it being 360 degrees?

2. In 1577664000 inches, how many miles? How many degrees of the equatorial circumference?

Questions. — ¶ 109. How many dimensions has extension? What are they? What is linear measure? Give the denominations, and the sign of each. Repeat the table. How is the inch usually divided? For what is the fathom used? For what is the geographical mile used? — its length? What is said of the degree of longitude, and its length? — of a degree of latitude? What causes the difference in the length of degrees? For what is the league used, and about what is its length in common miles?

3. How many inches from Boston to Washington, it being 482 miles?

4. In 30539520 inches, how many miles?

5. How many times will a wheel, 16 feet 6 inches in circumference, turn round in going from Boston to Providence, it being 40 miles?

6. If a wheel, 16 feet 6 inches in circumference, turn round 12800 times in going from Boston to Providence, what is the distance?

7. If a man step 2 feet 6 inches at once, how many steps will he take in walking 43 miles?

8. A man walked 90816 steps, of 2 feet 6 inches each, in a day; how many miles did he walk?

Cloth Measure.

¶ 110. Cloth Measure is a species of linear measure, being used to measure cloth and other goods sold by the yard in length, without regard to the width.

The denominations are ells, yards, quarters, and nails.

TABLE.

4 nails, (na.,) or 9 inches,	make	1 quarter,	marked qr
4 qrs., or 36 inches,		1 yard,	" yd.
3 qrs.		1 ell Flemish,	" E. Fl.
5 qrs.		1 ell English,	" E. E.
6 qrs.		1 ell French,	" E. Fr.

Note. — Eighths and sixteenths of a yard are now used instead of nails.

EXAMPLES FOR PRACTICE.

1. In 573 yds. 1 qr. 1 na., how many nails?

2. In 9173 nails, how many yards?

3. In 296 E. E. 3 qrs., how many nails?

4. In 5932 nails, how many E. E.?

5. In 151 E. E., how many yards?

6. In 188 yds. 3 qrs., how many E. E.?

7. In 29 pieces of cloth, each containing 36 E. Fl., how many yards?

8. In 783 yds., how many E. Fl.?

Questions. — ¶ 110. What is cloth measure? How used? Why a species of linear measure? Give the denominations, and the sign of each; the table. What is used instead of nails? How may yards be reduced to E. E.? to E. Fr.? to E. Fl.? How each of these to yards?

II. Land or Square Measure.

¶ 111. Square Measure is used in estimating the area of land, and other things wherein length and breadth are considered

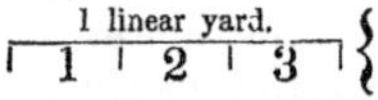

1 linear yard.
1 2 3
3 feet = 1 yard.

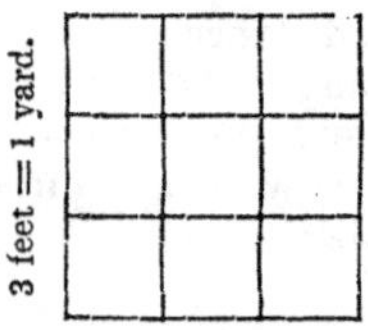

NOTE. — It takes 3 feet in length to make 1 linear yard.

But it requires a square, 3 feet = 1 linear yard in length, and 3 feet = 1 linear yard in breadth, to make 1 square yard. 3 feet in length and 1 foot in width, make 3 square feet, (3 squares in a row, ¶ 48.) 3 feet in length and 2 feet in width make $3 \times 2 = 6$ square feet, (2 rows of squares, ¶ 48.) 3 feet in length and 3 feet in width make $3 \times 3 = 9$ square feet, (3 rows of squares.)

3 feet = 1 yard.
9 sq. ft. = 1 sq. yd.

It is plain, also, that 1 square foot, that is, a square 12 inches in length and 12 inches in breadth, must contain $12 \times 12 = 144$ square inches, (12 rows, of 12 squares each.)

The denominations of square measure are miles, acres, roods, rods or poles, yards, feet, and inches.

TABLE.

144 square inches (sq. in.) make	1 square foot,	marked	sq. ft.
9 square feet	1 square yard,	"	sq. yd.
$30\frac{1}{4}$ sq. yds. $= 5\frac{1}{2} \times 5\frac{1}{2}$, or $272\frac{1}{4}$ sq. ft. $= 16\frac{1}{2} \times 16\frac{1}{2}$,	1 sq. rod, perch, or pole,	"	sq. rd. P
40 square rods	1 rood,	"	R.
4 roods, or 160 square rods	1 acre,	"	A.
640 acres	1 square mile,	"	M.

EXAMPLES FOR PRACTICE.

1. In 17 acres 3 roods 12 poles, how many square feet?

2. In 776457 square feet, how many acres?

3. Reduce 64 square miles to square feet.

4. In 1,784,217,600 square feet, how many square miles?

5. There is a town 6 miles square; how many square miles in that town? how many acres?

6. Reduce 23040 acres to square miles.

Questions. — ¶ 111. What is square measure? What is a square? Draw or describe a square inch; a square yard; a figure showing the square inches contained in a square foot. How many square inches in a row, and how many rows of square inches would it contain? *Multiply questions at pleasure.* What are the denominations of square measure? Repeat the table How does square measure differ from linear measure?

7. How many square feet on the surface of the globe, supposing it contain 197,663,000 square miles?

8. In 5510528179200000 square feet, how many square miles?

¶ 112. The *Surveyor's*, or what is called *Gunter's Chain*, is generally used in surveying land.

It is 4 rods, or 66 feet in length, and consists of 100 links

TABLE FOR LINEAR MEASURE.

$7\frac{92}{100}$ inches make	1 link,	marked	l.
25 links	1 rod,	"	rd.
4 rods, or 66 feet,	1 chain,	"	C.
80 chains	1 mile,	"	mi.

1. In 5 mi. 71 C., how many chains?

2. In 471 chains, how many miles?

3. Reduce 2 mi. 15 C. 3 rds. 18 l. to links.

4. Reduce 17593 links to miles.

5. In 75 C., how many feet?

6. In 4950 feet, how many chains?

TABLE FOR SQUARE MEASURE.

625 square links (sq. l.) make	1 square rod, perch, or pole,	marked	sq. rd. P.
16 square poles	1 square chain,	"	sq. C.
10 square chains	1 acre,	"	A.

NOTE.—Land is generally estimated in square miles, acres, roods and square poles or perches.

7. Reduce 8 A. 2 sq. C. 7 P. 456 sq. l. to square links.

8. In 824831 sq. l., how many acres?

9. In 80 A., how many square chains? how many square links?

10. In 8000000 sq. l., how many square chains? In 800 sq. C., how many acres?

III. CUBIC OR SOLID MEASURE.

¶ 113. Cubic or Solid Measure is used in measuring things that have length, breadth, and thickness; such as timber, wood, earth, stone, &c.

Questions.—¶ 112. What is generally used in surveying land What is its length? Of how many links does it consist? Repeat the table for linear measure for square measure. How is land generally estimated?

NOTE 1.—It has been shown (¶ 111,) that 1 square yard contains $3 \times 3 = 9$ square feet.

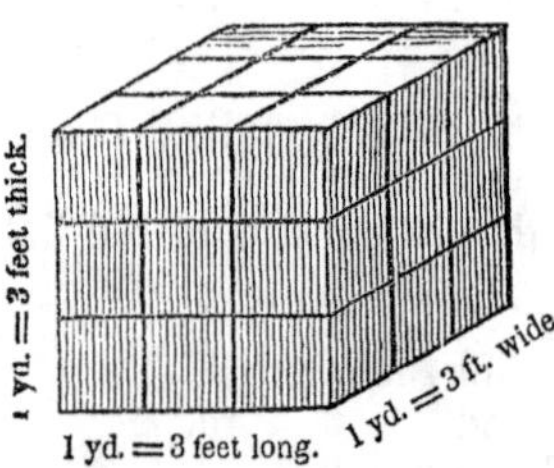

A block 3 feet long, 3 feet wide and 3 feet thick, is a cubic yard. The accompanying figure represents such a block.

Were a portion 1 foot in thickness cut off from the top of this block, the part cut off would be 3 feet long, 3 feet wide, and 1 foot thick, and would contain $3 \times 3 \times 1 = 9$ cubic feet.

The *bottom part* being 3 feet long, 3 feet wide, and 2 feet thick, would contain $3 \times 3 \times 2 = 18$ cubic feet.

But the entire block being 3 feet long, 3 feet wide, and 3 feet thick, contains $3 \times 3 \times 3 = 27$ cubic feet.

It is plain also, that a cubic foot, that is, a solid body, 12 inches long, 12 inches wide, and 12 inches thick, will contain $12 \times 12 \times 12 = 1728$ cubic or solid inches.

The denominations of cubic measure are cords, tons, yards, feet and inches.

TABLE.

1728 cubic inches, (cu. in.) $= 12 \times 12 \times 12$, that is, 12 inches in length, 12 in breadth, and 12 in thickness,	make 1 cubic foot,	marked	cu. ft
27 cubic feet, $3 \times 3 \times 3$,	1 cubic yard,	"	cu. yd.
40 feet of round timber, or 50 feet of hewn timber,	1 ton,	"	T.
42 cubic feet	1 ton of shipping, Used in measuring the capacity of ships,		T
16 cubic feet	1 cord foot, or 1 foot of wood,		C. ft.
8 cord feet, or 128 cubic feet,	1 cord of wood,		C.

Questions.—¶ 113. What is cubic measure? What distinctions do you make between a line, a surface and a solid? What is a cube? a cubic inch? a cubic foot? a cubic yard? For what is cubic or solid measure used? What are its denominations? Repeat the table. For what is the cubic ton used? What do you understand by a ton of round timber? What are the dimensions of a pile of wood containing 1 cord? What is a cord foot?

NOTE 2 — A cubic ton is used for estimating the cartage and transportation of timber. A ton of round timber is such a quantity (about 50 feet) as will make 40 feet when hewn square.

NOTE 3. — A pile or load of wood 8 feet long, 4 feet wide, and 4 feet high, contains 1 cord. $8 \times 4 \times 4 = 128$ cubic feet. A cord foot is 1 foot in length of such a pile.

1. Reduce 9 tons of round timber to cubic inches.

2. In 777600 cubic inches, how many tons of round timber?

3. In 37 cord feet of wood, how many solid feet?

4. In 592 solid feet of wood, how many cord feet?

5. Reduce 8 cords of wood to cord feet.

6. In 64 cord feet of wood, how many cords?

7. In 16 cords of wood, how many cord feet? how many solid feet?

8. 2048 solid feet of wood, how many cord feet? how many cords?

9. In 25 C., 5 C. ft., 9 cu. ft., 1575 cu. in. of wood, how many cubic inches?

10. In 5684967 cubic inches, how many cords?

Measures of Capacity.

I. WINE MEASURE.

¶ 114. Wine Measure is used in measuring all liquids except ale, beer, and milk.

The denominations are tuns, pipes, hogsheads, tierces, barrels, gallons, quarts, pints, and gills.

TABLE.

4 gills (gi.)	make 1 pint,	marked	pt.
2 pints	1 quart,	"	qt.
4 quarts	1 gallon,	"	gal.
31½ gallons	1 barrel,	"	bar.
42 gallons	1 tierce,	"	tier.
63 gallons, or 2 barrels,	1 hogshead,	"	hhd.
2 hogsheads	1 pipe,	"	P.
2 pipes, or 4 hogsheads,	1 tun,	"	T.

NOTE. — The wine gallon contains 231 cubic inches. A hogshead of molasses, &c , is no definite quantity, but is estimated by the gallon.

Questions. — ¶ 114. For what is wine measure used? What are its denominations? Repeat the table. How many cubic inches in a wine gallon? How many gallons in a hogshead of molasses? &c

1. Reduce 12 pipes of wine to pints.

2. In 12096 pints of wine how many pipes?

3. In 9 P. 1 hhd. 22 gals. 3 qts., how many gills?

4. Reduce 39032 gills to pipes.

5. In 25 tierces, how many gills?

6. In 33600 gills, how many tierces?

II. Beer Measure.

¶ 115. Beer Measure is used in measuring beer, ale and milk.

The denominations are hogsheads, barrels, gallons, quarts, and pints.

TABLE.

2 pints (pts.)	make	1 quart,	marked	qt.
4 quarts		1 gallon,	"	gal.
36 gallons		1 barrel,	"	bar.
54 gallons, or $1\frac{1}{2}$ barrels,		1 hogshead,	"	hhd.

Note. — The beer gallon contains 282 cubic inches.

1. Reduce 47 bar. 18 gal. of ale to pints.

2. In 13680 pints of ale, how many barrels?

3. In 29 hhds. of beer, how many pints?

4. Reduce 12528 pints to hogsheads.

III. Dry Measure.

¶ 116. Dry Measure is used in measuring all kinds of grain, fruit, roots, (such as potatoes and turnips,) salt, charcoal, &c.

The denominations are chaldrons, quarters, bushels, pecks, quarts, and pints.

TABLE.

2 pints (pts.)	make	1 quart,	marked	qt.
8 quarts		1 peck,	"	pk.
4 pecks		1 bushel,	"	bu.
8 bushels		1 quarter,	"	qr.
36 bushels		1 chaldron,	"	ch.

Note 1. — The dry gallon contains $268\frac{4}{5}$ cubic inches. The Winchester bushel, which is adopted as our standard, contains $2150\frac{2}{5}$ cubic inches. It is $18\frac{1}{2}$ inches in diameter, and 8 inches deep.

The quarter of 8 bushels is an English measure.

Questions. — ¶ **115.** What is the use of beer measure? What are its denominations? Repeat the table. How many cubic inches in a beer gallon?

NOTE 2. — The Imperial gallon, adopted in Great Britain in 1826, for all liquids and dry substances, contains $277\frac{274}{1000}$ cubic inches.

1. In 75 bushels of wheat, how many pints?

2. In 4800 pints, how many bushels?

3. Reduce 42 chaldrons of coal to pecks.

4. In 6048 pecks, how many chaldrons?

5. In 273 qrs. 6 bu. 3 pks. 7 qts. 1 pt. of wheat, how manv pints?

6. In 140223 pints, how many quarters?

Time.

¶ 117. Time is the measure of duration.

The denominations are years, months, weeks, days, hours, minutes, and seconds.

TABLE.

60 seconds (s.)	make 1 minute,	marked	m.
60 minutes	1 hour,	"	h.
24 hours	1 day,	"	d.
7 days	1 week,	"	w
52 weeks 1 day 5 hours 48 minutes 48 seconds, or 365 days 5 hours 48 minutes 48 seconds,	1 year,	"	yr.

NOTE 1. — As there is nearly ¼ of a day more than 365 days in a year, we add 1 day to February of certain years, thus giving them 366 days. If the excess was just ¼ of a day, we would add 1 day to every 4th year, thus making the years average 365 days 6 hours, the odd day being added to every year exactly divisible by 4.

But as the excess is not quite 6 hours, lacking about ¾ of a day in 100 years, a year divisible by 100, though divisible by 4, has only 365 days, unless it be divisible by 400, when it has 366 days. Thus, 1844 and 1600 had 366 days each, but 1845 and 1700 had only 365 days each.

A year of 366 days is called Bissextile, or Leap year.

The calendar months, into which the year is divided, are from 28 to 31 days in length.

Questions. — ¶ **116.** For what is dry measure used? What are its denominations? Repeat the table. How many cubic inches in a dry gallon? Describe the Winchester bushel. What measure is the quarter? What is said of the Imperial gallon? Which is the larger quantity, a quart of milk, or a quart of salt? — a quart of milk, or a quart of vinegar? — a quart of oats, or a quart of cider?

The number of days in each month may easily be remembered from the following lines:

Thirty days hath September,
April, June, and November;
All the rest have thirty-one,
Save February, which alone
Hath twenty-eight; and one day more
We add to it, one year in four.

EXAMPLES FOR PRACTICE.

1. How many seconds from Jan. 1, 1790, till March 1, 1804, including the two days named, and making allowance for leap years?

2. In 446947200 seconds how many weeks?

3. How many minutes from July 4th, M., to Sept. 29th, 6 o'clock, P. M.?

4. On what month, day, and hour, will 125640 minutes past 12 o'clock, M., July 4th, expire?

5. At Boston, on the longest days, the sun rises at 23 min. past 4 o'clock, and sets 40 min. past 7; how many seconds in such a day?

6. In 55020 seconds, how many hours?

7. How many minutes from the commencement of the war between America and England, April 19th, 1775, to the settlement of a general peace, which took place Jan. 20th, 1783?

NOTE 2. — The pupil will notice that the years 1776 and 1780 were leap years.

8. In 4079520 minutes how many years?

Circular Measure.

¶ 118. Circular Measure is used in computing latitude and longitude; also in measuring the motions of the earth, and other planets round the sun.

Questions. — ¶ 117. Of what is time the measure? What are the denominations? Repeat the table. Why is 1 day added to Feb. of certain years? Why is it not added to every 4th year? What years have 365 days, and what 366 days? What is a year of 366 days called? Name the calendar months, and the number of days in each.

The denominations are circles, signs, degrees, minutes, and seconds.

TABLE.

60 seconds (″)	make 1 minute, marked ′.
60 minutes	1 degree, " °.
30 degrees	1 sign, " S.
12 signs, or 360 degrees,	1 circle.

1. Reduce 9s. 13° 25 to seconds.

2. In 1020300″, how many degrees?

3. In 3 signs, how many minutes?

4. In 5400′, how many signs?

¶ 119. Miscellaneous Table.

20 units	make 1 score.	100 lbs. of raisins	make 1 cask.
5 score	1 hundred.	100 lbs. of fish	1 quintal.
12 units	1 dozen.	100 lbs.	1 hundred.
12 doz. = 144	1 gross.	18 inches	1 cubit.
12 gross = 144 doz.	1 great gross.	22 inches, nearly,	1 sacred cubit.
200 lbs. of beef, pork, or fish,	1 barrel.	1 gallon of train oil	7½ lbs.
		1 gallon of molasses	11 lbs.
196 lbs. of flour	1 barrel.	24 sheets of paper	1 quire
8 bushels of salt	1 hogshead.	20 quires	1 ream.
280 lbs. of salt at the salt works in N. Y.	1 barrel.	2 reams	1 bundle.
		5 bundles	1 bale.

A sheet folded in two leaves, or 4 pages, is called	a folio.
A sheet folded in four leaves, or 8 pages,	a quarto, or 4to.
A sheet folded in eight leaves, or 16 pages,	an octavo, or 8vo.
A sheet folded in twelve leaves, or 24 pages,	a duodecimo, or 12mo.
A sheet folded in 18 leaves, or 36 pages,	an 18mo.
A sheet folded in 24 leaves, or 48 pages,	a 24mo.

3 points make 1 line, } used in measuring the length of the rods of
12 lines 1 inch, } clock pendulums.
4 inches 1 hand, used in measuring the hight of horses.
6 feet 1 fathom, used in measuring depths at sea.

Reduction of Fractional Compound Numbers.

¶ 120. There are four particular cases in the reduction of fractional compound numbers. 1st, *To reduce a fraction of a higher denomination to one of a lower.* 2d, *To reduce a fraction of a lower denomination to one of a higher.* 3d, *To*

Questions.—¶ **118.** What are the uses of circula measure What are the denominations? Repeat the table.

reduce a fraction of a high denomination to integers of lower denominations. 4th, *To reduce integers of lower denominations to a fraction of a higher.* We will consider them in their order.

I. *To reduce a fraction of a higher denomination to one of a lower.*

1. Reduce $\frac{1}{280}$ of a pound to the fraction of a penny.

SOLUTION. — We must reduce $\frac{1}{280}$ of a pound to the fraction of a shilling by multiplying it by 20, since 20 shillings make £1. This done by ¶ 75, gives $\frac{1}{14}$ of a shilling, which multiplied by the composite number 12, (¶ 75, note 1,) is reduced to the fraction $\frac{6}{7}$ of a penny. Hence,

RULE.

Multiply as in the reduction of whole numbers, according to the rules for the multiplication of fractions.

II. *To reduce a fraction of a lower denomination to one of a higher.*

2. Reduce $\frac{6}{7}$ of a penny to the fraction of a pound

SOLUTION. — We must reduce $\frac{6}{7}$ of a penny to the fraction of a shilling, by dividing it by the composite number 12, since 12 pence make 1 shilling. This done by ¶ 81, note 2, gives $\frac{1}{14}$ of a shilling, which divided by 20, (multiplying the denominator,) is reduced to the fraction $\frac{1}{280}$ of a pound. Hence,

RULE.

Divide as in the reduction of whole numbers, according to the rules for the division of fractions.

EXAMPLES FOR PRACTICE.

3. Reduce $\frac{3}{760}$ of a pound of gold to the fraction of a grain.

4. Reduce $\frac{432}{19}$ of a grain of gold to the fraction of an ounce.

5. Reduce $\frac{1}{2740}$ of a hogshead of milk to the fraction of a pint.

6. Reduce $\frac{192}{685}$ of a pint of milk to the fraction of a hogshead.

7. Reduce $\frac{8}{45}$ of a hogshead of ale to the fraction of a barrel.

8. Reduce $\frac{8}{30}$ of a barrel of ale to the fraction of a hogshead.

9. Reduce $\frac{2}{22365}$ of a tun of oil to the fraction of a gill.

10. Reduce $\frac{16128}{22365}$ of a gill of oil to the fraction of a tun.

Questions. — ¶ **120.** How many cases in the reduction of fractional compound numbers? Give the first. How are integers reduced from a higher denomination to a lower? — from a lower denomination to a higher? How are fractions reduced from a higher denomination to a lower? Give the example and its explanation. Rule. How are they reduced from a lower denomination to a higher? Give Ex. 2 and the solution. Rule.

11. Reduce $\frac{1}{10000000000}$ of a square mile to the fraction of a square inch.

12. Reduce $\frac{156816}{390625}$ of a square inch to the fraction of a square mile.

13. Reduce $\frac{5}{2688}$ of a bushel to the fraction of a pint.

14. Reduce $\frac{5}{42}$ of a pint to the fraction of a bushel.

15. Reduce $\frac{11}{10080}$ of a week to the fraction of an hour.

16. Reduce $\frac{11}{60}$ of an hour to the fraction of a week.

17. A cucumber grew to the length of $\frac{1}{3960}$ of a mile; what part is that of a foot?

18. A cucumber grew to the length of 1 foot 4 inches $= \frac{16}{12} = \frac{4}{3}$ of a foot; what part is that of a mile?

19. Reduce $\frac{2}{9}$ of $\frac{1}{6}$ of a pound to the fraction of a shilling.

20. $\frac{20}{27}$ of a shilling is $\frac{2}{9}$ of what fraction of a pound?

21. Reduce $\frac{1}{8}$ of $\frac{2}{11}$ of 3 pounds to the fraction of a penny?

22. $\frac{180}{11}$ of a penny is $\frac{1}{8}$ of what fraction of 3 pounds? $\frac{180}{11}$ of a penny is $\frac{2}{11}$ of what part of 3 pounds? $\frac{180}{11}$ of a penny is $\frac{1}{8}$ of $\frac{2}{11}$ of how many pounds?

¶ 121. III. *To reduce a fraction of a high denomination to integers of lower denominations.*

1. How many shillings and pence in $\frac{2}{3}$ of a pound?

Solution. — Multiplying $\frac{2}{3}$ of a pound by 20, it is reduced to the fraction of a shilling, $\frac{40}{3}$. But as $\frac{40}{3}$ of a shilling is an improper fraction, (¶ 65,) it contains several shillings. The whole shillings we find, dividing the numerator by the denominator, to be 13, and a fraction, $\frac{1}{3}$, of a shilling remains, and this reduced to the fraction of a penny, is $\frac{12}{3}$ of a penny $=$ 4d. Hence 13s. 4d. is the *Ans.*

IV. *To reduce integers of lower denominations to a fraction of a higher denomination.*

2. What part of a pound is 13s. 4d.?

Solution. — In a whole pound there are 240 pence, and we wish to find what part of this number of pence is contained in 13s. 4d. 13s. 4d. reduced to pence, is 160 pence. Hence 13s. 4d. is 160 out of 240 pence contained in a whole pound, or $\frac{160}{240} = \frac{2}{3}$ of a pound, *Ans.*

Note 1. — The numerator and denominator of a fraction must be of the same denomination, since the former is a dividend, and the latter a divisor, both of which must be of one denomination, (¶ 33.) Hence, if there is a fractional part to the integer of the lowest denomination, for example, were it required to reduce 4d. $3\frac{2}{3}$qrs. to the fraction of a shilling, we should have to reduce 1s. to 3ds of a farthing for a denominator, and 4d. $3\frac{3}{5}$qrs. to 3ds of a farthing for a numerator. The former will then show the number of 3ds of a farthing in 1s., the latter how many of them are contained in 4d. $3\frac{3}{5}$qrs.

Hence,

RULE.

Reduce the given fraction to the next lower denomination, and, if it is then an improper fraction, reduce it to a whole or mixed number,—the integer is the number of this denomination. If a fraction remains, reduce it, as before, to the next lower denomination. So proceed, if a fraction continues to remain, to the lowest denomination.

Hence,

RULE.

Reduce the given sum to the lowest denomination contained in it for a numerator, and a unit of the required higher denomination to the same denomination for the denominator.

EXAMPLES FOR PRACTICE.

3. Reduce $\frac{3}{5}$ of 1 day to hours and minutes.

OPERATION.

```
 Numer. 3
        24
        ——
Denom. 5)72(14h. 24m. Ans.
        5
        —
        22
        20
        ——
         2
        60
        ———
        120
        10
        ———
         20
         20
```

4. Reduce 14h. 24min. to the fraction of a day.

OPERATION.

```
14h. 24 min.    1 da.
 60             24
 ——             ——
864 Numer.      24
                60
               ————
               1440 Denom.
```

$\frac{864}{1440} = \frac{3}{5}$ *day, Ans.*

Questions.—¶ **121.** What is case III.? Give the solution; the rule. Case IV.; the solution; the rule. Why must the numerator and denominator of a fraction be of the same denomination? What follows, then, in case of a fractional part?

5. What is the value of $\frac{3}{5}$ of a pound, Troy?

6. Reduce 7 oz. 4 pwt. to the fraction of a pound, Troy.

7. What is the value of $\frac{5}{8}$ of a pound, avoirdupois?

8. Reduce 8 oz. $14\frac{2}{5}$ dr. to the fraction of a pound, avoirdupois.

9. Reduce $\frac{4}{7}$ of a mile to its proper quantity.

10. Reduce 4 fur. 125 yds 2 ft. 1 in. $2\frac{1}{7}$ bar. to the fraction of a mile.

11. $\frac{4}{5}$ of a week is how many days, hours, and minutes?

12. 5 d. 14 h. 24 m. is what fraction of a week?

13. Reduce $\frac{7}{16}$ of an acre to its proper quantity.

14. Reduce 1 rood 30 poles to the fraction of an acre.

15. What is the value of $\frac{9}{10}$ of a yard?

16. Reduce 2 ft. 8 in. $1\frac{1}{5}$ b. to the fraction of a yard.

NOTE. — Let the pupil be required to reverse and prove the following examples:

17. Reduce 3 roods $17\frac{1}{2}$ poles to the fraction of an acre.

18. A man bought 27 gal. 3 qts. 1 pt. of molasses; what part is that of a hogshead?

19. A man purchased $\frac{5}{13}$ of 7 cwt. of sugar; how much sugar did he purchase?

20. 13 h. 42 m. $51\frac{3}{7}$ s. is what part or fraction of a day?

Reduction of Decimal Compound Numbers.

¶ 122. I. *To reduce the decimal of a higher denomination to integers of lower denominations.*

1. Reduce '375£ to integers of lower denominations

OPERATION.

```
'375£.
  20
------
7'500s.
  12
------
6'000d. Ans. 7s. 6d.
```

SOLUTION. — Multiplying '375 of a pound by 20, it is reduced to

II. *To reduce integers of lower denominations to a decimal of a higher denomination.*

2. Reduce 7s. 6d. to the decimal of a pound.

OPERATION.

```
12 | 6'0
   |------
20 | 7'500
   |------
     '375 of a pound, Ans
```

SOLUTION. — We divide 6d by 12, to reduce it to shillings, but

7'500s., observing the ordinary rule for pointing of decimals in the product, that is, 7 shillings, and '500 of a shilling. This decimal multiplied by 12 becomes 6'000d., that is, 6d. and no decimal. Hence '375 of a pound = 7s. 6d. *Ans.*

Hence,

RULE.

Multiply the given decimal by the number which will reduce it to the next lower denomination, pointing off decimals according to the ordinary rule; reduce this decimal to the next lower denomination, pointing off as before. So continue to do through all the denominations; the several integers will be those required.

as it will not make a whole shilling, we annex a cipher to reduce it to 10ths, (¶ 91;) then 12 in 60 tenths, 5 tenths of a shilling; annexing this to 7 shillings, we have 75 tenths of a shilling to reduce to the decimal of a pound, and dividing by 20, annexing ciphers to reduce the remainder to hundredths and thousandths, we have '375 of a pound.

Hence,

RULE.

Divide the lowest denomination, annexing ciphers as may be necessary, by the number which will reduce it to the next higher, and annexing the quotient to the number of this higher denomination, divide as before. So continue to do till the whole is brought to the required decimal.

EXAMPLES FOR PRACTICE.

3. Reduce '213 of a "long ton" to integers of lower denominations.

4. Reduce 4 cwt. 1 qr. 1 lb. 1 oz. 14'72 drs. to the decimal of a "long ton."

5. Reduce '6 of a lb. of emetic tartar to integers of lower denominations.

6. Reduce 7 ℥. 1 ʒ. 1 ℈. 16 grs. to the decimal fraction of 1 lb.

7. In '76754 of a square mile, how many integers of lower denominations?

8. What decimal of a square mile is 491 acres 36 square rods 26 square feet 19'584 square inches?

9. Reduce '3958 of a barrel of wine to integers of lower denominations.

10. Reduce 12 gal. 1 qt. 1 pt. 2'9664 gills of wine to the decimal of a barrel.

Questions.—¶ **122.** How many cases in the reduction of decimal compound numbers? What is case I.? Give the example and its solution. Give the rule. What is case II.? — the solution of the example? — the rule?

11. How many integers of lower denominations is ‘73 of a cord?

12. What decimal of a cord is 5 C. ft. 13 cu. ft. 760‘32 cu. inches?

13. In ‘648 of a quarter of wheat, how many integers of a less denomination?

14. In 5 bu. 5 qts. 1‘776 pts. of wheat, what fraction of a quarter?

15. Reduce ‘125 lbs. Troy to integers of lower denominations.

16. Reduce 1 oz. 10 pwt. to the fraction of a pound.

17. What is the value of ‘72 hhd. of beer?

18. Reduce 38 gals. 3‘52 qts. of beer, to the decimal of a hhd.

19. What is the value of ‘375 of a yard?

20. Reduce 1 qr. 2 na. to the decimal of a yard.

21. What is the value of ‘713 of a day?

22. Reduce 17 h. 6m. 43$\frac{1}{5}$ sec. to the decimal of a day.

Let the pupil be required to reverse and prove the following examples:

23. Reduce 4 poles to the decimal of an acre.

24. What is the value of ‘7 of a lb. of silver?

25. Reduce 18 hours 15 m. 50‘4 sec. to the decimal of a day.

26. Reduce 11 mi. 6 fur. 2 rods 3 yds. 2 ft. to the decimal of a degree on the equatorial circumference of the earth.

¶ 123. Review of Reduction of Compound Numbers.

Questions. — What are compound numbers? What is meant by the word denomination? What is reduction of compound numbers? What are the kinds, and how performed? Changing ells Eng. to yards is reduction — what kind? What is the use and what the denominations of Troy weight? Avoirdupois weight? Which is larger, 1 oz. Troy, or 1 oz. Avoirdupois? — 1 lb. Troy or 1 lb. Avoirdupois? What distinction do you make between the *long* and the short *ton*, and where are the two used? What distinctions do you make between linear, square, and cubic measure? What are the denominations in linear measure? — in square measure? — in cubic measure? How do you multiply by $\frac{1}{2}$? When the divisor contains a fraction, how do you proceed? How are the superficial contents of a square figure found? How is the solid contents of any body found in cubic measure? How many solid or cubic feet of wood make a cord? What is understood by a *cord foot?* How many such feet make a cord? How many rods in length is Gunter's chain? Of how many links does it consist? How many links make a rod? How many rods in a mile? How many square rods in an acre?

How many pounds make 1 cwt.? For what is circular measure used? Into how many parts is a *small* circle divided? — a *large* circle? — called what.

EXERCISES.

1. In 46£. 4s. sterling, how many dollars? (Consult ¶ 104, note 3.) *Ans.* $223'608.

2. How many rings, each weighing 5 pwt. 7 grs., may be made of 3 lb. 5 oz. 16 pwt. 2 grs. of gold? *Ans.* 158.

3. Suppose West Boston bridge to be 212 rods in length, how many times will a chaise wheel, 18 feet 6 inches in circumference, turn round in passing over it?

Ans. $189\frac{3}{37}$ times.

4. In 10 lb. of silver, how many spoons, each weighing 5 oz. 10 pwt.? *Ans.* $21\frac{9}{11}$ spoons.

5. How many shingles, each covering a space of 4 inches one way, and 6 inches the other, would it take to cover 1 square foot? How many to cover a roof 40 feet long, and 24 feet wide? (See ¶ 48.)

Ans. to the last, 5760 shingles.

6. How many cords of wood in a pile 26 feet long, 4 feet wide, and 6 feet high? *Ans.* 4 cords, and 7 cord feet.

7. There is a room 18 feet long, 16 feet wide, and 8 feet high; how many rolls of paper, 2 feet wide and 11 yards long, will it take to cover the walls? *Ans.* $8\frac{8}{33}$.

8. How many cord feet in a load of wood $6\frac{1}{2}$ feet long, 2 feet wide, and 5 feet high? *Ans.* $4\frac{1}{16}$ cord feet.

9. If a ship sail 7 miles an hour, how far will she sail in 3 w. 4 d. 16 h.?

10. A merchant sold 12 hhds. of brandy, at $2'75 a gallon; what did he receive for each hogshead, and to how much did the whole amount?

11. A goldsmith sold a tankard for £10 8s. at the rate of 5s. 4d. per ounce? how much did it weigh? *Ans.* 3 lbs. 3 oz.

12. An ingot of gold weighs 2 lb. 8 oz. 16 pwt.; how much is it worth at 3d. per pwt.?

13. If a cow give, on an average, 9 qts. of milk each day, how much will she give in a year, or 365 days?

Ans. 15 hhd. 11 gal. 1 qt.

14. Reduce 14445 ells Flemish to ells English.

15. There is a house, the roof of which is $44\frac{1}{2}$ feet in length, and 20 feet in width, on each of the two sides; if 3 shingles in width cover one foot in length, how many shingles will it take to lay one course? If 3 courses make one

foot, how many courses will there be on one side of the roof? How many shingles will it take to cover one side? —— to cover both sides? *Ans.* 16020 shingles.

16. How many steps, of 30 inches each, must a man take in traveling $54\frac{1}{2}$ miles?

17. How many seconds of time would a person redeem in 40 years, by rising each morning $\frac{1}{2}$ hour earlier than he now does?

18. If a man lay up 4 shillings each day, Sundays excepted, how many dollars would he lay up in 45 years?

19. If 9 candles are made from 1 pound of tallow, how many dozen can be made from 24 pounds?

20. If one pound of wool make 60 knots of yarn, how many skeins, of ten knots each, may be spun from 4 pounds 6 ounces of wool? *Ans.* $26\frac{1}{4}$ skeins.

21. How many hours from the commencement of the common Christian era till Dec. 10, 1847, 12 o'clock, noon, allowance being made for leap years? How many weeks?

* *Ans.* { 16189932 hours. / $96368\frac{9}{14}$ weeks. }

22. What part of a pwt. is $\frac{7}{1920}$ of a pound Troy?
Ans. $\frac{7}{8}$ pwt.

23. What fraction of a pound is $\frac{3}{4}$ of a farthing?
Ans. £$\frac{1}{1280}$.

24. What fraction of an ell English is 4 qrs. $1\frac{1}{2}$ na.?
Ans. $\frac{7}{8}$ E. E.

25. What fraction of a yard is 2 qrs. $2\frac{2}{3}$ na.?
Ans. $\frac{2}{3}$ yd.

26. What fraction of a day is 16 h. 36 m. $55\frac{5}{13}$ s.?
Ans. $\frac{9}{13}$ d.

27. What fraction of a mile is 6 fur. 26 r. 11 ft.?
Ans. $\frac{5}{6}$ mi.

28. What is the value of $\frac{3}{13}$ of a "long ton?"
Ans. 4 cwt. 2 qrs. 12 lb. 14 oz. $12\frac{4}{13}$ drs.

29. What decimal of a day is 55 m. 37 sec.?
Ans. '03862 + d.

30. What decimal of a pound Troy is 10 oz. 13 pwt. 9 gr.?
Ans. '8890625.

31. What is the value of '397 of a yard?
Ans. 1 qr. 2 na. + lb.

* Consult ¶ 117, Note 1.

Addition of Compound Numbers.

¶ 124. 1. A boy bought a knife for 1 shilling 9 pence, and a comb for 1 shilling 6 pence; how much did he give for both? *Ans.* 3s. 3d.

2. A grocer sold at one time 2 qts. of molasses, at another time 3 qts., at another 1 qt., at another 3 qts., and at another 2 qts.; how many gallons did he sell?

SOLUTION. — 3 qts. + 2 qts. + 1 qt. + 3 qts. + 2 qts. = 11 qts., and 11 qts. = 2 gal. 3 qts. *Ans.* 2 gal. 3 qts.

3. A boy had 30 rods to walk; he walked the first 10 rods in 30 seconds, the next 10 rods in 45 sec., and the last 10 rods in 20 sec., how many minutes was he in walking the 30 rods? *Ans.* 1 min. 35 sec.

4. What is the amount of 1 yd. 2 ft. 6 in + 2 yds. 1 ft. 8 in.? *Ans.* 4 yds. 1 ft. 2 in.

5. A man has two bottles which he wishes to fill with wine; one will contain 2 gal. 3 qts. 1 pt., and the other 3 qts.; how much wine can he put in them? *Ans.* 3 gal. 2 qts. 1 pt.

The uniting together in one sum of several compound numbers is called Compound Addition.

6. A man bought a horse for £15 14s. 6d., a pair of oxen for £20 2s. 8d., and a cow for £5 6s. 4d.; what did he pay for all?

SOLUTION. — As the numbers are large, we write them down, placing those of the same denomination under each other, and, beginning with those of the least value, add up each kind separately; thus:—

OPERATION.

	£	*s.*	*d.*
	15	14	6
	20	2	8
	5	6	4
Ans.	41	3	6

Then, adding up the pence, we find the amount to be 18, which we divide by 12 to reduce to shillings; the remainder, which is pence, we write under the column of pence, and add the 1 shilling to the column of shillings. Adding up the shillings, we reduce them to pounds; setting the remainder, 3 shillings, under the column of shillings, we carry the one pound to the column of pounds, which we add up, setting down the whole amount, as we do the amount of the last column in simple addition.

NOTE. — The operations in compound numbers differ from those of simple numbers but in one particular. In *simple* numbers we are governed by the law that figures increase in a tenfold proportion from right to left, (¶ 9.) *Compound* numbers having no regular system of units, we are governed by the *relations* between the *different denominations* in which the several quantities are expressed.

The above process is sufficient to establish the following

RULE

For Addition of Compound Numbers.

I. Write the numbers so that those of the same denomination may stand under each other.

II. Add together the numbers in the column of the lowest denomination, and carry for that number which it takes of the same to make one of the next *higher* denomination. Proceed in this manner with all the denominations, till you come to the last, whose amount is written as in simple numbers.

PROOF. — The same as in addition of simple numbers.

EXAMPLES FOR PRACTICE.

7.

y.	w.	d.	h.	m.	s.
57	7	6	23	55	11
84	8	0	16	42	18
32	24	5	5	18	5

8.

T.	cwt.	qr.	lb.	oz.	dr.
14	11	1	16	5	10
25	0	2	11	9	15
7	18	0	25	11	9

9. Bought a silver tankard, weighing 2 lb. 3 oz., a silver cup, weighing 3 oz. 10 pwt., and a silver thimble, weighing 2 pwt. 13 grs.; what was the weight of the whole?

Ans. 2 lb. 6 oz. 12 pwt. 13 grs.

10. A ship landed at N. York the following invoice of English goods, viz., 78 tons 3 cwt. 2 qrs. 26 lbs. of cotton goods, 135 tons 15 cwt. 1 qr. 9 lbs. of iron, 90 tons 12 cwt. 2 qrs. 20 lbs. of woollen goods, 225 tons 9 cwt. 17 lbs. of coal, and 106 tons 1 qr. of earthen ware; what was the whole amount? *Ans.* 636 tons 1 cwt. 16 lbs.

Questions. — ¶ **124.** What is compound addition? How are lower denominations reduced to higher? (¶ 105.) Repeat the rule for compound addition. How do you carry from farthings to pence? — from pence to shillings? — from shillings to pounds? How do operations in compound numbers differ from those in simple numbers? How do you carry through the several denominations of avoirdupois weight? — long measure? &c., &c.

NOTE.—It will be recollected, (¶ 106,) that what is called the "long ton" is used in invoices of English goods, and of coal from Pennsylvania.

11. A boat took in freight as follows: at one place, 9576 lbs. of butter; at another, 11 tons of pork; at a third, 7 T. 18 cwt. 27 lbs. of coal; what was the entire freight in "short tons?" *Ans.* 24 tons 1299 lbs.

12. A merchant bought 3 pieces of linen, measuring as follows: 41 E. Fl. 1 qr. 2 na., 18 E. Fl. 2 qr. 3 na., 57 E. Fl. 1 na; how many Flemish ells in the whole? *Ans.* 117 E. Fl. 1 qr. 2 na.

13. A draper bought 3 pieces of English broadcloth measuring as follows, viz., 75 E. E. 4 qr. 2 na., 31 E. E. 1 qr., 28 E. E. 1 na.; how many English ells in the whole? *Ans.* 135 E. E. 3 na.

14. There are four pieces of cloth, which measure as follows, viz., 36 yds. 2 qrs. 1 na., 18 yds. 1 qr. 2 na., 46 yds. 3 qrs. 3 na., 12 yds. 0 qr. 2 na.; how many yards in the whole? *Ans.* 114 yards.

15. A man travelled as follows, viz., the 1st day, 35 mi. 7 fur. 38 rd.; 2d day, 4 mi. 2 rd.; 3d day, 37 mi. 3 fur. 19 rd.; 4th day, 44 mi.; what was the length of his journey? *Ans.* 121 mi. 3 fur. 19 rd.

16. Bought of Williams and Brother, London, 1 copy of Shakspeare for £7 14s. 6d., 1 copy of Arnold's Works for £19 10s. 9d., 1 copy of the Edinburgh Encyclopedia for £27 6s., 1 quarto Bible for £8 6d., 1 copy of Johnson's Works for £15 2s.; what did the whole cost? *Ans.* £77 13s. 9d.

17. Bought at Liverpool 1 bale of cotton goods for £9 10s. 3d., 1 box of jewelry for £227 4s., 1 gross of buttons for £6 9s. 8d.; what did I pay for the whole? *Ans.* £243 3s. 11d.

18. There are 3 fields, which measure as follows, viz., 17 A. 3 R. 16 P., 28 A. 5 R. 18 P., 11 A. 25 P.; how much land in the three fields? *Ans.* 58 A. 1 R. 19 P.

19. A raft of hewn timber consisted of 3 cribs; the 1st crib contained 29 T. 36 cu. ft. 1229 cu. in.; the 2d, 12 T 19 cu. ft. 64 cu. in.; the 3d, 8 T. 11 cu. ft. 917 cu. in.; how much timber did the raft contain? *Ans.* 50 T. 27 cu. ft. 482 cu. in.

20. A man removed 79 cu. yds. 22 cu. ft. of earth in digging a cellar, 9 cu. yds. 26 cu. ft. in digging a drain, and 22 cu. yds. 17 cu. ft. in digging a cistern; how much earth did he remove? *Ans.* 112 cu. yds. 11 cu. ft

21. In one pile of wood are 37 cords 1.9 cu. ft. 76 cu. in.; in another, 9 cords 104 cu. ft.; in a 3d, 48 cords 7 cu. ft. 127 cu. in.; in a 4th, 61 cords 139 cu. in.; how much wood in the four piles? *Ans.* 156 C. 102 ft. 342 in.

22. A vintner sold in one week 51 hhd. 53 gal. 1 qt. 1 pt. of wine; in another week, 27 hhd. 39 gal. 3 qts.; and in another week, 9 hhd. 13 gal. 3 qts.; how much did he sell in the three weeks? *Ans.* 88 hhd. 43 gal. 3 qts. 1 pt.

23. A milk-man sold in one week, 70 gal. 3 qts. of milk; in another week, 67 gal. 1 qt.; how many hogsheads did he sell? *Ans.* 2 hhds. 30 gal.

24. A farmer sowed 36 bush. 2 pks. 5 qts. 1 pt. of wheat, and 19 bush. 3 pks. 7 qts. of barley; how many bushels did he sow? *Ans.* 56 bush. 2 pks. 4 qts. 1 pt.

25. A printer used in one week 6 bales, 7 reams, 9 quires and 9 sheets of paper, and in another week 14 bales, 9 reams, 19 quires and 15 sheets; how much paper did he use in the two weeks? *Ans.* 21 bales 7 reams 9 quires.

26. A ship sailed in one week as follows, viz., on Monday, 3° 8′ 45″ south, 1° 51′ east; on Tuesday, 2° 36′ south, 2° 1′ 15″ east; on Wednesday, 4° 52″ south, 1° east; on Thursday, 1° 48′ 52″ south, 3° 16′ 22″ east; on Friday, 1° 19′ south, 48′ 29″ east; and on Saturday, 59′ 30″ south, 3° 52′ 11″ east; what was her distance south and east from the place of starting? *Ans.* { South 13° 52′ 59″. East 12° 49′ 17″.

27. A man plastered a church of the following dimensions, viz., the end walls contained 116 sq. yds. 7 sq. ft. 96 sq. in. each; the side walls, 178 sq. yds. 138 sq. in. each, and the ceiling 439 sq. yds. 6 sq. ft. 78 sq. in.; what was the whole amount of plastering in the church?

Ans. 1029 sq. yds. 5 sq. ft. 114 sq. in.

28. What is the amount of 40 weeks 3 d. 1 h. 5 m. + 16 w. 6 d 4 m. + 27 w. 5 d. 2 h.? *Ans.* 85 wk. 3 h. 9 m.

29. A miller sold flour as follows, viz., 4 bar. 176 lbs. 8 oz., 18 bar. 40½ lbs., 1 bar. 104 lbs. 7 oz., 181¾ bar., how much did he sell in all? *Ans.* 206 bar. 76 lbs. 7 oz.

30. Five bags of wheat weighed as follows, viz., 2¼ bush. 2 bush. 21 lb. 7 oz., 1½ bush. 18 lbs., 2 bush. 50 lbs., 1 bush. 58¾ lbs.; what was their entire weight, calling 60 lbs. a bushel? *Ans.* 11 bush. 13 lbs. 3 oz.

NOTE.—If the four following examples be correctly wrought, the results will be the same as hose here given.

31. What is the sum of 35 bar. 27 gal. 3 qts. + 19 bar 5 gal. 1 qt. + 7 bar. 13 gal. 3 qts.?

Ans. 62 bar. 15 gal. 1 qt.

32. What is the sum of 12 rd. 9 ft. 4 in. + 15 rd. 7 ft. 8 in. + 6 rd. 4 ft. 5 in.? *Ans.* 34 rd. 4 ft. 11 in.

33. What is the sum of the following distances on the equatorial circumference of the earth, viz., 59 deg. 46 mi. 6 fur. 39 rds. 15 ft. 10 in.; 216 deg. 39 mi. 7 fur. 39 rds. 4 ft. 7 in.; 78 deg. 53 mi. 7 fur. 38 rds. 9 ft. 8 in.?

Ans. 355 deg. 2 mi. 4 fur. 11 rds. 2 ft. 7 in.

34. What is the sum of 2 A. 75 P. 248 sq. ft. 72 sq. in. + 3 A. 120 P. 177 sq. ft. 85 sq. in. + 15 A. 17 P. 84 sq. ft. 80 sq. in.? *Ans.* 21 A. 53 P. 238 sq. ft. 57 sq. in.

35. A hardware merchant sold several bills of screws, as follows: 25 great gross 9 gross 7 doz. 11 screws; 15 great gross 7 gross 8 doz.; 40 great gross 4 doz.; what was the whole amount sold?

Ans. 81 gr. gr. 5 gr. 7 doz. 11 screws.

Addition of Fractional Compound Numbers.

¶ 125. 1. To $\frac{7}{8}$ of an hour, add $\frac{7}{8}$ of a minute.

SOLUTION. — First reduce each fraction to its proper quantity; see ¶ 121. $\frac{7}{8}$ of an hour = 52 min. 30 sec.; $\frac{7}{8}$ of a minute = $52\frac{1}{2}$ sec.; and 52 min. 30 sec. + $52\frac{1}{2}$ sec. = 53 min. $22\frac{1}{2}$ sec.

Ans. 53 min. $22\frac{1}{2}$ sec.

NOTE. — It may sometimes be more convenient to reduce the fractions to the same denomination, add them together, and reduce their fractional sum to its proper quantity.

2. To $\frac{7}{8}$ of a pound, add $\frac{3}{4}$ of a shilling. *Ans.* 18s. 3d.

3. To $\frac{5}{6}$ of a gallon, add $\frac{3}{4}$ of a pint. *Ans.* 3 qts. $1\frac{5}{12}$ pts.

4. To $\frac{1}{2}$ lb. Troy, add $\frac{7}{12}$ of an oz.

Ans. 6 oz. 11 pwt. 16 gr.

5. To $\frac{3}{4}$ of a mile, add $47\frac{3}{11}$ rods. *Ans.* 239 rds. 4 ft. 6 in.

6. To $\frac{3}{8}$ of $20\frac{1}{2}$ yds., add $\frac{5}{6}$ of $9\frac{1}{4}$ yds.

Ans. 15 yds. 1 qr. $2\frac{1}{3}$ na.

Subtraction of Compound Numbers.

¶ 126. 1. From a piece of tape, containing 9 yds. 3 qrs., sold 4 yds. 1 qr.; how much remained?

2. A woman having 6 lbs. of butter, sold 3 lbs. 10 oz.; how much had she left?

SOLUTION. — Since there are no oz. in the first number or minuend, we take from 6 lbs. 1 lb. = 16 oz.; 16 oz. — 10 oz. = 6 oz., and 5 lb. — 3 lbs. = 2 lbs. *Ans.* 2 lbs. 6 oz.

3. How much is 1 ft. — (less) 6 in.? 1 ft. — 8 in.? 6 ft. 3 in. — 1 ft. 6 in.? 7 ft. 8 in. — 4 ft. 2 in.? 7 ft. 8 in. — 5 ft. 10 in.?

4. How much is 4 weeks 3d. — 3 w. 4 d.? 3 w. 1 d. — 2 w. 5 d.?

Finding the difference between two compound numbers, is called Compound Subtraction.

5. From 9 days 15 h. 30 m., take 4 d. 9 h. 40 min.

OPERATION.

	d.	*h.*	*m.*
Minuend,	9	15	30
Subtrahend,	4	9	40
Ans.	5	5	50

SOLUTION. — As the quantities are large, it will be more convenient to write them down, the less under the greater, minutes under minutes, hours under hours, &c., since we must subtract those of the same denomination from each other. We cannot take 40 m. from 30 m., but we may, as in simple numbers, borrow from the 15 h. in the minuend, 1 hour = 60 minutes, which added to 30 m. in the minuend, makes 90 m., and 40 m. from 90 m. leaves 50 m., which we set down.

Proceeding to the hours, having borrowed 1 from the 15 h. in the minuend, we must make this number 1 less, calling it 14 h., and say, 9 (hours) from 14 (hours) leaves 5, (hours,) which we set down.

Lastly, proceeding to the days, 4 d. from 9 d. leaves 5 d., which we set down, and the work is done.

6. From £27, take £15 12s. 6d.

OPERATION.

27£.	*s.*	*d.*
15	12	6
11	7	6

SOLUTION. — We have no pence from which to take the 6d., but we must go to the pounds, and borrow £1 = 20s., and from the 20s. borrow 1s. = 12d. Then 6d. from 12d. leave 6d.; 12s. from 19s. (which remain of the £1) leave 7s.; and £15 from £26 remaining, leave £11. *Ans.* £11 7s. 6d.

The process in the foregoing examples may be presented in form of a

RULE

For Subtracting Compound Numbers.

I. Write the less quantity under the greater, placing similar denominations under each other.

II. Beginning with the least denomination, take the lower

number in each from the upper, and write the remainder underneath.

III. If the lower number of any denomination be greater than the upper, borrow one from the next higher denomination of the minuend, reduce it to this lower denomination, subtract the lower number therefrom, and to the remainder add the upper number, remembering to call the denomination from which you borrowed 1 less.

Proof. — The same as in simple subtraction.

EXAMPLES FOR PRACTICE.

7. A London merchant sold goods to a New York house to the amount of £136 7s. 6½d., and received in payment £50 10s. 4¾d.; how much remained due?

Ans. £85 17s. 1¾d.

8. A man bought a farm in Canada West for £1256 10s., and, in selling it, lost £87 10s. 6d.; how much did he sell it for? *Ans.* £1168 19s. 6d.

9. A hogshead of molasses, containing 118 gal., sprang a leak, when it was found only 97 gal. 3 qts. 1 pt. remained in the hogshead; how much was the leakage?

Ans. 20 gal. 0 qt. 1 pt.

10. There was a silver tankard which weighed 3 lb. 4 oz.; the lid alone weighed 5 oz. 7 pwt. 13 grs.; how much did the tankard weigh without the lid?

Ans. 2 lb. 10 oz. 12 pwt. 11 grs.

11. From 256 A. 1 R. 10 P., take 87 A. 6 P. 10 sq. yd.

Ans. 169 A. 1 R. 3 P. 20 sq. yds. 2 sq. ft. 36 sq. in.

12. From 15 lb. 2 oz. 5 pwt., take 9 oz. 8 pwt. 10 grs.

13. Bought a piece of black broadcloth, containing 36 yds 2 qrs.; two pieces of blue, one containing 10 yds. 3 qrs. 2 na., the other, 18 yds. 3 qrs. 3 na.; how much more was there of the black than of the blue?

14. A farmer has two mowing fields; one containing 13 acres 3 roods, the other, 14 acres 3 roods; he has two pastures, also; one containing 26 A. 2 R. 27 P., the other, 45 A. 2 R. 33 P.; how much more has he of pasture than of mowing?

Questions. — ¶ **126.** What is compound subtraction? How do you write the quantities? — the denominations? Why do you so write them? When the lower number of any denomination exceeds the upper, how do you proceed? Why? Repeat the rule for compound subtraction. Proof.

NOTE.—If the five following examples be correctly wrought, the results will be the same as those here given.

15. From 28 mi. 5 fur. 16 rd., take 15 mi. 6 fur. 26 rd. 12 ft. *Ans.* 12 mi. 6 fur. 29 rd. 4 ft. 6 in.

16. From 27 P. 16 sq. ft. 71 sq. in., take 11 P. 110 sq. ft. 60 sq. in. *Ans.* 15 P. 178 sq. ft. 47 sq. in.

17. From 19 P. 55 sq. ft. 126 sq. in., take 7 P. 92 sq. ft 11 sq. in. *Ans.* 11 P. 236 sq. ft. 7 sq. in.

18. From 64 A. 2 R. 11 P. 29 sq. ft., take 26 A. 7 R. 34 P. 132 sq. ft. *Ans.* 36 A. 2 R. 16 P. 169 sq. ft. 36 sq. in.

19. From 9 rd. 5 yds. 2 ft. 11 in., take 10 rd. 0 yd. 1 ft. 2 in.? *Ans.* 3 in.

20. From a pile of wood, containing 21 cords, was sold, at one time, 8 cords 76 cubic feet; at another time, 5 cords 7 cord feet; what was the quantity of wood left? *Ans.* 6 cords 68 cu. ft.

21. London is 51° 32′, and Boston 42° 23′ N. latitude; what is the difference of latitude between the two places? *Ans.* 9° 9′.

22. Boston is 71° 3′, and the city of Washington is 77° 43′ W. longitude; what is the difference of longitude between the two places? *Ans.* 6° 40′.

23. The moon is 8 signs 12° 25′ 45″ east of the sun, and Mars is 11 signs 4° 50′ 28″ east of the sun; how far is Mars east of the moon? *Ans.* 2 s. 22° 24′ 43″.

24. An apothecary had 9 ℔ 8 ℥ 2 ʒ 1 ℈ 13 grs. of jalap, but has used, in various mixtures, 4 ℔ 7 ℥ 5 ʒ 2 ℈ 17 grs. what quantity has he left? *Ans.* 5 ℔ 4 ʒ 1 ℈ 16 grs.

25. From 124 lbs. 14 oz. 6 drs. of Epsom salts, sold 116 lbs. 7 oz. 13 drs.; what quantity remained unsold? *Ans.* 8 lbs. 6 oz. 9 drs.

26. Shipped to London 725 qrs. 3 bu. 2 pecks of Indian corn, but unfortunately 218 qrs. 5 bu. 3 pks. became damaged; what quantity remained uninjured? *Ans.* 506 qrs. 5 bu. 3 pks.

27. A ship loaded with 615 T. 7 cwt. of coal for Boston, encountering a tempest, a part is thrown overboard; there were weighed out on landing, 409 T. 13 cwt. 2 qrs. 27 lbs.; how much was lost? *Ans.* 205 T. 13 cwt. 1 qr. 1 lb.

¶ 127. *Distance of* TIME *from one date to another.*

The distance of time from one date to another may be found by subtracting the first date from the last, observing to number the months according to their order.

1. A note, bearing date Dec. 28th, 1846, was paid Jan. 2d, 1847; how long was it at interest?

OPERATION.

A. D. 1847.	1st m.		2d day.
1846.	12	"	28 "
Ans. 0	0		4

NOTE.—In casting interest, and in finding the difference of time between dates, *each* month is reckoned 30 days.

2. A note, bearing date Oct. 20th, 1823, was paid April 25th, 1825; how long was the note at interest?

3. What is the difference of time from Sept. 29th, 1844, to April 2d, 1847? *Ans.* 2 y. 6 m. 3 d.

4. A man bought a farm April 14th, 1842, and was to pay for it Sept. 1st, 1847, paying interest after Oct. 30th, 1843; how much time had he in which to pay for the farm? How much time without interest? For how long a time was he to pay interest?

Ans. 5 yr. 4 m. 17 d.; 1 " 6 " 16 "; 3 " 10 " 1 "

Subtraction of Fractional Compound Numbers

¶ 128. 1. From $\frac{2}{5}$ of a week, take $\frac{11}{16}$ of a day.

SOLUTION.—First reduce each fraction to its proper quantity. $\frac{2}{5}$ of a week = 2 da. 19 h. 12 m.; $\frac{11}{16}$ of a day = 16 h. 30 m.; and 2 da 19 h. 12 m. — 16 h. 30 m. = 2 da. 2 h. 42 m. *Ans.* 2 da. 2 h. 42 m

NOTE.—We may, if we please, reduce the fractions to the same denomination, subtract them, and reduce their fractional difference to its proper quantity.

2. From $\frac{3}{5}$ of an ounce, take $\frac{7}{8}$ of a pwt. *Ans.* 11 pwt. 3 grs.

3. From $\frac{29}{32}$ of a bushel, take $\frac{15}{16}$ of a peck.
Ans. 2 pk. 5 qts. 1 pt.

4. From $\frac{1}{24}$ of a mile, take $\frac{1}{6}$ of a furlong. *Ans.* 6 rds. 11 ft.

5. From $\frac{4}{5}$ of $19\frac{1}{4}$ gallons, take $\frac{1}{2}$ of $3\frac{4}{5}$ quarts.
Ans. 14 gal. 3 qts. 1 pt. $1\frac{3}{5}$ gi

Questions.—¶ 127. How is the distance of time from one date to another found? 1 month is reckoned how many days?

Multiplication and Division of Compound Numbers.

¶ **129.** *To multiply and divide by* 12 *or less.*

1. A man has 3 pieces of cloth, each measuring 10 yds. 3 qrs.; how many yds. in the whole?

2. If 3 pieces of cloth contain 32 yds. 1 qr., how many yards in one piece?

3. A man has 5 bottles, each containing 2 gal. 1 qt. 1 pt.; what do they all contain?

4. A man would put 11 gal. 3 qts. 1 pt. of vinegar into 5 bottles of the same size, what does each contain?

Hence, Compound Multiplication is, repeating a compound number as many times as there are units in the multiplier.

Hence, Compound Division is, dividing a compound number into as many parts as are indicated by the divisor.

5. At 1£. 5s. 8$\frac{3}{4}$d. per yard, what will 6 yards of cloth cost?

6. If 6 yards of cloth cost 7£. 14s. 4$\frac{1}{2}$d., what is the price per yard?

As the numbers are large, we write them down before multiplying and dividing.

OPERATION.

£.	s.	d.	qrs.	
1	5	8	3	*price of* 1 *yard.*
			6	*number of yds.*
Ans. 7	14	4	2	*cost of* 6 *yards.*

SOLUTION. — 6 times 3qrs. are 18qrs. = 4d. and 2qrs. over; we write down the 2qrs.; then, 6 times 8d. are 48d., and 4 to carry makes 52d. = 4s. and 4d. over; we write down the 4d.; again, 6 times 5s. are 30s. and 4 to carry makes 34s. = 1£. and 14s. over; 6 times 1£. are 6£., and 1 to carry makes 7£., which we write down; and it is plain, that the united products arising from the several denominations is the real product arising from the whole compound number.

OPERATION.

	£.	s.	d.	qrs.	
6)	7.	14	4	2	*cost of* 6 *yards.*
	1	5	8	3	*price of* 1 *yard.*

SOLUTION. — We divide 7£. by 6, to see how many £ each yard will cost, and find it to be 1£.; but 6 yards at 1£. per yard would cost only 6£. Hence, the 6 yards cost 1£. 14s. 4d. 2 qrs. more. Reducing 1£. to shillings, and adding 14 shillings, we have 34 shillings, which, divided by 6, will give 5 shillings more as the price of each yard, and 4 shillings more, which, reduced to pence and added to 4d. will make 52d., and dividing by 6, we have 8d. and 4 remainder: this remainder reduced to qrs. is 16 and 2 are 18qrs., dividing by 6, the quotient is 3qrs. Hence, each yard will cost 1£. 5s. 8d. 3qrs.

The processes in the foregoing examples may now be presented in form of a

RULE.

For multiplying a Compound Number when the multiplier does not exceed 12.

Multiply each denomination separately, beginning at the least, as in multiplication of simple numbers, and carry as in addition of compound numbers, setting down the whole product of the highest denomination.

For dividing a Compound Number when the divisor does not exceed 12.

By short division, find how many times the divisor is contained in the highest denomination, under which write the quotient, and if there be a remainder reduce it to the next less denomination, adding thereto the number given, if any, of that denomination, and divide as before.

Proceed in this manner through all the denominations, and the several quotients will be the answer required.

EXAMPLES FOR PRACTICE.

7. What will be the cost of 5 pairs of shoes, at 10s. 6d. a pair?

8. At 2£. 12s. 6d. for 5 pairs of shoes, what is that a pair?

9. In 5 barrels of wheat, each containing 2 bu. 3 pks. 6 qts., how many bushels?

10. If 14 bu. 2 pks. 6 qts. of wheat be equally divided into five barrels, how many bushels will each contain?

11. How many yards of cloth will be required for 9 coats, allowing 4 yds. 1 qr. 3 na. to each?

12. If 9 coats contain 39 yds. 3 qrs. 3 na., what does 1 coat contain?

13. In 7 bottles of wine, each containing 2 qts. 1 pt. 3 gills, how many gallons?

14. If 5 gal. 1 gill of wine be divided equally into 7 bottles, how much will each contain?

Questions. — ¶ **129.** What is compound multiplication? Where do you begin to multiply? How do you proceed when the multiplier does not exceed 12? How do you carry? Repeat the rule, when the multiplier is 12 or less. What is compound division? How do you write the numbers when the divisor does not exceed 12? Where do you begin the division? Where write the quotient? If there be a remainder, how do you proceed? Repeat the rule.

15. What will be the weight of 8 silver cups, each weighing 5 oz. 12 pwt. 17 grs.?

16. If 8 silver cups weigh 3 lb. 9 oz. 1 pwt. 16 grs., what is the weight of each?

17. How much sugar in 12 hogsheads, each containing 9 cwt. 3 qrs. 21 lb.?

18. If 119 cwt. 1 qr. of sugar be divided into 12 hogsheads, how much will each hogshead contain?

19. How much beer in 9 casks, each containing 1 bar. 7 gal. 3 qts. 1 pt.?

20. If 9 equal casks of beer contain 10 bar. 34 gal. 3 qts. 1 pt., what quantity in each?

21. A house has 7 rooms, averaging 1 sq. rod 57 sq. ft. 55 sq. inches; what do they all contain?

22. If 7 rooms contain 8 sq. rods 129 sq. ft. 61 sq. in., what is the average?

23. A boat on the Erie canal averages 21 m. 65 rods 13 ft. a day; what is that for 5 days?

24. A boat moves 106 m. 8 rods 15 ft. 6 in. in 5 days, what is the average of each day?

25. What quantity of land in 6 fields, each containing 17 A. 7 sq. C. 12 sq. rd. 133 sq. l.?

26. If 6 equal fields contain 106 A. 6 sq. C. 9 sq. rds. 173 sq. l.; how much land in each?

27. How much wood in 12 piles, each containing 7 C. 5 c. ft. 12 cu. ft.?

28. In 12 piles of wood are 92 C. 5 c. ft., how much in each?

¶ 130. *To multiply and divide by a composite number.*

1. In 15 loads of oats, each measuring 42 bu. 3 pk. 2 qts., how many bushels?

SOLUTION. — Multiplying the quantity in 1 load by 3, we have the quantity in 3 loads, and multiplying the quantity in 3 loads by 5, we have the quantity in 5 times 3, or 15 loads.

2. If 15 loads of oats measure 642 bu. 0 pk. 6 qts., how many bushels in each load?

SOLUTION. — Dividing the quantity in 15 loads by 5, we have the quantity in $\frac{1}{5}$ of 15, or 3 loads, and dividing the quantity in 3 loads by 3, we have the quantity in 1 load.

Questions. — ¶ 130. When the multiplier, or divisor, is a composite number, how may the operation be contracted in multiplication? — in division?

OPERATION.	OPERATION.
bu. pk. qts.	*bu. pk. qts.*
42 3 2 *in* 1 *load.*	One factor, { 5) 642 0 6 *in* 15 *loads.*
3 *one factor.*	The other factor, { 3) 128 1 6 *in* 3 *loads.*
128 1 6 *in* 3 *loads.*	*Ans.* 42 3 2 *in* 1 *load.*
5 *the other factor.*	
642 0 6 *in* 15 *loads, Ans.*	

Hence, *When the multiplier or divisor exceeds* 12 *and is a composite number,*

RULE.

Multiply by each of the component parts of the multiplier. The last product will be the answer.

Divide by each of the component parts of the divisor. The last quotient will be the answer.

3. What will 24 barrels of flour cost, at 2£. 12s. 4d. a barrel?

4. Bought 24 barrels of flour, for 62£. 16s.; how much was that per barrel?

5. How many bushels of apples in 112 barrels, each barrel containing 2 bu. 1 pk.?

NOTE. — 8, 7, and 2, are factors of 112.

6. In 112 barrels are 252 bushels of apples; how many bushels in 1 barrel?

7. How much molasses in 84 hogsheads, each hogshead containing 112 gal. 2 qts. 1 pt. 3 gi.?

8. Bought 84 hogsheads of molasses, containing 9468 gal. 1 qt. 1 pt.; how much in a hogshead?

9. How many bushels of wheat in 135 bags, each containing 2 bu. 3 pecks?

$3 \times 9 \times 5 = 135.$

10. 371 bu. 1 pk. of wheat are equally divided into 135 bags; how much in each?

11. Sold 25 pieces of cloth, each containing 32 yds. 2 qr. 1 na.; how much in the whole?

12. Sold 814 yds. 1 na. of cloth in 25 equal pieces; how much in each piece?

¶ 131. *To multiply and divide by any number greater than* 12, *which is not a composite number.*

1. How many yards of sheeting in 139 pieces, each piece containing 31 yds. 3 qrs. 3 na.?

2. Bought 139 pieces of sheeting, containing 4439 yds. 1 qr. 1 na., how many yards in 1 piece?

Solution. — 139 is not a composite number. We may, however, decompose this number thus, 139 = 100 + 30 + 9.

We may now multiply the number of yards in 1 piece by 10, which will give the number of yards in 10 pieces, and this product again by 10, which will give the number of yards in 100 pieces.

We may next multiply the number of yards in 10 pieces by 3, which will give the number of yards in 30 pieces, and the number of yards in 1 piece by 9, which will give the number of yards in 9 pieces, and these three products, added together, will evidently give the number of yards in 139 pieces; thus:

yds.	qrs.	na.	
31	3	3	*in* 1 *piece.*
		10	
319	1	2	*in* 10 *pieces.*
		10	
3193	3	0	*in* 100 *pieces.*
958	0	2	*in* 30 *pieces.*
287	1	3	*in* 9 *pieces.*
4439	1	1	*in* 139 *pieces.*

Note. — In multiplying the number of yards in 10 pieces, (319 yds. 1 qr. 2 na.,) by 3, to get the number of yards in 30 pieces, and in multiplying the number of yards in 1 piece, (31 yds. 3 qrs. 3 na.,) by 9, to get the number of yards in 9 pieces, the multipliers, 3 and 9, need not be written down.

Solution. — When the divisor cannot be produced by the multiplication of small numbers, the better way is to divide after the manner of long division, setting down the work of dividing and reducing as follows:

```
     yds.  qr. na. yds. qr. na.
139) 4439  1   1 ( 31   3   3
     417
     ---
      269
      139
      ---
      130
        4
      ---
      521 ( 3 qrs.
      417
      ---
      104
        4
      ---
      417 ( 3 na.
      417
```

We divide 4439 yds. by 139, to ascertain the number of yards in each piece, which we find to be 31, and a remainder of 130 yds., which will not be a whole yard to each piece; but reducing them to quarters, adding in the 1 qr., we have 521 qrs., which we divide by 139 to ascertain the number of quarters in each piece, and find it to be 3, and a remainder of 104 qrs.; we reduce this remainder to nails, adding in the 1 nail, and have 417 na., which we divide by 139 for the number of nails in each piece, and find it to be 3, without a remainder. Hence, there are 31 yds. 3 qrs. 3 na. in each piece, *Ans.*

Questions. — **¶ 131.** When the multiplier or divisor exceeds 12, how is the multiplication performed? — the division?

Hence, *When the multiplier or divisor exceeds* 12, *and is not a composite number,*

RULE.

Multiply first by 10, and this product by 10, which will give the product for 100; and if the hundreds in the multiplier be more than one, multiply the product of 100 by the *number* of hundreds; for the *tens*, multiply the product of 10 by the number of tens; for the *units*, multiply the *multiplicand;* and the sum of these several products will be the product required.

Divide after the manner of long division, setting down the work of dividing and reducing.

EXAMPLES FOR PRACTICE.

3. How many acres in 241 wild lots, each containing 75 acres 2 roods 25 rods?

4. There are surveyed in an unsettled district, 18233 acres 25 rods of land, which is divided into 241 equal lots; how many acres in each lot?

5. How many pounds of tea in 23 chests, each containing 78 lbs. 9 oz.?

NOTE. — The pupil will easily perceive the method of operation, when the multiplier is less than 100.

6. If 1806 lbs. 15 oz. of tea be divided equally into 23 chests, how much will be in each chest?

7. Bought 375 bales of English goods at 9£. 11s. 6d. per bale; what did the whole cost?

8. Bought 375 bales of English goods for 3590£. 12s. 6d.; what did each bale cost?

9. How many bushels of wheat are raised on 125 acres, averaging 22 bush. 3 pecks 5 quarts to the acre?

10. A wealthy farmer harvested 125 acres of wheat, which yielded 2863 bush. 1 peck 1 qt.; what was the average per acre?

¶ 132. *Difference in longitude and time between different places.*

Every circle, whether great or small, is supposed to be divided into 360 equal parts, called degrees.

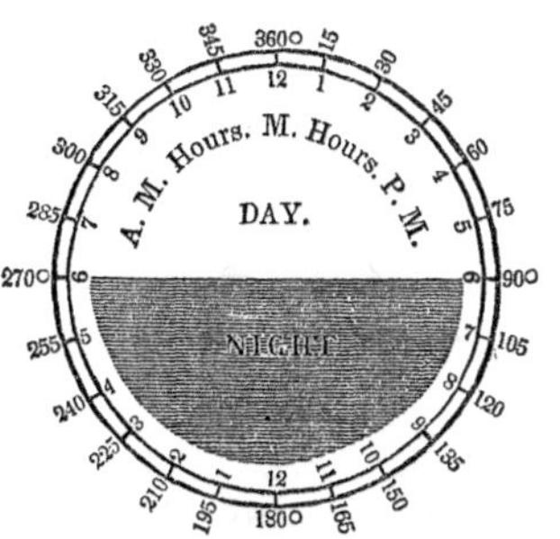

Let the accompanying diagram represent the great circle of the earth, called the equator, divided, as you here see, into 24 equal parts of 15 degrees each, ($360° \div 24 = 15°$.)

As the sun apparently passes round the earth in 24 hours, it will pass through one of these divisions, or 15° in 1 hour $=$ 60 minutes of time, and of course it will pass 1° of motion in $\frac{1}{15}$ of $60 = 4$ minutes of time, and 1′ of motion in $\frac{1}{60}$ of 4 minutes ($= 240$ seconds $\div 60) = 4$ seconds of time. Hence it follows, that the apparent motion of the sun round the earth, from east to west, is

15° of motion in 1 hour of time,
1° of motion in 4 minutes of time, and
1′ of motion in 4 seconds of time.

From these premises, it follows that when

there is a difference in longitude between two places there will be a corresponding difference in the hour, or time of the day. The difference in longitude being **15°**, the difference in time will be 1 hour; the place *easterly* having noon, or any other specified time, 1 hour *sooner* than the place *westerly*.

Hence, if the difference in longitude, in degrees, and minutes, be multiplied by **4**, the product will be the difference in time in minutes and seconds, which may be reduced to hours

there is a difference in time between two places, there is a corresponding difference in their longitude. If the difference in time be **1** hour, the difference in longitude will be **15°**; if 4 minutes, the difference in longitude will be **1°**, &c.

Hence, if the difference in *time* (in minutes and seconds) between two places be *divided* by **4**, the quotient will be the difference in *longitude*, in degrees and minutes.

1. What is the difference in time between London and Washington, the difference in longitude being 77°?

SOLUTION. — The sun apparently moves 1° in 4 minutes of time, and it will move 77° in 77 times 4 minutes = 308 minutes. Or, we may multiply 77 by 4, since either factor may be the multiplicand. And 308 min. = 5 h. 8 min, *Ans.*

2. What is the difference in longitude between London and Washington, the difference in time being 5 h. 8 m.?

SOLUTION. — The sun's apparent motion is 1° in 4 minutes of time, and we wish to know how far it will move in 308 minutes. It will move as many degrees as 4 min. is contained times in (can be subtracted from) 308 min And 308 ÷ 4 = 77°, *Ans.*

3. When it is 12 o'clock at the most easterly extremity of the island of Cuba, what will be the hour at the most westerly extremity, the difference in longitude being 11°?

4. When it is 12 o'clock at the most easterly extremity of the island of Cuba, it is 16 minutes past 11 o'clock at the western extremity; what is the difference in longitude between the two points?

5. Supposing a meteor should appear so high that it could be seen at once by the inhabitants of Boston, 71° 3′, of Washington, 77° 43′, and of the Sandwich Islands, 155° west longitude; if the time be 47 minutes past 11 o'clock of Dec. 31, 1847, at Washington, what will be the time at Boston, and at the Sandwich Islands?

Ans. At Boston, 13 min. 40 sec. A. M. (morning) of Jan. 1st, 1848; at the Sandwich Islands, 37 min. 52 sec. past 6 o'clock, P. M., of Dec. 31, 1847.

¶ 133. Review of Compound Numbers.

Questions. — What distinction do you make between simple and compound numbers? (¶ 102.) What is the rule for adding compound numbers? In ¶ 124, Ex. 33, what difficulty is met with in carrying from miles to degrees? How is it obviated? Rule for subtracting com-

---

Questions. — ¶ **132.** What circle is the diagram intended to represent? Into how many divisions is it divided? how many degrees in each division; and what does it represent? In what time does the sun move 1°, and why? — 1′, and why? What does M. signify? A. M.? P. M.? What motion causes the apparent motion of the sun? When the difference of longitude between places is known, how may the difference in time be calculated? When it is noon at any place, is it before noon or after noon at places easterly? — at places westerly? Why? When there is a difference in time between places, what follows? When the difference in time is given, how may the difference in longitude be found?

pound numbers? — for multiplying when the multiplier does not exceed 12? — when it does exceed 12, and is a composite number? — when *not* a composite number? — for dividing compound numbers, when the divisor does not exceed 12? — when it exceeds 12 and is a composite number? — when *not* a composite number? How is the distance of time from one date to another found? How many degrees does the earth revolve from west to east in 1 hour? In what time does it revolve 1°? Where is the time or hour of the day sooner — at the place most easterly or most westerly? The difference in longitude between two places being known, how is the difference in time calculated? The difference in time being known, how is the difference in longitude calculated?

EXERCISES.

1. A gentleman is possessed of 1½ dozen of silver spoons each weighing 3 oz. 5 pwt.; 2 doz. of tea-spoons, each weighing 15 pwt. 14 gr.; 3 silver cans, each 9 oz. 7 pwt.; 2 silver tankards, each 21 oz. 15 pwt.; and 6 silver porringers, each 11 oz. 18 pwt.; what is the weight of the whole?
Ans. 18 lb. 4 oz. 3 pwt.

2. In 35 pieces of cloth, each measuring 27 yds. 3 qrs., how many yards? *Ans.* 971 yds. 1 qr.

3. How much wine in 9 casks, each containing 45 gal. 3 qts. 1 pt.?

4. If a horse travel a mile in 12 min. 16 sec., in what time would he travel 176 miles? *Ans.* 1 d. 11 h. 58 m. 56 sec.

5. If 8 horses consume 889 bu. 2 pks. 6 qts. of oats in 365 days, how much will one horse consume in 1 day?
Ans. 1 pk. 1 qt. 1 pt. 2 gills.

6. I hold an obligation against George Brown, of London, of 735£. 11s. 6d., on which are two endorsements, viz., 61£. 5s. and 195£. 13s. 11d.; what remains unpaid?
Ans. 478£. 12s. 7d.

7. Liverpool, Jan. 1, 1848.

Isaac Derwent, of Boston, U. S.

Bought of Shipley & Co.

10	boy's hats, No.	1,	4s. 6d.
12	do.	2,	5s.
4	do.	3,	5s. 6d.
4	do.	9,	10s.
4	do.	10,	11s.
6	do.	11,	12s.
6	men's hats,		14s.
1	trunk for packing,		1£. 4s.
			19£. 11s.

Received payment, Shipley & Co.
by G. Williams.

NOTE. — If the three following examples be wrought correctly, the answers will be as here given.

8. A man divides 16 bar. 23 gal. 3 qts. of oil into 5 large vessels; how much does he put in each?

Ans. 3 bar. 11 gal. $0\frac{1}{5}$ qt.

9. On an acre of ground were erected 21 buildings, occupying on an average 3 sq. rds. 112 ft. 81 in.; how much remained unoccupied? *Ans.* 2 roods 8 rods 86 ft. 63 in.

10. If a man build 3 rds. 9 ft. 7 in. in length of wall in 1 day, how many rods can he build in 15 days?

Ans. 53 rds. 11 ft. 9 in.

11. If a ship sail 3° 18′ 45″ in one day, how far will it sail in the month of June? *Ans.* 99° 22′ 30″.

12. If a druggist sell 1 gross 7 doz. bottles of sarsaparilla in 1 week, how many gross will he sell in the months of April, May and June, at the same rate? *Ans.* 20 gr. 7 doz.

13. If a man, employed in counting money from a heap, count 100 silver dollars in a minute, and continue at the work 10 hours each day, how many days will it take him to count a million? *Ans.* $16\frac{2}{3}$ days.

14. At the same rate, how many years, reckoning 365 days to a year, would it take him to count a billion?

Ans. 45 years $241\frac{2}{3}$ days.

15. Were 1000 men employed at this same business, each one counting at the same rate, 10 hours each day, how many years would it take them to count a quadrillion?

Ans. 45662 years $36\frac{2}{3}$ days.

ANALYSIS.

¶ 134. In most examples in arithmetic, two things are given to find a third. Thus, in the relations of the price and quantity, the quantity and the price of a unit may be given to find the price of the quantity, or the quantity and its price to find the price of a unit, or the price of a unit and of a quantity to find the quantity. The same principles may readily be applied to other calculations.

This method of operating is called Analysis. Analysis, therefore, may be defined, *the solving of questions on general principles.*

We have presented (¶ 46) these rules in connection, as applied to whole numbers, and separately as applied to any quantities in ¶¶ 77, 85, and 83. We will now present them together as applied to any quantities, with examples which mutually prove each other, except where there are some fractional losses.

I. *The price of unity, and the quantity being given, to find the price of the quantity,*

RULE.

Multiply the price by the quantity.

II. *The quantity, and the price of the quantity being given, to find the price of unity,*

RULE.

Divide the cost by the quantity.

III. *The price of unity, and the price of a quantity being given, to find the quantity,*

RULE.

Divide the price of the quantity by the price of unity.

EXAMPLES FOR PRACTICE.

1. At $302‘40 per tun, what will 1 hhd. 15 gal. 3 qts. of wine cost?

2. At $94‘50 for 1 hhd. 15 gal. 3 qts. of wine, what is that per tun?

3. At $302‘40 per tun, how much wine may be bought for $94‘50.

4. At $2‘215 per gal., what cost $3\frac{1}{4}$ qts.?

5. At $1‘80 for $3\frac{1}{4}$ qts. of wine, what is that per gal.?

6. At $2‘215 per gal., how much wine may be bought for $1‘80?

7. At $96‘72 per ton for pot-ashes, what will $\frac{5}{8}$ of a ton cost?

8. If $\frac{5}{8}$ of a ton of pot-ashes cost $60‘45, what is that per ton?

9. At $96‘72 per ton, how much pot-ashes may be bought for $60‘45?

10. If a yard of cloth cost $2‘5, what will ‘8 of a yard cost?

11. If ‘8 of a yard of cloth cost $2, what is that per yard?

12. At $2‘5 per yard, how much cloth may be purchased for $2?

13. If a ton of pot-ashes cost £27 10s., what will 14 cwt. cost?

14. If 14 cwt. of pot-ashes cost £19 5s., what is that per ton?

15. At £27 10s. a ton for pot-ashes, what quantity may be bought for £19 5s.?

16. If a bushel of wheat cost $1 92, what will 1 pk. 4 qts. cost?

17. If 1 pk. 4 qts. of wheat cost $‘72, what is that per bushel?

18. At $1‘92 per bushel, how much wheat may be bought for $‘72?

19. If a yard of broadcloth cost $6, what will 16 yds. 2 qrs. 3 na. cost?

20. If 16 yds. 2 qrs. 3 na. of broadcloth cost $100'125, what is that per yard?

21. At $6 per yard, how much broadcloth may be bought for $100'125?

22. If a ton of hay cost $13, what will 1850 lbs. cost?

23. If 1850 lbs. of hay cost $12'025, what is that per ton?

24. At $13 per ton, how many lbs. of hay may be bought for $12'025?

25. If an eagle weigh 11 pwt. 6 grs., what will be the weight of 665 eagles?

26. If 665 eagles weigh 31 lbs. 2 oz. 1 pwt. 6 grs., what is the weight of each?

27. How many eagles, each weighing 11 pwt. 6 grs., may be coined from 31 lbs. 2 oz. 1 pwt. 6 grs. of standard gold?

NOTE.—After the same manner let the pupil reverse the following examples:

28. At $1'75 a bushel for wheat, how many quarters can be bought for $200? *Ans.* 14 qrs. 2 bu. 1 pk. 1'1428$\frac{4}{7}$ qts.

29. What will 3 qrs. 2 na. of broadcloth cost, at $6 per yard?

30. At $22'10 for the transportation of 6500 lbs. 46 miles what is that per ton? $6'80.

31. Bought a silver cup, weighing 9 oz. 4 pwt. 16 grs., for $11'08; what was that per ounce?

32. A lady purchased a gold ring, giving at the rate of $20 per ounce; she paid for the ring $1'25; how much did it weigh? *Ans.* 1 pwt. 6 grs.

Examples requiring several operations.

33. If 6 bushels of wheat cost $7'50, what will 28 bushels cost?

NOTE 1. It was customary formerly to perform all examples simi-

---

Questions.—¶ **134.** What do you say of every example in arithmetic? What severally are given and required in the relations of the price and quantity? Answer the questions in ¶ 46. (*The teacher will ask them.*) What is given and what required under I.?—the rule?—under II.?—the rule?—under III.?—the rule? How were such examples as the 33d formerly wrought? What method is preferable? What do we do by this method? Explain the solution of this example;—the first solution of example 34;—the second method. What is this called? What is an analytic solution?

lar to the above by a rule called *the Rule of Three*, or, on account of its supposed importance, *the golden rule*. But they are more intelligibly solved by analysis; that is, from the things given in the question find the price of unity, and having the price of unity, find the price of the quantity, the price of which is required. Thus,

OPERATION.

6) \$7'50, *price of* 6 *bu.*

\$1'25, " 1 *bu.*
28

1000
250

\$35'00, *price of* 28 *bu.*

SOLUTION. — Dividing the price of 6 bu. by 6 will give the price of 1 bu., and multiplying the price of 1 bu. by 28 will give the price of 28 bu.

34. If $\frac{11}{13}$ of a bushel of corn cost $\frac{7}{15}$ of a dollar, what will $\frac{32}{43}$ of a bushel cost?

SOLUTION. — We divide the price of the quantity, $\frac{7}{15}$, by the quantity, $\frac{11}{13}$, to get the price of unity: $\frac{7}{15} \div \frac{11}{13} = \frac{91}{165}$ of a dollar, price of 1 bu. And multiplying the price of 1 bu. by $\frac{32}{43}$ of a bushel, will give the price of the quantity required: $\frac{91}{165} \times \frac{32}{43} = \frac{2912}{7095}$ of a dollar.

Or, to give a more full analysis, if $\frac{11}{13}$ of a bushel cost $\frac{7}{15}$ of a dollar, $\frac{1}{13}$ would cost $\frac{1}{11}$ as much: $\frac{1}{11}$ of $\frac{7}{15} = \frac{7}{165}$; and $\frac{13}{13} =$ 1 bu., would cost 13 times as much as $\frac{1}{13}$. $\frac{7}{165} \times 13 = \frac{91}{165}$ of a dollar, price of 1 bushel. Then $\frac{1}{43}$ of $\frac{91}{165} = \frac{91}{7095}$, is the price of $\frac{1}{43}$ of a bushel, and 32 times the price of $\frac{1}{43}$, that is, $\frac{91}{7095} \times 32 = \frac{2912}{7095}$ of a dollar, is the price of $\frac{32}{43}$ of a bushel.

NOTE 2. This process is called an analytical solution, or *solving the question by analysis.*

35. If 16 men will finish a piece of work in $28\frac{1}{3}$ days, how long will it take 12 men to do the same work?

NOTE 3. Find in what time 1 man would do it, and 12 men would do it in $\frac{1}{12}$ of the time. *Ans.* $37\frac{7}{9}$ days.

36. How many yards of alpacca, which is $1\frac{1}{4}$ yards wide, will be required to line 20 yards of cassimere $\frac{3}{4}$ of a yard wide?

NOTE 4. Find the square contents of the cassimere, which is the same as the square contents of the alpacca. We have, then, the square contents and the width of the alpacca given to find the length? ¶ 49.
Ans. 12 yds. of alpacca.

37 If 7 horses consume $2\frac{3}{4}$ tons of hay in 6 weeks, how much will 12 horses consume in 8 weeks? *Ans.* $6\frac{2}{7}$ tons.

38. If 5 persons drink $7\frac{4}{5}$ gallons of beer in 1 week, how much will 8 persons drink in $22\frac{1}{2}$ weeks?
Ans. $280\frac{4}{5}$ gallons.

39. A merchant bought several bales of velvet, each containing $129\frac{17}{27}$ yards, at the rate of \$7 for 5 yards, and sold them at \$11 for 7 yards, gaining \$200 on them; how many bales were there? *Ans.* 9 bales.

40. If $\frac{1}{3}$ lb. less by $\frac{1}{6}$ costs $13\frac{1}{5}$d., what cost 14 lb. less by $\frac{1}{6}$ of 2 lb.? *Ans.* £4 9s. $9\frac{3}{25}$d.

41. If 5 acres 1 rood produce 26 quarters 2 bushels of wheat, how many acres will be required to produce 47 quarters 4 bushels? *Ans.* 9 A. 2 R.

42. If 9 students spend £$10\frac{7}{5}$ in 18 days, how much will 20 students spend in 30 days? *Ans.* £39 18s. $4\frac{8}{81}$d.

43. If $\frac{3}{5}$ yd. cost \$$\frac{7}{8}$, what will $40\frac{1}{2}$ yds. cost?
Ans. \$59·062 +.

44. If $\frac{7}{16}$ of a ship costs \$251, what is $\frac{3}{32}$ of it worth?
Ans. \$53·785 +.

45. At £$3\frac{5}{8}$ per cwt., what will $9\frac{2}{3}$ lbs. cost?
Ans. 6s. $3\frac{5}{56}$d.

46. A merchant, owning $\frac{4}{5}$ of a vessel, sold $\frac{2}{3}$ of his share for \$957; what was the vessel worth? *Ans.* \$1794·375.

47. If $\frac{5}{8}$ yd. cost £$\frac{5}{7}$, what will $\frac{9}{15}$ of an ell Eng. cost?
Ans. 17s. 1d. $2\frac{6}{7}$q.

PRACTICE.

¶ 135. I. *When the price is an aliquot part of a dollar.*

NOTE.—For the definition of aliquot part, see ¶ 55.

ALIQUOT PARTS OF 1 DOLLAR.

Cents.		Cents.	
50	$= \frac{1}{2}$ of 1 dollar.	$12\frac{1}{2}$	$= \frac{1}{8}$ of 1 dollar.
$33\frac{1}{3}$	$= \frac{1}{3}$ of 1 dollar.	10	$= \frac{1}{10}$ of 1 dollar.
25	$= \frac{1}{4}$ of 1 dollar.	$6\frac{1}{4}$	$= \frac{1}{16}$ of 1 dollar.
20	$= \frac{1}{5}$ of 1 dollar.	5	$= \frac{1}{20}$ of 1 dollar.

1. What will be the cost of **4857** yards of calico at **25** cents (= $\frac{1}{4}$ of 1 dollar) per yard?

OPERATION.
4)4857 dollars,

$1214'25, *Ans.*

SOLUTION. — At $1 a yard, the cost would be as many dollars as there are yards, that is, $4857; and at $\frac{1}{4}$ of a dollar a yard, it is plain, that the cost will be $\frac{1}{4}$ as many dollars as there are yards, that is, $\frac{\$4857}{4}$ = $1214'25.

After dividing the unit figure 7, there is a remainder of 1, (dollar, which we reduce to cents by annexing ciphers, and continue the division.

This manner of computing the cost of articles, *by taking aliquot parts*, is called PRACTICE, from its daily use among merchants and tradesmen.

Hence, when the price is an aliquot part of a dollar, this general

RULE OF PRACTICE.

Divide the price at $1 per pound, yard, &c., by the number expressing the aliquot part, the quotient will be the answer in dollars.

NOTE. — If there be a remainder, it may be reduced to cents and mills, by annexing ciphers, and the division continued.

EXAMPLES FOR PRACTICE.

2. What is the value of 14756 yards of cotton cloth, at $12\frac{1}{2}$ cents, or $\frac{1}{8}$ of a dollar per yard?

By practice.
8)14756

Ans. $1844'50

By multiplication.
```
   14756
    '125
   -----
   73780
  29512
 14756
 -------
$1844'500
```
Ans. as before.

NOTE. — By comparing the two operations, it will be seen that the operation by practice is much shorter than the one by multiplication.

3. What is the cost of 18745 pounds of tea, at $'50, = $\frac{1}{2}$ dollar, per pound? *Ans.* $9372'50.

4. What is the value of 9366 bushels of potatoes, at $33\frac{1}{3}$ cents, or $\frac{1}{3}$ of a dollar, per bushel? $\frac{9366}{3}$ = $3122, *Ans.*

Questions. — **¶ 135.** What do you understand by aliquot parts? What are the aliquot parts of a dollar? When the price of 1 yard, 1 pound, &c., is an aliquot part of a dollar, how may the cost of any quantity of that article be found? What is this manner of computing called? Why? Repeat the rule. What two things are given, and what one is required, in example 3d? — in example 4th? 5th? 6th? 7th? &c.

5. What is the value of 48240 pounds of cheese, at $'06¼ = $\frac{1}{16}$ of a dollar, per pound? *Ans.* $3015.

6. What cost 4870 oranges, at 5 cents, = $\frac{1}{20}$ of a dollar apiece? *Ans.* $243'50.

7. What is the value of 151020 bushels of apples, at 20 cents, = $\frac{1}{5}$ of a dollar, per bushel? *Ans.* $30204.

8. What will 264 pounds of butter cost, at 12½ cents per pound? *Ans.* $33.

9. What cost 3740 yards of cloth, at $1'25 per yard?

4) $3740 = cost at $1' per yard.
935 = cost at $ '25 per yard.

Ans. $4675 = cost at $1'25 yer pard.

10. What is the cost of 8460 hats, at $1'12½ apiece? —— at $1'50 apiece? —— at $3'20 apiece? —— at $4'06¼ apiece? *Ans.* $9517'50. $12690. $27072. $34368'75

¶ 136. II. *To find the value of articles sold by the* 100 *or* 1000.

1. What is the value of 865 feet of timber, at $5 per hundred?

OPERATION.

865
5

$4325 = *value at $5 per foot.*

Were the price $5 per *foot*, it is plain the value would be 865 × $5 = $4325; but the price is $5 for 100 feet; consequently, $4325 is 100 times the true value of the timber; and therefore, if we divide this number ($4325) by 100, we shall obtain the true value; which we do by cutting off two right hand figures. *Ans.* $43'25.

Were the price so much per *thousand*, the same remarks would apply, with the exception of cutting off *three* figures, instead of two. Hence,

To find the value of articles sold by the 100 or 1000,

RULE.

I. Multiply the number and price together.

II. If the price be by the 100, cut off *two* figures at the right; if by the 1000, cut off *three* figures at the right; the *product* will be the answer, in the same denomination as the price, which, if cents or mills, may be reduced to dollars

2. What is the cost of 4250 bricks, at $5'75 per 1000? *Ans.* $24'43¾.

OPERATION.

```
 $5'75
  4250
 ------
  2875
 1150
2300
----------
$24'43|750
```

In his example, we cut off three figures from the right hand of the product, because the bricks were sold by the 1000. The remaining figures at the left express the cost of the bricks in the lowest denomination of the price, viz., cents, which we reduce to dollars by pointing off two places for cents

3. What will 3460 feet of timber cost, at $4 per hundred?

4. What will 24650 bricks cost, at 5 dollars per 1000?

5. What will 4750 feet of boards cost, at $12'25 per 1000? *Ans.* $58'187+.

6. What will 38600 bricks cost, at $4'75 per 1000?

7. What will 46590 feet of boards cost, at $10'625 per 1000?

8. What will 75 feet of timber cost, at $4 per 100?

9. What is the value of 4000 bricks, at 3 dollars per 1000?

10. Wilderness, February 8, 1847

Mr. Peter Carpenter,

Bought of Asa Falltree,

5682	feet	Boards,	at $6 per M.
2000	"	"	" 8'34 "
800	"	Thick Stuff,	" 12'64 "
1500	"	Lathing,	" 4 "
650	"	Plank,	" 10 "
879	"	Timber,	" 2'50 per C.
236	"	"	" 2'75 "

Received payment, $101'849.

Asa Falltree.

NOTE.—M. stands for the Latin *mille*, which signifies 1000, and C. for the Latin word *centum*, which signifies 100.

¶ 137. III. *To find the cost of articles by the ton of* 2000 *lbs*

1. What cost 3[illegible] lbs. of [illegible]y, at $12'40 a ton?

Questions.—¶ 136. To find the cost of articles sold by 100, or 1000, what is the first step proposed by the rule?—the second? In what denomination will the product be? How will you find the cost of 725 bricks, at $4'25 a thousand? How many figures in all do we point off?

OPERATION.

$12'40 ÷ 2 = $6'20 = *price of* 1000 *lbs*
$6'20
3684
————
$22'84 | 080

SOLUTION. — A ton equals 2000 lbs. Dividing the price of one ton by 2, we have the price of 1000 lbs., which is $6'20. Multiplying this cost by the number of pounds, 3684, it will give the cost at $6'20 per pound, a result which is 1000 times too large. ¶ 136. We therefore divide this product by 1000, cutting off three right hand figures, and have the cost at $6'20 per 1000 lbs., or $12'40 per ton.

Ans. $22'84 +.

Hence,

RULE.

Multiply 1 half the cost of 1 ton by the number of pounds, and point off three figures from the right hand. The remaining figures will be the price, in the denomination of the price of 1 ton, which, if cents or mills, may be reduced to dollars.

NOTE. — At $12'40 per ton, or $6'20 per 1000 lbs.
100 lbs. will cost $'62 removing the separatrix 1 figure to the left.
10 " " $'062 " " 2 figures "
1 " " $'0062 " " 3 " "

2. What is the cost of 15742 lbs. of Anthracite coal, at $7'50 per ton? *Ans.* $59'032+.

3. What will be the transportation on 49826 lbs. of iron, from New York to Chicago, at $11 per ton? *Ans.* $274'043.

4. What will be the storage on 13991 lbs. of goods, at $2'50 per ton? *Ans.* $17'488+.

5. What will be the cost of 658 lbs. of hay, at $7'38 per ton? —— at $5'25? —— at $8'50? —— at $9'00? —— at $9'50? —— at $11? —— at $12?
Ans. $2'428. $1'727. $2'796½. $2'961. $3'125½. $3'619 $3'948.

6. At $7'00 per ton, what will be the cost of 424 lbs. of hay? —— 530 lbs.? —— 658 lbs.? —— 750 lbs.? —— 896 lbs.? —— 918 lbs.? —— 1024 lbs.? —— 1216 lbs.? —— 1350 lbs.? —— 1600 lbs.? —— 1890 lbs.?
Ans. $1'484. $1'85½. $2'303. $2'62½. $3'136. $3'213. $3'584. $4'256. $4'72½. $5'60. $6'61½.

Questions. — ¶ **137.** When the cost of 1 ton is given, how do you find the cost of 1000 pounds? How do you find the cost of any number of pounds?

¶ 138. IV. *When the price is the aliquot part of a £*

ALIQUOT PARTS OF A £.

10s. = ½	of 1£.	6s. 8d. = ⅓	of 1£.
5s. = ¼	"	3s. 4d. = ⅙	"
4s. = ⅕	"	2s. 6d. = ⅛	"
2s. = $\frac{1}{10}$	"	1s. 8d. = $\frac{1}{12}$	"

1. Bought of John Smith, Liverpool, 7685 yards of black broadcloth at 10s. per yard; what did it cost?

OPERATION.

2)7685£. = *price at* 1£. *per yard.*

3842£. 10*s.* = " 10*s.* "

SOLUTION.—At 1£. per yard the price is 7685£., and one half of this is the price at 10s. per yard. The remainder of 1£. may be reduced to shillings and then divided.

Hence, when the price is the aliquot part of 1£.,

RULE.

Divide the price of the quantity at 1£. per yard, bushel, &c., by the number expressing the aliquot part. The answer will be in pounds.

EXAMPLES FOR PRACTICE.

2. What cost 1873 reams of paper, at 6s. 8d. per ream? *Ans.* 624£. 6s. 8d.

3. What cost 10416 bushels of salt, at 3s. 4d. per bushel? *Ans.* 1736£.

4. Bought 640 lbs. colored thread, at 7s. 6d. per pound; what was the whole cost?

OPERATION.

8|4)640£. = 1£. *per lb.*

2| 160£. = 5*s.* "

80£. = 2*s.* 6*d.* "

240£. = 7*s.* 6*d.* "

SOLUTION.—7s. 6d. is not an aliquot part of a pound, but it is equal to 5s. + 2s. 6d., and taking ¼ of 640£. we have the price at 5s. per lb., and ⅛ of 640£., or ½ of 160£., is the price at 2s. 6d. Then, adding together the prices at 5s. and 2s. 6d. per lb., we have the price at 7s. 6d. per lb.

5. What cost 866 yards of black silk, at 14s. per yd.? *Ans.* 606£. 4s.

6 What cost 7 T. 8 cwt. of iron, at 16s. 8d. per cwt.? *Ans.* 123£. 6s. 8d.

Questions.—¶ 138. Give the aliquot parts of 1£. Explain the principle on which the first example is performed?—the fourth example? Rule.

¶ 139. *. When the price is an aliquot part of a shilling*

ALIQUOT PARTS OF A SHILLING.

6d. = $\frac{1}{2}$ of 1s.	2d. = $\frac{1}{6}$ of 1s.
4d. = $\frac{1}{3}$ "	$1\frac{1}{2}$d. = $\frac{1}{8}$ "
3d. = $\frac{1}{4}$ "	1d. = $\frac{1}{12}$ "

Reasoning as above, we have this

RULE.

Divide the price of the quantity at 1s. per lb., yd., &c., by the number expressing the aliquot part; the answer will be in shillings.

EXAMPLES FOR PRACTICE.

1. Sold 348216 lbs. of cotton, for 4d. per lb.; what did I receive? *Ans.* 116072s. = 5803£. 12s.

2. Bought 2490 yds. of calico, at 9d. per yard; what did it all cost? 9d. = 6d. + 3d. *Ans.* 93£. 7s. 6d.

3. Bought 4000 papers of pins, at $4\frac{1}{2}$d. per paper; what was the cost? $4\frac{1}{2}$d. = 3d. + $1\frac{1}{2}$d.

4. What cost 7430 lbs. of sugar, at 6d. per lb.? —— at 4d.? —— at 3d.? —— at 2d.? —— at $1\frac{1}{2}$d.? —— at 1d.?

¶ 140. *To find the price of a quantity less than unity, when it is an aliquot part, or parts, of* 1.

1. At $1'50 per bushel for wheat, what will 2 pecks and 4 quarts cost?

OPERATION.

```
8|2) 1'50
    ------
 4)  '75
     '18¾
    ------
     '93¾
```

SOLUTION. — We divide the price of 1 bushel by 2 which gives the price of 2 pecks, or $\frac{1}{2}$ a bushel. Then as 4 quarts is $\frac{1}{8}$ of 1 bushel, we divide $1'50 by 8, or as it is $\frac{1}{4}$ of half a bushel, we divide $'75 by 4, for the price of 4 quarts, and the quotient, added to the price of 2 pecks, gives the price of 2 pecks and 4 quarts. *Ans.* $'93¾.

Hence,

RULE.

Take such part, or parts, of the price of unity as the quantity is of 1; the part, or sum of the parts taken, will be the price of the quantity.

Questions. — **¶ 139.** Give the aliquot parts of 1 shilling. Rule. Show how the 2d example can be performed by Practice; the 3d.

¶ 140. What is the subject of this ¶? Why divide $1'50 by 8, to get the price of 4 qts. of wheat? Why divide $'75 by 4 for the same purpose? Give the rule.

EXAMPLES FOR PRACTICE.

2. What costs 3 qts. of oil, at $‘94 per gal.? *Ans.* $‘70½.

3. What shall I receive for building 90 rods of road, at $1200 per mile? *Ans.* $337‘50.

4. Bought 65 lbs. of pork, at $17‘25 per barrel; what did I pay? *Ans.* $5‘60 +.

5. What will 14 quires of paper cost, at $3‘00 per ream? *Ans.* $2‘10.

6. At $8‘50 per month of 30 days for the rent of a house, what will be the rent for 18 days?

```
2) $8‘50
   ------
5) 4‘25 for 15 days.
    ‘85 for  3 days.
   ------
 $5‘10 Ans.
```

SOLUTION.—Take half of the rent for 30 days, which will be the rent for 15 days, and one fifth of the rent for 15 days will be the rent for 3 days, and add together the rent for 15 and 3 days, the sum will be the rent for 18 days.

7. What will a man's salary amount to in 7 months, at the rate of $500 a year?

```
2) $500
   ------
6)  250 = for 6 months.
    41‘66 +, 1 mo.
   ------
   291‘66 +, 7 mo.
```

SOLUTION.—One half the year's salary will be the salary for 6 months, and one sixth of this the salary for 1 month

8. What will be a man's salary for 8 months and 21 days, at $400 per annum, that is, by the year? *Ans.* $290.

9. What will 5 cord feet and 12 solid feet of wood cost, at $2‘50 per cord? *Ans.* $1‘80, nearly.

10. What will 11 oz. of sugar cost, at 12 cents per pound? *Ans.* $‘082½.

11. What wil. 3⅝ yards of broadcloth cost, at $4‘00 per yard? *Ans.* $14‘50.

¶ 141. *To reduce shillings, pence, and farthings, to the decimal of a pound, by inspection.*

There is a simple and concise method of reducing shillings, pence, and farthings to the decimal of a pound, by *inspection*. The reasoning in relation to it is as follows:

$\frac{1}{10}$ of 20s. is 2s.; therefore every 2s. is $\frac{1}{10}$, or ‘1£. Every shilling is $\frac{1}{20} = \frac{5}{100}$, or ‘05£. Pence are readily reduced to farthings. Every farthing is $\frac{1}{960}$£. Had it so happened that 1000 farthings, instead of 960, had made a pound, then every farthing would have been $\frac{1}{1000}$, or '001£. But 960 increased by $\frac{1}{24}$ part of itself is 1000; consequently, 24 farthings are exactly $\frac{25}{1000}$, or '025£., and 48 farthings are exactly $\frac{50}{1000}$, or '050£. For, add $\frac{1}{24}$ of any number of farthings to the number, and it will be reduced to thousandths of a pound.

If the farthings are 20, they will equal '020$\frac{20}{24}$, which we will call '021, since $\frac{20}{24}$ is more than $\frac{1}{2}$ of a thousandth. If the farthings are 14, they will equal '014$\frac{14}{24}$, which we will call '015 for the same reason. But if the farthings are only 10, they will equal '010$\frac{10}{24}$, which we call '010, since $\frac{10}{24}$ is less than $\frac{1}{2}$ a thousandth. If the farthings are 31 = '031$\frac{31}{24}$ = 32$\frac{7}{24}$, we call them '032, for the same reason. And if the farthings be 42 = 42$\frac{42}{24}$ = 43$\frac{18}{24}$, we call them '044. The result will always be nearer than $\frac{1}{2}$ of 1 thousandth of a pound. Thus, 17s. 5$\frac{3}{4}$d. is reduced to the decimal of a pound as follows: 16s. = '8£. and 1s. = '05£. Then 5$\frac{3}{4}$d. = 23 farthings, which, increased by 1, (the number being more than 12, but not exceeding 36,) is '024£., and the whole is '874£., the *Ans.*

Wherefore, *to reduce shillings, pence, and farthings to the decimal of a pound, by inspection,— Call every two shillings one tenth of a pound; every odd shilling, five hundredths; and the number of farthings, in the given pence and farthings, so many thousandths, adding one, if the number be more than twelve and not exceeding thirty-six, and two, if the number be more than thirty-six.*

NOTE.—If the farthings be just 12 = '012$\frac{12}{24}$, they are equal to '0125; if 36, they are equal to '0375. 48 farthings = 12d. = 1s. equal '05 of a pound, as above.

EXAMPLES FOR PRACTICE.

1. Find, by inspection, the decimal expressions of 9s. 7d., and 12s. 0$\frac{3}{4}$d. *Ans.* '479£., and '603£.

2. Reduce to decimals, by inspection, the following sums, and find their amount, viz.: 15s. 3d.; 8s. 11$\frac{1}{2}$d.; 10s. 6$\frac{1}{4}$d.; 1s. 8$\frac{1}{2}$d.; $\frac{1}{2}$d., and 2$\frac{1}{4}$d. *Amount,* £1'833.

Questions.—¶ 141. What is the rule for reducing shillings, pence, and farthings, to the decimal of a pound, by inspection? What is the reasoning in relation to this rule?

¶ **142.** *To reduce the decimal of a pound to shillings pence, and farthings, by inspection.*

Reasoning as above, (¶ 141,) the first three figures in any decimal of a pound may readily be reduced to shillings, pence, and farthings, by *inspection.* Double the *first* figure, or *tenths*, for shillings, and if the second figure, or hundredths, be *five*, or *more* than five, reckon *another* shilling; then, after the five is deducted, call the figures in the second and third places so many farthings, abating *one* when they are above twelve, and *two* when above thirty-six, and the result will be the answer, within ½ a farthing. Thus, to find the value of '876£. by inspection :—

'8 tenths of a pound	=16 shillings.
'05 hundredths of a pound	= 1 shilling.
'026 thousandths, abating 1, = 25 farthings,	= 0s. 6¼d.
'876 of a pound,	=17s. 6¼d. *Ans.*

1. Find, by inspection, the value of £'523, and £'694.
Ans. 10s. 5½d., and 13s. 10½d.

2. Find the value of £'47.

Note. — When the decimal has but *two* figures, after taking out the shillings, the remainder, to be reduced to *thousandths*, will require a cipher to be annexed to the right hand. *Ans.* 9s. 4¾d.

3. Value the following decimals by inspection, and find their amount, viz.: £'785, £'357, £'916, £'74, £'5, £'25, £'09, and £'008. *Ans.* £3 12s. 11d.

PERCENTAGE.

¶ **143.** 1. A man owns a farm of 320 acres, 5 per cent. of which is marsh; how many acres are marsh?

OPERATION.

```
 320
 '05
-----
16'00
```

Solution. — Per cent. signifies hundredth part. The number placed before *per cent.* signifies how many hundredths are taken, being really the second figure of a decimal fraction; thus, 5 per cent. of 320 acres is '05 of that quantity, and since *of* implies

---

Questions. — ¶ 142. How may the first three figures of any decimal of a pound be reduced to shillings, pence, and farthings, by inspection? Explain the reasons for this operation

multiplication, we multiply 320 by '05, pointing off as in decimal fractions, and get the *Ans.* 16 acres.

The finding of a certain per cent., or a certain number of hundredths of a quantity, is called percentage; and it is performed by the following

RULE.

Multiply the quantity by the rate per cent., written decimally as hundredths.

NOTE. — **Per cent. is from the Latin, which signifies by the hundred.**

7 per cent. is '07. 25 per cent. is '25. 50 per cent. is '50.
100 per cent. is 1'00 ($\frac{100}{100}$, or the whole.)
125 per cent. is 1'25 ($\frac{125}{100}$, more than the whole.)
1 per cent. is . . . . '01
$\frac{1}{2}$ per cent. is a half of 1 per cent., ($\frac{1}{2}$ of 1 hundredth, or $\frac{5}{1000}$ of the whole,) . . . . . '005
$\frac{1}{4}$ per cent. is a fourth of 1 per cent, that is, $\frac{1}{4}$ of $\frac{1}{100}$, = '0025
$\frac{3}{4}$ per cent. is 3 times $\frac{1}{4}$ per cent., that is, $\frac{3}{4}$ of $\frac{1}{100}$, = '0075
$\frac{1}{8}$ per cent., ($\frac{1}{8}$ of a hundredth, that is, $\frac{1}{8}$ of $\frac{1}{100}$,) = '00125
4$\frac{1}{2}$ per cent. is '04$\frac{1}{2}$ = '045, (the 5 expressing 10ths of 100ths,) . . . . . '045

EXAMPLES.

Write 2$\frac{1}{2}$ per cent. as a decimal fraction.

2 per cent. is '02, and $\frac{1}{2}$ per cent. is '005. *Ans.* '025.

Write 4 per cent. as a decimal fraction. —— 4$\frac{1}{2}$ per cent. —— 4$\frac{3}{4}$ per cent. —— 5 per cent. —— 7$\frac{1}{4}$ per cent. —— 8 per cent. —— 8$\frac{3}{4}$ per cent. —— 9 per cent. —— 9$\frac{1}{2}$ per cent. —— 10 per cent. —— 10$\frac{1}{2}$ per cent. —— 12$\frac{1}{2}$ per cent. —— 121 per cent. —— 133$\frac{1}{3}$ per cent.

EXAMPLES FOR PRACTICE.

2. A farmer gives 10 per cent. of 460 bushels of wheat for threshing; how many bushels does he give?

Ans. 46 bushels.

3. A farmer rented ground on which 409 bushels of oats were raised, receiving 30 per cent. for the rent; how many bushels did he receive? *Ans.* 122'7 bushels.

Questions. — ¶ **143.** What does per cent. signify? — percentage? — the number before *per cent.*? How is '05 of a quantity obtained, and why? Give the rule for percentage. How is any per cent. from 1 to 99 expressed? 100 per cent. 125 per cent.? $\frac{1}{2}$ per cent.? 4$\frac{1}{2}$ per cent.

4. A beef weighs 895 lbs., of which 9 per cent. is bone; what does the meat weigh? *Ans.* 814'45 lbs.

5. A schooner, freighted with 725 barrels of flour, encountered a storm, when it was found necessary to throw 28 per cent. of the cargo overboard; how many barrels were thrown overboard, and how many were saved?

Ans. to the last, 522 barrels.

6. A forwarding merchant agreed to transport 2000 bushels of corn, worth $692'75, from Buffalo to Albany for 12½ per cent. on its value; what was the cost of transportation?

Ans. $86'59⅜.

7. A farmer had a flock of 639 sheep, which increased 33⅓ per cent. in 1 year; how many sheep had he at the expiration of the year? *Ans.* 852 sheep.

8. A man owing a debt of $1942'71½, pays 16⅜ per cent. of it; how much of it remains due? *Ans.* $1624'595 +.

9. A man, worth $4861, lost 28½ per cent. of it by endorsing with his neighbor; how much of it did he lose?

Ans. $1385'385.

10. What is ¾ per cent. of $115? *Ans.* $'862½.

11. What is ⅞ per cent. of $376? *Ans.* $3'29.

12. A gentleman, worth $4280, spent 15½ per cent. of his property in educating his son; how much did the son's education cost his father? *Ans.* $663'40.

13. A merchant has outstanding accounts to the amount of $1960; 22 per cent. of which is due in 3 months, and the remainder in 6 months. What is the amount due in 3 months? —— in 6 months? *Ans.* to the last, $1528'80.

14. A merchant who fails in business pays 63 per cent. on his debts; what does a man receive whose demands are $2465? *Ans.* $1552'95.

15. What does another man lose, whose demands are $3615 against the same merchant? *Ans.* $1337'55.

16. A young man is left with $5000, and loses 15 per cent. in paying too high a price for a farm, 15 per cent. of the remainder in selling the farm for less than its value; he expends 15 per cent. of what is left in an excursion to the west, 15 per cent. of what he has when he gets back in an unfortunate investment in railroad stocks, and 15 per cent. of the residue in trade; what has he then left? *Ans.* $2218'526 +.

NOTE. — Under the general subject of Percentage will be considered Insurance, Stocks, Brokerage, Profit and Loss, Interest, Discount, Commission, Bankruptcy, Partnership Banking, Taxes and Duties.

Insurance.

¶ **144.** Insurance is security to individuals against loss of property from fire, storms at sea, &c.

Companies incorporated for the purpose, having a certain capital to secure their responsibility, insure property at so much per cent. a year. When any property insured is destroyed by the agent insured against, the company pays to the owner the sum for which it is insured. The sums paid by the several individuals insured, make up the losses, and pay the company for doing the business.

Premium is the sum paid for insurance.

Policy is the writing of agreement.

An *Underwriter* is an insurer, whether it be an incorporated company or an individual.

Insurance at sea, called *Marine insurance,* is usually for a certain voyage. It is sometimes effected by an individual; it is then called out-door insurance.

The rate per cent. of insurance is in proportion to the risk. Property is not insured for its entire value, lest it should be fraudulently destroyed.

EXAMPLES FOR PRACTICE.

1. What is the annual insurance of $1000 on an academy, at $\frac{1}{2}$ per cent? *Ans.* $5.

2. Insured $14500 on a factory at $1\frac{3}{4}$ per cent. per annum; what was the premium? *Ans.* $253'75.

3. What is the premium for insuring $6000 on a store and goods at $\frac{3}{4}$ per cent.?

SOLUTION. — At 1 per cent. the sum is as many cents as there are dollars, or 6000 cents, which reduced is $60'00, and $\frac{3}{4}$ per cent. is $\frac{3}{4}$ of this, or, *Ans.* $45.

NOTE. — In this manner the percentage on any sum at 1 per cent. or less may be calculated with ease.

4. What must be paid for insuring $800 on a farm house, at $\frac{1}{4}$ per cent.? *Ans.* $2.

Questions. — ¶ **144.** What is insurance? Who insure? How is insurance estimated? Who pays, if the property be destroyed? How much? What remunerates the company? What is premium? — policy? — an underwriter? — marine insurance? How is marine in surance often effected? What is it then called? What is life insurance, and for what purpose is it effected? What is said of health insurance? Example. Why is not property insured for its entire value?

5. The Marine Insurance Company insures $17500 on the cargo of the ship Minerva, from Boston to Constantinople, at 2 per cent.; what is the premium? *Ans.* $350.

6. Amos Lawrence insures $34000 on the ship Washington and cargo from Canton to Boston, at $8\frac{1}{4}$ per cent.; what does he receive? *Ans.* $2805.

7. The New England Life Insurance Company insures $2000 on a person's life for one year at a premium of $1\frac{1}{20}$ per cent.; what is the sum? *Ans.* $21.

NOTE 1. — Life insurance is effected that the heirs of the individual, in case of his death, may receive the sum on which the premium is paid. The insurance is usually for one year, for seven years, or for life, and the annual rate per cent. is determined by a careful estimate made from bills of mortality of the probable chances of death with persons of different ages.

NOTE 2. — The premiums of health insurance companies, which have lately been organized, are specified sums to be paid annually in proportion to what is received weekly in case of sickness. Thus in the Massachusetts Company, $5'00 a year is paid by a person at the age of 30, to secure $4'00 a week in sickness. The premium is determined from a careful estimate of the probabilities of health.

Mutual Insurance.

¶ 145. The rate at which companies can afford to insure is estimated from the probable losses that will occur. But when the losses are small, large profits are made by the company. Or the losses at some time may be greater than the means at the command of the company, whereby its capital will be annihilated, while the losses of the insured will not be fully made up.

Hence, mutual insurance companies have been formed, to average the losses that may actually occur, which it is the aim of all insurance to do. Each one gives a premium note of so much per cent. on the property which he wishes to insure, the rate being determined by the risk of the property. The amount of these notes are the capital of the company, and a per cent. is paid down on them, to furnish money for

---

Questions. — **¶ 145.** How is the rate of insurance in ordinary companies estimated? What objections to this method? What is the aim of all insurance? Describe the premium note. How is the rate determined? How is money procured for use? What is the capital of the company? Why, and on what, are assessments made? Apply these principles to Ex. 1, and its solution.

immediate use. Any losses that occur more than this are averaged on the premium notes.

EXAMPLES FOR PRACTICE.

1. What sum is paid by a farmer for insuring $1500 on his buildings for five years in the Cheshire Mutual Insurance Company, the premium note being 7 per cent., of which 3 per cent. is paid down, and assessments paid afterwards of 2, $1\frac{1}{2}$, $3\frac{1}{4}$, and $\frac{3}{4}$ per cent.?

OPERATION.

$1500
'07
———
105'00
'$10\frac{1}{2}$
———
$52\frac{1}{2}$
10 50
———
$$11'02\frac{1}{2}$

SOLUTION. — First find the amount of the premium note, which is 7 per cent. of $1500 = $105, and $3 + 2 + 1\frac{1}{2} + 3\frac{1}{4} + \frac{3}{4} = 10\frac{1}{2}$ per cent. of $105 = $11'02\frac{1}{2}$, *Ans.*

2. What sum must be paid for insuring $2845 on a store for the same time, and with the same assessments, the premium note being 12 per cent.? *Ans.* $35'847.

3. What must be paid as above, premium note 15 per cent.? *Ans.* $44'808$\frac{3}{4}$.

4. Insured $5000 on a flouring mill for five years, in the Tompkins Co. Mutual, premium note 22 per cent. of which $4\frac{1}{2}$ per cent. was paid down, and $2\frac{1}{2}$ per cent. in assessments; what did it cost per year? *Ans.* $15'40.

5. Insured for five years $3200 on a house of worship in the Vt. Mutual, premium note 11 per cent., on which $3\frac{1}{2}$ per cent. was paid down, and four assessments were made respectively of $1\frac{1}{4}$, $2\frac{1}{2}$, 2, and $\frac{5}{6}$ per cent.; what is the whole sum paid? *Ans.* $35'49$\frac{1}{3}$.

6. How much more would it cost to insure the same property for the same time in the Ætna Insurance Company at $\frac{1}{2}$ per cent. each year? *Ans.* $44'50$\frac{2}{3}$.

7. What must be paid annually to insure $750 for five years on a library, premium note 6 per cent., paid down 4 per cent., sum of assessments $9\frac{1}{2}$ per cent.? *Ans.* $1'21$\frac{1}{2}$.

8. Insured for five years $900 on furniture, premium note 5 per cent. sum of payments on it 6 per cent.; how much is paid? *Ans.* $2'70.

Stocks.

¶ 146. In the construction of a railroad, which costs say $200000, the sum is divided into shares usually, of $100, each individual paying the amount of a certain number of shares, which are called his stock in the road. The one thus paying towards the road is called a stockholder, and is remunerated by a proportional share of the profits. It follows, of course, that the road may be so profitable that each share will be worth more than $100; the stock is then said to be above par. If the road is unprofitable, a share will be worth less than $100, and the stock is said to be below par. When a share is worth just $100, the stock is said to be at par. The stockholders together constitute the railroad company, and the sum of the shares is the capital of the company.

Manufactories, too large for individual enterprise, banks, &c., are conducted in a similar manner.

When governments borrow money, the sum each lends is said to be his stock in what are called the government funds.

EXAMPLES FOR PRACTICE.

1. What is the value of 35 shares in the Fitchburg railroad, at 120 per cent.? $\frac{120}{100}$ of $3500 = how much? *Ans.* $4200.

2. Sold 15 shares of the Eastern railroad at $7\frac{1}{2}$ per cent. advance; what sum did I receive? *Ans.* $1612'50.

3. What do I pay for 20 shares in the Old Colony railroad, at $1\frac{1}{4}$ per cent. below par? *Ans.* $1975.

4. What are 28 shares in the Vt. Central railroad worth, at $11\frac{1}{4}$ per cent. below par? *Ans.* $2485.

5. What are 45 shares in the Exchange Bank worth, at 8 per cent. below par? *Ans.* $4140.

6. What are 30 shares in the Western railroad worth, at $9\frac{1}{2}$ per cent. above par? *Ans.* $3285.

7. For what must I sell $5000 U. S. 6 per cent. stock, that is, stock on which 6 per cent. per annum is to be paid, at $1\frac{1}{2}$ per cent. discount? *Ans.* $4925.

8. What is $3200 in the Amoskeag Cotton Manufacturing Co. worth, at 17 per cent. above par? *Ans.* $3744.

Questions.—¶ 146. How is a railroad built? What is a share? —a stockholder, and how remunerated? When, and why, is stock above par? —below par? —at par? What is the company? —the capital? What else are established on the same plan? What are government funds?

9. What is $2000 in the Ocean Steam Navigation Co. worth, at 2 per cent. advance? *Ans.* $2040.

10. Bought 9 shares in the Western Transportation Co., at 4 per cent. below par; what did I pay? $864.

Brokerage.

¶ 147. Brokerage is an allowance made to a dealer in money, stocks, &c., who is called a broker. The allowance is generally a certain per cent. of the money paid out or received.

EXAMPLES FOR PRACTICE.

1. A western merchant procures $3500 in bills on N. E banks of a Boston broker, for Ohio money, the broker charging 2 per cent. for the accommodation; what brokerage does he pay? *Ans* $70.

2. A drover exchanges $2240 of country money for city bills, paying $\frac{1}{8}$ per cent. on his country money; what does he receive? *Ans.* $2237'20.

3. A broker is directed to buy 150 shares of the N. Y. and Erie railroad stock. He pays $92 per share, and receives $\frac{3}{4}$ per cent. on the money advanced; what does he receive on the whole? *Ans.* $103'50.

4. What must I pay a New York broker for $5000 of city bank bills, in bills on eastern banks, at $\frac{1}{4}$ per cent.? *Ans.* $5012'50.

5. A broker sells for an individual 90 shares of the Fitchburg railroad for $125 a share, receiving 1 per cent. on what money he gets; what does he receive? *Ans.* $112'50.

6. Bought $6000 in gold coin, paying the broker 1 per cent. for it; what does he receive? *Ans.* $60.

7. Sold $5200 in gold sovereigns, at a discount of $\frac{1}{2}$ per cent., for good bank bills, which are more convenient for me to carry; how much in bills do I receive? *Ans.* $5174.

Profit and Loss.

¶ 148. 1. Bought cloth at 40 cents a yard; how must I sell it to gain 25 per cent.?

Questions. — ¶ 147. What is brokerage? — a broker? **How is brokerage calculated?**

SOLUTION. — When the price at which goods are bought is given to find the price for which they must be sold, in order to gain or lose a certain per cent., the calculation is by the general rule for percentage. *Ans.* $'50.

NOTE. — The profit or loss must be added to or subtracted from the price of purchase.

EXAMPLES FOR PRACTICE.

2. Bought a hogshead of molasses for $60; for how much must I sell it to gain 20 per cent.? *Ans.* $72.

3. Bought broadcloth at $2,50 per yard; but, it being damaged, I am willing to sell it so as to lose 12 per cent.; how much will it be per yard? *Ans.* $2'20.

4. Bought calico at 20 cents per yard; how must I sell it to gain 5 per cent.? —— 10 per cent.? —— 15 per cent.? —— to lose 20 per cent.?

Ans. to the last, 16 cents. per yard.

Interest.

¶ 149. Interest is an allowance made by a debtor to a creditor for the use of money.

Per annum signifies *for a year.*

The rate per cent. per annum is the number of dollars paid for the use of 100 dollars, or the number of cents for the use of 100 cents for 1 year.

NOTE. — Percentage has hitherto been computed at a certain per cent., usually *without regard to time; interest* is computed at a certain per cent. for *one year*, and in the same proportion for a longer or shorter time.

Principal is the money due, for which interest is paid.

Amount is the sum of the principal and interest.

Legal interest is the rate per cent. established by law.

Usury is any rate per cent. *higher* than the legal rate

The *legal rate per cent.* varies in different countries, and in the different States. It is

---

Questions. — ¶ **148.** What do you understand by profit? — by loss? What are given, and what required, in this ¶? How found?

¶ **149.** What is interest? How is it computed? What does per annum signify? — rate per cent. per annum? What distinction do you make between percentage and interest? What is the principal? — the amount? — legal interest? — usury? What is the legal rate per cent. in each of the states? When no rate per cent. is mentioned, what rate per cent. is understood? How is interest for one year computed? — for more than a year?

6 *per cent.* in all the New England States, in Pennsylvania Delaware, Maryland, Virginia, N. Carolina, Tennessee Kentucky, Ohio, Indiana, Illinois, Missouri, Arkansas, New Jersey, District of Columbia, and on debts or judgments in favor of the U. States.

7 *per cent.* in New York, S. Carolina, Michigan, Wisconsin, and Iowa.

8 *per cent.* in Georgia, Alabama, Florida, Texas, and Mississippi.

5 *per cent.* in Louisiana.

When no rate per cent. is named, the *legal* rate per cent. of the *state* where the business is transacted, is always understood.

At 6 per cent., a sum equal to $\frac{6}{100}$ of the principal lent or due is paid for the use of it one year; at 7 per cent., a sum equal to $\frac{7}{100}$ of it, and so of any other rate per cent. Hence,

To find the interest of any sum for 1 year, is to take such a fractional part of the principal as is indicated by the rate per cent., as in percentage, and by the same rule; that is, we *multiply the principal by the rate per cent.*

For *more* years than 1, multiply the interest for 1 year by the number of years.

Interest is computed not only for one or more years, but in "the same proportion" for *months and days.*

¶ 150. *To find the interest on any sum for months at any rate per cent.*

1. What is the interest of $216'80, at 7 per cent., for 1 month? —— for 2 months? —— 3 mo.? —— 4 mo.? —— 5 mo.? —— 6 mo.? —— 7 mo.? —— 8 mo.? —— 9 mo.? 10 mo.? —— 11 mo.?

First, find the interest for 1 year, and then for the months take fractional parts of the interest for 1 year.

OPERATION.

$216'80 *Principal.*
'07 *rate per ct.*
————
$15'1760 *Int.* 12 *mo.*

NOTE. — The pupil will readily perceive methods of simplifying and shortening the operation, according to ¶ 140. Thus, he may take ½ of the interest for 1 year, that is, the int. for 6 months, and ⅓ the int. for 6 months, = the interest for 2 months, and add these together for the interest 8 months.

Questions. — ¶ **150.** How is interest calculated for months? What part or parts do we take for the interest 3 months? — 4 months? — 5 months — 6 months? — 7 months? — 8 months? — 9 months? &c.

$15'1760 *Int.* 12 *mo*

For 1 mo. take $\frac{1}{12}$ of the int. for 12 mo.			= $1'264 +	*Ans.*
" 2 mo.	" $\frac{1}{6}$	"	= 2'529 +	"
" 3 mo.	" $\frac{1}{4}$	"	= 3'794	"
" 4 mo.	" $\frac{1}{3}$	"	= 5'058 +	"
" 5 mo.	" $\frac{5}{12}$	"	= 6'323 +	"
" 6 mo.	" $\frac{1}{2}$	"	= 7'588	"
" 7 mo.	" $\frac{1}{2}+\frac{1}{12}$	"	= 8'852 +	"
" 8 mo.	" $\frac{1}{2}+\frac{1}{6}$	"	= 10'117 +	"
" 9 mo.	" $\frac{1}{2}+\frac{1}{4}$	"	= 11'382	"
" 10 mo.	" $\frac{1}{2}+\frac{1}{4}+\frac{1}{12}$	"	= 12'646 +	"
" 11 mo.	" $\frac{1}{2}+\frac{1}{4}+\frac{1}{6}$	"	= 13'911 +	"

EXAMPLES FOR PRACTICE.

2. What is the interest of $450 for 9 months, at the rate in Louisiana? What is the amount?

Ans. to the last, $466'87½.

NOTE. — The amount, which is the sum due, is found by adding the principal and interest together.

3. What is the amount of $87'50 on interest 7 months, at the rate in Georgia? *Ans.* $91'583.

4. What will be the interest of $163, for 4 months, at the rate in S. Carolina? *Ans.* $3'803.

5. What will be the interest of $850, for 10 months, at the rate in Kentucky? *Ans.* $42'50.

¶ 151. *To find the interest on any sum for days.*

1. What is the interest of $216'80, at 7 per cent., for 1 day? —— for 2 days? —— for 3 days, and so on, to 29 days?

NOTE 1. — In computing interest, 1 month is reckoned 30 days.

First, find the interest for 1 year, then for 1 month, as in Ex. 1, last ¶. These operations we need not repeat here, but take the interest as there found, $1'264 for 1 month, of which we may take parts, thus:

Questions. — ¶ 151. How many days are called a month in computing interest? How is the interest for days found? — for 3 days? — 5 days? — 8 days? — 10 days? — 12 days? — 18 days? — 21 days? — 25 days? — 28 days? Explain the principle of the direction how to take $\frac{1}{30}$ of a number. What is the amount, and how found?

			OPERATION	
			3)$1'264	*Int.* 1 *mo.*
For 1 day take	$\frac{1}{30}$ of Int. for 1 mo.		= $'042 +	" 1 *day.*
" 5 days "	$\frac{1}{6}$	"	= '210 +	" 5 *days*
" 6 " "	$\frac{1}{5}$	"	= '252 +	" 6 "
" 10 " "	$\frac{1}{3}$	"	= '421 +	" 10 "
" 15 " "	$\frac{1}{2}$	"	= '632	" 15 "
" 18 " "	3 times $\frac{1}{5}$, or $\frac{1}{2} + \frac{1}{10}$		= '758	" 18 "
" 20 " "	2 times $\frac{1}{3}$		= '842	" 20 "

NOTE 2.—To get $\frac{1}{30}$ of a number, divide it by 3, setting the first figure of the quotient 1 place towards the right, as in the operation. For 2 days, take 2 times the interest for 1 day. For 9 days, $\frac{1}{5}$ of int. for 30 days plus $\frac{1}{2}$ of $\frac{1}{5}$ of int. for 30 days. For 11 days, $\frac{1}{5}$ plus $\frac{1}{6}$. For 16 days, take 4 times the interest for 4 days, or $\frac{1}{2}$ the int. for 1 month, + the int. for 1 day. For 20 days, we may take 2 times the int. for $\frac{1}{3}$ of 1 month. For 25 days, $\frac{1}{2} + \frac{1}{3}$ the int. for 1 month, varying after this manner as may suit our convenience.

EXAMPLES FOR PRACTICE.

2. What is the interest of $400, at the rate in Alabama, for 9 days? *Ans.* $'80.

3. What is the interest of $75, for 19 days, at the rate in Florida? *Ans.* $'31$\frac{2}{3}$.

4. What is the interest of $500, for 25 days, at the rate in Texas? *Ans.* $2'777.

¶ 152. *When the time is expressed in more than one denomination, as in days and months, or years, months, and days.*

1. What is the interest of $32'25, for 1 year 7 months 19 days, at 4$\frac{1}{2}$ per cent.?

OPERATION.

	$32'25	*Principal.*
	'045	*rate per cent.*
	16125	
	12900	
	$1'45125	*Int. for* 1 *year.*
$\frac{1}{2}$ *Int. of* 12 *mo.* =	'7256 +	" 6 *months.*
$\frac{1}{6}$ " 6 " =	'1209 +	" 1 *month.*
$\frac{1}{2}$ " 30 *da.* =	'0604 +	" 15 *days.*
$\frac{1}{5}$ " 15 " =	'0120 +	" 3 "
$\frac{1}{3}$ " 3 " =	'0040 +	" 1 *day.*
Ans.	$2'3741 +	*Int. for* 1 *yr.* 7 *mo.* 19 *da.*

Hence, *To find the interest on any sum, for any time, at any rate per cent.*, *this*

GENERAL RULE.

I. *For* 1 *year.* Multiply the principal by the rate per cent., pointing off as in decimal fractions, and the product will be the interest for 1 year.

II. For more years than 1, multiply the interest for 1 year by the number of years.

III. *For months.* First find the interest for 1 year, of which take such fractional part as is denoted by the given number of months.

IV. *For days.* Take such part of the interest for 1 month as is denoted by the given number of days.

V. *For years, months and days, or for any two of these denominations of time.* Find the interest of each separately, and add the results together.

EXAMPLES FOR PRACTICE.

1. What is the interest of **$84**, for 1 year 9 months 20 days, at the legal rate in Alabama? *Ans.* $12‘133.

2. What is the interest of **$147**, for 2 years 8 months 12 days, at the rate in Michigan? *Ans.* $27‘783.

3. What is the interest of **$248**, for 2 years 6 months 20 days, at 9 per cent.? *Ans.* $57‘04.

4. What is the interest of **$161‘08**, for 11 months 19 days, at the rate in N. Y.? *Ans.* $10‘931 +.

5. What is the interest of **$73‘25**, for 1 year 9 months 12 days, at 8 per cent.? *Ans.* $10‘45 +.

6. What is the interest of **$910‘50**, for 3 years 9 months 26 days, at 7 per cent.? *Ans.* $243‘609 +.

7. What is the amount of **$185‘26**, in 2 years 3 months 11 days, at $7\frac{1}{2}$ per cent.? *Ans.* $216‘947 +.

8. What is the interest of **$656** from Jan. 9 to Oct. 9 following, at $\frac{1}{2}$ per cent.?

$656 *Principal.*
2) 6‘56 = *Int.* 1 *yr. at* 1 *per cent.*
2) 3‘28 = " " $\frac{1}{2}$ "
2) 1‘64 = " 6 *mo* " "
‘82 = " 3 *mo.* " "
Ans. $2‘46 *Int.* 9 *mo. at* $\frac{1}{2}$ *per cent.*

SOLUTION. — Remove the separatrix two places to the *left*, and the sum itself will express the interest for 1 year at 1 per cent., ¶ 149, $\frac{1}{2}$ of which will be the interest for 1 year at $\frac{1}{2}$ per cent., of which take fractional parts for 9 months.

Questions. — **¶ 152.** Explain the operation, Ex. 1. What is the general rule?

9. What is the interest of \$46‘28, for 2 years 3 months 23 days, at 5 per cent.? *Ans.* \$5‘354 +.

10. What will be the amount of \$175‘25, in 5 years 8 months and 21 days, at 6 per cent.? *Ans.* \$235‘448 +.

11. What will be the amount of \$96‘50, for 1 year 11 months 29 days, at 12½ per cent.? *Ans.* \$120‘591 +.

NOTE. — At 12½ per cent., ⅛ of the principal will be the interest for 1 year.

12. What is the interest of \$54‘81, for 1 year and 6 months, at the rate in Louisiana? *Ans.* \$4‘11.

13. What is the interest of \$500, for 9 months 9 days, at the rate in Georgia? *Ans.* \$31.

14. What is the interest of \$62‘12, for 1 month 20 days, at 4 per cent.? *Ans.* \$‘345.

15. What is the interest of \$85, for 10 months 15 days, at 12½ per cent.? *Ans.* \$9‘296.

16. What is the amount of \$53, at 10 per cent., for 7 months? *Ans.* \$56‘091.

17. What is the interest of \$327‘825, at the rate in Florida, for 1 year? *Ans.* \$26‘226.

18. What is the interest of \$325, for 3 years, at the rate in Pennsylvania? *Ans.* \$58‘50.

19. What is the interest of \$187‘25, for 1 year 4 months, at the rate in Delaware? *Ans.* \$14‘98.

20. What is the interest of \$694‘84, for 9 months, at 10 per cent.? *Ans.* \$52‘113.

21. What is the interest of \$32‘15, for 1 year, at 4½ per cent.? *Ans.* \$1‘446 +.

22. What is the amount of \$600, in 2 years, at the rate in New England? *Ans.* \$672.

23. What is the interest of \$57‘78, for 1 year 4 months 17 days, at 4 per cent.? *Ans.* \$3‘19.

24. What is the amount of \$298‘59, from May 19th, 1847, till Aug. 11th, 1848, at the rate in Texas? (See ¶ 127.)

25. What is the amount of \$196, from June 14, 1847, till April 29, 1848, at 5¾ per cent.? *Ans.* \$205‘861 +.

To compute interest for months and days, when the rate is 6 per cent.

I. Of One Dollar.

¶ 153. A method brought out in the "Scholar's Arithmetic," 1801, and revised in "Adams' New Arithmetic,"

1827, will be preferred by many to the foregoing, in those states where the legal rate is 6 per cent.

Induction. The interest on $1 for 1 year, at 6 per cent., being

'06 cents, is
'01 cent for 2 months,
'005 mills (or $\frac{1}{2}$ a cent) for 1 month of 30 days, and
001 mill for every 6 days; 6 being contained 5 times in 30.

Hence, it is very easy to find, by *inspection*, the interest of 1 dollar, at 6 per cent., for *any given time.*

The *cents* will be equal to *half* the greatest even number of months.

The *mills* will be 5 for the odd month, if there be one, and 1 for every time 6 is contained in the given number of days, with such

Part of 1 *mill*, as the days *less than* 6, are *part* of 6 days.

1. What is the interest of $1, at 6 per cent., for 9 months 18 days?

Solution. — The greatest even number of months is 8, the interest for which will be $'04; the mills, reckoning 5 for the odd month, and 3 for the 18 (3 times 6 = 18) days, will be $'008, which, added to the cents, give 4 cents 8 mills for the interest of $1 for 9 months and 18 days. *Ans.* $'048.

2. What will be the interest of $1 for 5 months 6 days? —— 6 months 12 days? —— 7 months? —— 8 months 24 days? —— 9 months 12 days? —— 10 months? —— 11 months 6 days? —— 12 months 18 days? —— 15 months 6 days? —— 16 months?

3. What is the interest of $1 for 13 months 16 days?

Solution. — The cents will be 6, and the mills 5, for the odd month, and 2 for 2 times 6 = 12 days, and there is a remainder of 4 days, the interest for which will be such part of 1 mill as 4 days is part of 6 days that is, $\frac{4}{6} = \frac{2}{3}$ of a mill. *Ans.* $'067$\frac{2}{3}$.

4. What is the interest of $1 for 12 months 3 days?

Questions. — ¶ 153. At 6 per cent., what is the interest of $1 for 1 year? — for 2 months? — for 1 month of 30 days? — for every 6 days? How, then, may the interest of $1, at 6 per cent., for any given time, be found by inspection? If there is no odd month, and the number of days be less than 6, what is to be done? Why? The interest on $1 for a number of days less than 6, is what? How do you find it written in the examples which have been given?

NOTE.—*If there is no odd month, and the number of days be less than* 6, *so that there are no mills, a cipher* must be put in the place of mills; thus, for 12 months 3 days, the cents will be '06, the mills 0 the 3 days ½ a mill. *Ans.* $'060½.

5. What will be the interest of $1, for 2 months 1 day? —— 4 months 2 days? —— 6 months 3 days? —— 8 months 4 days? —— 10 months 5 days? —— for 3 days? —— for 1 day? —— for 2 days? —— for 4 days? —— for 5 days? *Ans.* to the last, $'000⅚.

II. OF ANY SUM.

¶ 154. 1. What is the interest of $75, for 10 months 12 days?

FIRST OPERATION.

```
$'052 Int. on $1.
   75
 ----
  260
 364
 ----
$3'900 Int. of $75.
```

SECOND OPERATION.

```
 $75
 '052
 ----
  150
 375
 ----
$3'900 Int. of $75.
```

SOLUTION.—We find the interest of $1, by the last ¶, which is $'052, and 75 times this sum, as in the first operation, will be the interest of $75; or, since either factor may be made the multiplicand, (¶ 21,) we multiply $75 by '052, thus taking 52 thousandths of the principal, for that is the part taken, when the interest of each dollar is $'052.

2. What is the interest of $56'13, for 8 months 5 days?

OPERATION

```
        3|2) $56'13 Principal.
              '040⅚
             ------
            224520  Int. for 8 mo.
½ multip. =   2806    "    3 days.
⅓    "    =   1871    "    2   "
             ------
         $2'29197  Int. for 8 mo. 5 days. Ans.
```

SOLUTION.—We find the interest of $1 for the time to be $'040⅚, and we multiply the principal by '040⅚. As 1 thousandth of the multiplicand is taken (1 mill or thousandth being the unit of the multiplier) for every 6 days, for the days *less* than 6, we take such fractional part of the mul tiplicand as the odd day or days is of 6. Thus, for 3 days, we take ½ of the multiplicand, which will be the interest for that time in mills For 2 days, we take ⅓ of the multiplicand. Adding together the interest for 8 months, 3 days, and 2 days, the sum will be the interest for 8 months and 5 days.

NOTE 1. — As the interest of $1 for 6 days is 1 mill, that of $10 for the same time will be 10 mills = 1 cent. Hence, if the sum on which interest is to be cast be *less* than $10, the interest, for any number of days *less* than 6, will be *less* than 1 cent; consequently, *in business* transactions, if the sum be less than $10, such days need not be regarded.

Hence, *To find the interest of any given sum, in Federal Money, for any length of time, at* 6 *per cent.,*

RULE.

I. Find the interest on $1 for the given time by inspection.

II. Multiply the principal by this sum, written as a decimal, and point off the result as in multiplication of decimals.

EXAMPLES FOR PRACTICE.

3. What is the interest of $194, for 4 months 12 days? *Ans.* $4‘268.
4. Interest of $263‘48, for 2 mo. 21 days? $3‘556.
5. Amount of $985, for 5 years 8 months? $1319‘90.
6. Interest of $87‘19, for 1 year 3 months? $6‘539.
7. " of $116‘08, for 11 mo. 19 days? $6‘751.
8. " of $200, for 8 mo. 4 days? $8‘133.
9. " of $0‘85, for 19 mo.? $‘08.
10. " of $8‘50, for 1 year 9 mo. 12 days? $‘909.
11. " of $675, for 1 mo. 21 days? $5‘737.
12. " of $8673, for 10 days? $14‘455.
13. " of $0‘73, for 10 mo.? $‘036.
14. " of $126‘46, for 9 mo.? $5‘69.
15. " of $318‘, for 10 mo. 16 days? $16‘748.
16. " of $418‘, for 1 year 7 mo. 17 days? $40‘894.
17. " of $268‘44, for 3 yrs. 5 mo. 26 ds.? $56‘193
18. " of $658, from Jan. 9 to Oct. 9 following? $29‘61.

Questions. — ¶ **154.** After the interest of $1 is found, how is the interest of $75 found, by the first operation, Ex. 1? — by the second operation? Why? In what denomination is the ⅚ of the multiplicand taken in Ex. 2, and why? What is said of the interest of $10 for less than 6 days, and why? Give the rule. How may the interest of any sum be found for 6 days? — for less than 6 days?

19. Interest of $96, for 3 days?

NOTE 2. — The interest of $1 for 6 days being 1 mill, the dollars *themselves* express the interest in *mills for six days*, of which we may take parts.

20. Interest of $73‘50, for 2 days?
21. " of $180‘75, for 5 days?
22. " of $15000, for 1 day?

¶ 155. When 6 per cent. interest is required for a large number of years, it will be more convenient to find the interest for one year, and multiply it by the number of years; after which, find the interest for the months and days, if any, as usual.

1. What is the interest of $520‘04, for 30 years and 6 months?

OPERATION.

```
     $520‘04 Principal.
         ‘06
     -------
2)$31‘2024 Int. 1 year.
        30
    --------
  $936‘0720 "  30 years.
  $ 15‘6012 "   6 mo.
    --------
  $951‘6732 "  30 years. 6 mo.
```

Ans. $951‘673.

2. What is the interest of $1000, for 120 years? *Ans.* $7200.

3. What is the interest on $400 for 10 years 3 months and 6 days? *Ans.* $246‘40.

4. What is the interest of $220, for 5 years? —— for 12 years? —— 50 years? *Ans.* to the last, $660.

5. What is the amount of $86, at interest 7 years? *Ans.* $122‘12.

6. What is the amount of $750, on interest 9 years 4 mo. 14 days? *Ans.* $1171‘75.

Questions. — ¶ 155. How do we get 6 per cent. interest for a large number of years? How, when there are also months and days?

To find the interest on pounds, shillings, and pence.

¶ 156. 1. What is the interest of £36 9s. 6½d., for 1 year, at 6 per cent.?

Reduce the shillings, pence, &c., to the decimal of a pound, by inspection, (¶ 141,) then proceed in all respects as in federal money Having found the interest, reverse the operation, and reduce the first three decimals to shillings, &c., by inspection, (¶ 142.)

Ans. £2 3s. 9d.

2. Interest of £36 10s., for 18 mo. 20 days, at 6 per cent.? *Ans.* £3 8s. 1½d.

3. Interest of £95, for 9 mo.? *Ans.* £4 5s. 6d.

4. What is the amount of £18 12s., at 6 per cent. interest, for 10 months 3 days? *Ans.* £19 10s. 9¼d.

5. What is the amount of £100, for 8 years, at 6 per cent.? *Ans.* £148.

6. What is the amount of £400 10s., for 18 months, at 6 per cent.? *Ans.* £436 10s. 10d. 3qr.

7. What is the amount of £640 8s., at interest for 1 year at 6 per cent.? —— for 2 years 6 months? —— for 10 years? *Ans.* to the last, £1024 12s. 9½d.

8. What is the amount of £391 17s., for 3 years 3 mo., at 4½ per cent.?

9. What is the amount of £235 3s. 9d., from March 5, 1846, till Nov. 23, 1846, at 5¼ per cent.? *Ans.* £244 8¾d.

To calculate interest on notes, &c., when partial payments have been made.

¶ 157. Payment of *part* of a note or other obligation, is called a *partial payment.*

It has been settled in the Supreme Court of the U. States, and their practice adopted by nearly all the states in the Union, that payments shall be applied to keep down the interest, and that neither *interest* nor *payment* shall ever *draw interest.* Hence, if the payment at any time *exceed* the interest computed to the *same* time, that excess is taken from the principal; but if the payment be *less* than the interest, the principal remains unaltered. Hence, the

---

Questions.—¶ 156. How do we proceed, when the principal is pounds, shillings, and pence?

RULE.

Compute the interest on the principal to the time when the payment, or payments, (if the first be less than the interest,) shall equal or exceed the interest due; subtract the interest from the payment, or *sum* of the payments, made within the time for which interest was computed, and deduct the excess from the principal.

The remainder will form a new principal, with which proceed as with the first.

1. $116'666.

Boston, May 1st, 1842.

For value received, I promise to pay James Conant, or order, one hundred and sixteen dollars sixty-six cents and six mills, on demand, with interest. Samuel Rood.

On this note were the following endorsements:

Dec. 25, 1842, received	$ 16'666	
July 10, 1843, "	$ 1'666	
Sept. 1, 1844, "	$ 5'000	
June 14, 1845, "	$ 33'333	
April 15, 1846, "	$ 62'000	

NOTE. — In finding the *times* for computing the interest, consult ¶ 127.

What was due August 3, 1847? *Ans.* $23'775.

NOTE 1. — The transaction being in Massachusetts, the rate of interest will be 6 per cent.

The first principal on interest, from May 1, 1842,		$116'666
Payment, Dec. 25, 1842, (exceeding interest due,)	$16'666	
Interest to time of 1st payment, . .	4'549	
		12'117
Remainder for a new principal, ...		$104'549
Payment, July 10, 1843, less than interest then due, ...	$1'666	
Payment, Sept. 1, 1844, less than interest then due, ...	$5'000	
Amount carried forward,	$6'666	

Questions. — **¶ 157.** What is a partial payment of a note or obligation? What court has established a rule for computing interest on notes, &c., on which partial payments have been made? How extensively is this rule adopted? Repeat the rule. What is the great fundamental principle on which this rule is based? What is customary when notes with endorsements are paid within one year of the time they are given?

Amount brought forward,	$6'666	
Payment, June 14, 1845, . .	$33'333	
Amount, (exceeding interest due,)	$39'999	
Interest from Dec. 25, 1842, to June 14, 1845, (29 mo. 19 days,) . . .	15'490	
		24'509
		$80'040
Payment, April 15, 1846, (exceeding interest due,)	$62'000	
Interest from June 14, 1845, to April 15, 1846, (10 mo. 1 day,) . . .	4'015	
		57'985
Remainder for a new principal, ...		$22'055
Interest due August 3, 1847, from April 15, 1846, (15 months 18 days,)		1'720
Balance due Aug. 3, 1847. . . *Ans.*		$23'775

2. $867'33.

Buffalo, Dec. 8th, 1842.

On demand, for value received, I promise to pay James Hadley, or bearer, eight hundred and sixty-seven dollars and thirty-three cents, with interest after three months.

Wm. R. Dodge.

On this note were the following endorsements, viz.:

April 16, 1843, received $ 136'44.
April 16, 1845, received $ 319'.
Jan. 1, 1846, received $ 518'68.

What remained due July 11, 1847? *Ans.* $31'765 +.

3. $1000.

Boston, Jan. 1, 1840.

For value received, I promise to pay George A. Curtis, or order, on demand, one thousand dollars, with interest.

Caleb Nelson.

On this note were the following endorsements: — April 1, 1840, $24; Aug. 1, 1840, $4; Dec. 1, 1840, $6; Feb. 1, 1841, $60; July 1, 1841 $40; June 1, 1844, $300; Sept. 1, 1844, $12; Jan. 1, 1845, $15; and Oct. 1 1845, $50.

What remained due, June 1, 1846? *Ans.* $843'083 +.

Note. 2 — When notes are paid within one year from the time they were given, and have endorsements, it is common to subtract from the amount of the principal for the whole time the amount of each endorsement from its date till the day of settlement.

4. $300.

Mobile, (Alabama,) June 10 1846.

For value received, we jointly and severally promise to Reuben Washburn to pay him, or order, on demand, three hundred dollars, with interest.

Louis P. Legg.
Sanford Comstock.

On this note were endorsements: Jan. 20, 1847, $116; March 2, 1847, $49'50; April 26, 1847, $85.

What remained due June 2, 1847?

NOTE. — The rate in Alabama is 8 per cent. *Ans.* $67'894 +.

CONNECTICUT METHOD.

¶ 158. The Supreme Court of Connecticut have established a method somewhat different from the U. S. Court rule, which may be practised by those belonging to the state. The substance of the method is presented in the following

RULE.

I. Payments a year, or more than a year after the date of the note, or from each other, and those less than the interest due, are treated according to the U. S. Court rule.

II. Find the amount of any other payment from date till one year from the time the note was given, or of a former cast, and subtract it from the amount of the principal for one year. The remainder is a new principal.

NOTE. — If the note be settled in less than a year from the time of a cast, find the amount of each subsequent payment that has been made till the time of settlement, and subtract it from the amount of the principal found till the same time.

$1100.

Woodstock, Ct., Jan. 1, 1841.

For value received, I promise to pay Henry Bowen, or order, eleven hundred dollars, on demand, with interest.

James Marshall.

On this note are the following endorsements: — Sept. 1, 1841, $30; April 1, 1842, $200; Dec. 1, 1842, $180; March 1, 1844, $195; Sept 16, 1844, $250; May 16, 1846, $100; July 16, 1846, $170.

What remains due Jan. 16, 1847? *Ans.* $234'134 +.

VERMONT COURT RULE.

Regulating the mode of casting, and the allowance of interest in certain cases.

¶ 159. I. When the contract is for a sum with interest, payments made *before* the debt falls due are to be applied exclusively to the principal; payments made *after* the debt falls due are to be applied to the interest till the same is extinguished.

II. Interest accruing after the debt falls due is to be extinguished annually, if the payments are sufficient for that purpose. But interest upon interest is not allowable.

III. Interest, upon a contract for the payment of interest *annually*, is to be cast upon that interest after the same falls due, up to the time of payment.

NOTE. — See ¶ 162, Note 2, and Example; also Note 3, and Example.

COMPOUND INTEREST.

¶ 160. A promises to pay B $256 in 3 years, with interest annually; but at the end of 1 year, not finding it convenient to pay the interest, he consents to pay interest on the interest from that time, the same as on the principal.

Simple interest is that which is allowed for the principal only.

Compound interest is that which is allowed for both principal and interest, when the latter is not paid at the time it becomes due.

To calculate Compound Interest,

RULE.

Add together the interest and principal at the end of each year, and make the amount the principal for the next succeeding year. From the last amount subtract the principal.

1. What is the compound interest of $256, for 3 years, at 6 per cent.?

Questions. — **¶ 160.** What is simple interest? — compound interest? What is the method of computing compound interest?

$256	given sum, or first principal.
'06	
15'36	interest, } to be added together.
256'00	principal, }
271'36	amount, or principal for 2d year.
'06	
16'2816	compound interest, 2d year, } added together.
271'36	principal, do. }
287'6416	amount, or principal for 3d year.
'06	
17'25846	compound interest, 3d year, } added together.
287'641	principal, do. }
304'899	amount.
256	first principal subtracted.

Ans. $48'899, compound interest for 3 years.

2. At 6 per cent., what will be the compound interest, and what the amount, of $1 for 2 years? —— what the amount for 3 years? —— for 4 years? —— for 5 years? —— for 6 years? —— for 7 years? —— for 8 years?

Ans. to the last, $1'593 +.

¶ 161. It is plain that the amount of $2, for any given time, will be 2 times as much as the amount of $1; the amount of $3 will be 3 times as much, &c.

Hence, we may form the amounts of $1, for several years, into a table of *multiplicands* for finding the amount of *any sum* for the same time.

Questions.—¶ **161.** How do we compute by the table? When there are months and days, how do you proceed? In what time will any sum, at 6 per cent., double itself?

TABLE,

Showing the amount of $1, or £1, for any number of years from 1 to 40, at 5 and 6 per cent., and from 1 to 20 at 7 and 8 per cent.

Years.	5 per cent.	6 per cent.	Years.	5 per cent.	6 per cent.
1	1·05	1·06	21	2·785963	3·399564
2	1·1025	1·1236	22	2·925261	3·603537
3	1·15762 +	1·19101 +	23	3·071524	3·819750
4	1·21550 +	1·26247 +	24	3·225100	4·048935
5	1·27628 +	1·33822 +	25	3·386355	4·291871
6	1·34009 +	1·41851 +	26	3·555673	4·549383
7	1·40710 +	1·50363 +	27	3·733456	4·822346
8	1·47745 +	1·59384 +	28	3·920129	5·111687
9	1·55132 +	1·68947 +	29	4·116136	5·418388
10	1·62889 +	1·79084 +	30	4·321942	5·743491
11	1·71033 +	1·89829 +	31	4·538039	6·088101
12	1·79585 +	2·01219 +	32	4·764941	6·453387
13	1·88564 +	2·13292 +	33	5·003189	6·840590
14	1·97993 +	2·26090 +	34	5·253348	7·251025
15	2·07892 +	2·39655 +	35	5·516015	7·686087
16	2·18287 +	2·54035 +	36	5·791810	8·147252
17	2·29201 +	2·69277 +	37	6·081407	8·636087
18	2·40661 +	2·85433 +	38	6·385477	9·154252
19	2·52695	3·02559 +	39	6·704751	9·703507
20	2·65329 +	3·20713 +	40	7·039989	10·285718

Years.	7 per cent.	8 per cent.	Years.	7 per cent.	8 per cent.
1	1·07	1·08	11	2·10485 +	2·331638 +
2	1·1449	1·1664	12	2·25219 +	2·518170 +
3	1·225043	1·259712	13	2·40984 +	2·719623 +
4	1·310796 +	1·360488 +	14	2·57853 +	2·917193 +
5	1·402551 +	1·469328 +	15	2·75903 +	3·150569 +
6	1·50073 +	1·586874 +	16	2·95216 +	3·402614 +
7	1·60578 +	1·713824 +	17	3·15881 +	3·674823 +
8	1·71818 +	1·850930 +	18	3·37993 +	3·968809 +
9	1·83845 +	1·999004 +	19	3·61652 +	4·286314 +
10	1·96715 +	2·158924 +	20	3·86968 +	4·629219 +

Note 1.—When there are months and days, first find the amount for the *years*, and on that amount cast the interest for the months and days; this, added to the amount, will give the answer.

1. What is the amount of $600·50, for 20 years, at 5 per cent. compound interest? —— at 6 per cent.?

SOLUTION. — $ 1 at 5 per cent., by the table, is $ 2'65329; therefore, 2'65329 × 600'50 = $ 1593'30 + *Ans.* at 5 per cent.; and 3'20713 × 600'50 = $ 1925'881 + *Ans.* at 6 per cent.

2. What is the amount of $40'20, at 6 per cent. compound interest, for 4 years? —— for 10 years? —— for 18 years? —— for 12 years? —— for 3 years and 4 months? —— for 20 years 6 months and 18 days?

Ans. to the last, $133'181 +.

NOTE 2. — Any sum at 6 per cent. compound interest, will double itself in 11 years 10 months and 22 days.

3. What is the amount of $750, at 7 per cent., compound interest, for 16 years? *Ans.* $2214'12.

4. What is the amount of $150, at 8 per cent., compound interest, for 20 years?

¶ 162. ANNUAL INTEREST.

1. $500'00.

Keene, N. H., Feb. 2d, 1843.

For value received, I promise to pay George Hooper, or order, five hundred dollars, with interest annually till paid.

Henry Truman.

What was due Aug. 2, 1847, no payment having been made?

SOLUTION. — A note like the above, with the promise to pay interest annually, is not considered, in courts of law, a contract for anything more than simple interest on the principal. Interest does not become principal by operation of law. If the annual interest be not paid, the creditor can bring his *action* for it at the end of each year. If he neglects to do this in Massachusetts, and in some other states, he is considered as waiving his claim, and must be contented to receive simple interest.

In New Hampshire, interest is allowed on the *annual* interest, in the nature of *damages* for its detention and use, from the time it becomes payable till paid. Hence, when a note is written " with interest annually," the following is the N. H.

COURT RULE.

Compute separately the interest on the principal, from the time the note is given till the time of payment, and the interest on each year's interest from the time it should be paid, till the time of payment. The sum of the interests thus obtained will be the interest sought, to which add the principal for the amount due.

Applying this rule to a cast on the above note, the first year's interest of $30, is on interest 3 years and 6 months; the 2d year's interest is on interest 2 years and 6 months, &c.

The operation may be written down as follows:

Interest on the principal,	$500'00,	4 years 6 months,	$135'00
" 1st year's int.	($30)	3 years 6 months,	6'30
" 2d "	"	2 years 6 months,	4'50
" 3d "	"	1 year 6 months,	2'70
" 4th "	"	6 months,	'90
		Amount of interest,	$149'40

Then $500 + $149'40 = *Ans.* $649'40, amount due.

NOTE 1.—Among business men the mutual understanding and practice, oftentimes, is compound interest, when the note is written with interest annually; but compound interest cannot be *legally* enforced, unless it be so expressed in the note. The interest, by the method presented in the rule, is due at the end of each year, but as it is not paid then, it is on simple interest, just as any other debt would be, if not paid when due.

NOTE 2.—The same method is practised in Vermont when no payments have been made, and there is no time specified when the note is due.

2. $1000'00.

Brattleboro', Vt., June 10, 1842.

For value received, we, jointly and severally, promise to pay Joseph Steen, or order, one thousand dollars, on demand, with interest annually.

Samuel W. Ford,
Stephen Wise.

What was due Sept. 10, 1847? *Ans.* $1355'50.

NOTE 3.—But when payments have been made, we find the amount of the principal for one year, and having found the amount of

Questions.—¶ **162.** When a note is written with the promise to pay interest annually, how is it considered in courts of law? If the annual interest be not paid, what may the creditor do? If he neglects to do this, how is it considered by the courts in Massachusetts, and in some other states? What is the New Hampshire court practice? On what principle is simple interest allowed on the annual interest? Repeat the court rule. What mutual understanding among business men? How can the interest on interest be regarded as interest on any debt? What other state practises the same method? When payments have been made, what is done?

each payment made during that year from its date till the end of the same year, we subtract it from that amount, and the remainder will be a new principal, with which we proceed as before.

3. $400.

Windsor, Aug. 2, 1844.

For value received, I promise to Bishop and Tracy, to pay to them, or order, four hundred dollars, on demand, with annual interest. Asa H. Truman.

On this note are endorsements as follows: April 2, 1845, $50 June 2, 1845, $30; Jan. 2, 1846, $100; May 17, 1847, $80.

What remained due Aug. 2, 1847?

OPERATION.

Principal,	. . . .	$400'00
Interest 1 year,	. . . . . .	24'00
1st payment,	$50 + int. 4 mo. $1 = $51'00,	
2d "	$30 + int. 2 mo. $'30 = 30'30.	Am't, 424'00
		81'30
		New prin. 342'70 &c.
		Ans. $194'34.

The time, rate per cent., and amount being given, to find the principal.

¶ 163. 1. What sum of money, put at interest 1 year and 4 months, at 6 per cent., will amount to $61'02?

SOLUTION. — $1'08 is the amount of $1 for 1 year and 4 months, and $61'02 is the amount of as many dollars as the number of times $1'08 is contained in $61'02. $61'02 ÷ $1'08 = $56'50, the principal required. Hence,

RULE.

Divide the given amount by the amount of 1 dollar, at the given rate and time.

2. What principal, at 8 per cent., in 1 year 6 months, will amount to $85'12? *Ans.* $76.

3. What principal, at 6 per cent., in 11 months 9 days will amount to $99'311?

Questions. — ¶ 163. What is the subject of this paragraph Repeat the first example. Why do you divide $61'02 by $1'08? What is the rule for finding the principal, when the time, rate per cent., and amount are given?

NOTE. — The interest of $ 1, for the given time, is '056½; but, when there are odd days, instead of writing the parts of a mill as a common fraction, it will be more convenient to write them as a *decimal*, thus, '0565; that is, extend the decimal to four places.

Ans. $ 94.

4. A produce buyer purchased 1000 bushels of wheat, on credit, and agreed to pay 15 per cent. on the purchase money; at the expiration of 4 months he paid the debt and interest, which together amounted to **$1500**; what was the value of the wheat? *Ans.* $1428'571 --.

Discount.

¶ 164. 1. I purchase a horse, agreeing to pay for him $106, one year from the time he comes into my possession, without interest. When I take the horse, I propose to pay down, for a just allowance; what must I pay, money being worth 6 per cent.?

SOLUTION. — I should not pay the whole sum, $106; for the one who received it might loan it to a third person and receive $112'36 for it at the end of the year, when he *should* receive only $106. Consequently, I must pay a sum, which, if loaned, will amount to $106 in a year, which is $100, *Ans.*

An allowance made by a creditor to a debtor, for paying money due at some future time without interest, before the time agreed on for payment, is called *Discount*, and the sum paid is called the *Present Worth.*

2. I sell a piece of wild land in Wisconsin, for $868, to be paid in 4 years, without interest, since the purchaser is to receive no profit from his purchase. Wishing ready money, I transfer the debt to a third person for a sum, which, put at 6 per cent. interest, for the time, would amount to $868; what do I receive for my debt?

SOLUTION. — Since $1 in 4 years will amount to $1'24, for every time $1'24 can be subtracted from, or is contained in, $868, I shall receive $1. $868 ÷ $1'24 = $700, *Ans.*

The rule is evidently the same as in the last paragraph.

Questions. — ¶ **164.** What is the subject of this paragraph? Repeat the first example. Solve it. What is discount? — present worth? How is it found? Proof? How is discount found?

PROOF. — Find the amount of the result at the given rate and time this amount will be equal to the given sum.

3. How much ready money must be paid for a note of $ 18, due 15 months hence, discounting at the rate of 6 per cent.?

Ans. $ 16'744+.

4. What is the present worth of $ 56'20, payable in 1 year 8 months, discounting at 6 per cent.? —— at $4\frac{1}{2}$ per cent.? —— at 5 per cent.? —— at 7 per cent.? —— at $7\frac{1}{2}$ per cent.? —— at 9 per cent.? *Ans* to the last, $ 48'869.

5. What is the present worth of $ 834, payable in 1 year 7 months and 6 days, discounting at the rate of 7 per cent.? *Ans.* $ 750+.

6. What is the discount on $ 321'63, due 4 years hence, discounting at the rate of 6 per cent.?

NOTE. — If the present worth be subtracted from the given amount, the remainder will be the discount.

Ans. $ 62'25+.

7. Sold goods for $ 650, payable one half in 4 months, and the other half in 8 months; what must be discounted for present payment, at 6 per cent.? *Ans.* $ 18'872+.

8. A merchant purchases in New York city, goods to the amount of $ 5378, on 6 months credit, paying 6 per cent. more than if he had paid down. What would he have saved if he had borrowed money at 7 per cent. per annum, with which to make his purchase? What would he save in 20 years, averaging $2\frac{1}{2}$ such purchases each year?

Ans. to the last, $ 6342.

Commission.

¶ 165. 1. A merchant in Utica receives $988 from a house in New York, with which to purchase butter, after deducting for his services 4 per cent. on what money he shall lay out; how much will he pay for butter?

SOLUTION. — Of every $1'04 which he receives he must lay out 1 dollar, and retain 4 cents, thus having 4 cents for each dollar which he pays for butter. Then, as many times as $1'04 is contained in $988, so many dollars he must pay out. $988 ÷ $1'04 = $950, *Ans.* $950.

NOTE. — Had we multiplied $988 by '04 to ascertain how much he received for his services, it would have given him 4 cents on each dollar which he received. This would have given him 4 cents for laying out 96 cents, instead of 100 cents, as required by the question.

Questions. — ¶ 165. What is the first example? Give the solution. Show the error, if performed as supposed in the note. What is commission? What are persons, who buy and sell goods for others, called? When they receive money to disburse, how is their commission estimated? When the value of the goods bought or sold is known, how is the commission estimated?

The per cent. or amount allowed to persons for their services in assisting merchants and others in purchasing and selling goods, and for transacting other business, is called *Commission.*

Persons who buy and sell goods for others, are called *Agents*, *Commission Merchants*, *Correspondents*, and *Factors.*

When agents, &c., receive money to disburse, their commission is estimated in the same manner as discount, by the rule ¶ 163.

2. Received $2475, with which to purchase wool, after deducting my commission of 5 per cent.; how many dollars will I pay out? What will be the amount of my commission?

Ans. to the last, $117'857.

3. Sent my agent $4820, with which to purchase wheat, after deducting his commission of 7½ per cent.; how much money will he expend, and what will be the amount of his commission?

Ans. He will expend $4483'72 +.

NOTE.—When the value of the goods bought or sold is known, the commission is estimated upon that value, in the same manner as percentage. (¶ 143.)

4. A commission merchant sold goods to the amount of $1422, at 5 per cent. commission; how much did he receive for his services?

Ans. $71'10.

5. My correspondent sends me word that he has purchased goods to the value of $1286, on my account; what will be his commission, at 2½ per cent.? *Ans.* $32'15.

6. What must I allow my correspondent for selling goods to the amount of $2317'46, at a commission of 3¼ per cent.?

Ans. $75'317.

7. A tax on a certain town is $1627'18, on which the collector is to receive 2½ per cent. for collecting; what will he receive for collecting the whole tax? *Ans.* $40'679.

The time, rate per cent., and interest being given, to find the principal.

¶ 166. 1. What sum of money, put at interest 16 months, will gain $10'50, at 6 per cent.?

SOLUTION.—$1 in 16 months, at 6 per cent., will gain $'08; and since $'08 is the interest of $1 at the given rate and time, $10'50 is the interest of as many dollars as the number of times $'08 is contained in (can be subtracted from) $10'50. $10'50 ÷ $'08 = $131'25, the principal required.

Hence, the

RULE.

Find the interest of **$1**, at the given rate and **time, by**

which divide the given interest; the quotient will be the principal required.

EXAMPLES FOR PRACTICE.

2. A man paid $4'52 interest, at the rate of 6 per cent., at the end of 1 year 4 months; what was the principal? *Ans.* $56'50.

3. A man received, for interest on a certain note, at the end of 1 year, $20; what was the principal, allowing the rate to have been 6 per cent.? *Ans.* $333'333⅓

4. A man leases a farm for $562, which sum is 10 per cent. of the value of the farm; how much is the farm worth?

The principal, interest, and time being given, to find the rate per cent.

¶ 167. 1. If I pay $3'78 interest, for the use of $36 for 1 year and 6 months, what is that per cent.?

Solution. — The interest of $36, at 1 per cent., for 1 year and 6 months, is $'54, and consequently $3'78 is as many per cent. as the times $'54 is contained in $3'78. $3'78 ÷ '54 = 7 per cent.

Hence, the

RULE.

Find the interest on the given sum, at 1 per cent., for the given time, by which divide the given interest; the quotient will be the rate at which interest is paid.

EXAMPLES FOR PRACTICE.

2. If I pay $2'34 for the use of $468, 1 month, what is the rate per cent.? *Ans.* 6 per cent.

3. At $46'80 for the use of $520, 2 years, what is that per cent.? *Ans.* 4½ per cent.

4. A stockholder, who owned 10 shares, of $100 each, of the Tonawanda Railroad company stock, received a dividend of $50 every 6 months, what per cent. was that on the money invested? *Ans.* 10 per cent.

5. A widow lady, whose expenses are $324 a year, has $5400 in money; at what rate per cent. must she loan it, that the interest may pay her expenses?

The principal, rate per cent., and interest being given, to find the time.

¶ 168. 1. The interest on a note of $36, at 7 per cent., was $3'78; what was the time?

Questions. — **¶ 166.** When the time, rate per cent., and interest are given, how may the principal be found? Explain the principles of the rule.

¶ 167. When the principal, interest, and time are given how do you find the rate per cent.? Explain the principles of the rule.

SOLUTION. — The interest of $36, 1 year, at 7 per cent., is $2'52. Since $2'52 will pay for the use of $36 1 year, $3'78 will pay for the use of it as many years as the times $2'52 is contained in $3'78. $3'78 ÷ 2'52 = 1'5 years, = 1 year 6 months, the time required.

Hence, the

RULE.

Find the interest for 1 year on the principal given, at the given rate, by which divide the given interest; the quotient will be the time required, in years and decimal parts of a year.

EXAMPLES FOR PRACTICE.

2. If $31'71 interest be paid on a note of $226'50, what was the time, the rate being 6 per cent.?

Ans. 2'33⅓ = 2 years 4 months.

3. On a note of $600, paid interest $20, at 8 per cent.; what was the time?

Ans. '416 + yr. = 5 mo., nearly. It would be exactly 5 but for the fraction lost.

4. The interest on a note of $217'25, at 4 per cent., was $28'242; what was the time? *Ans.* 3 years 3 months.

NOTE. — When the rate is 6 per cent., we may divide the interest by ½ the principal, and the *quotient*, removing the separatrix *two* places to the left, will be the answer required, in months and decimals of a month. ¶ 153.

The percentage of any number of dollars being given, to find the rate.

¶ 169. 1. A merchant purchases a piece of broadcloth for $60; what will be the per cent. of gain, if he sells it for $67'20?

```
$67'20
 60'00
 -----
60')7'20('12
    60
    --
    120
    120
    ---
      0
```

SOLUTION. — Subtracting the price which he gave from the price for which he sells the cloth, we have left $7'20, the gain on $60, of which we must take $\frac{1}{60}$ for the gain on $1. We get 12 cents as the gain on $1, or 100 cents. Hence the gain is 12 hundredths of the sum paid, or, *Ans.* 12 per cent.

Hence, the

RULE.

Divide the percentage of the number of dollars by the num-

Questions. — ¶ 168. When the principal, rate per cent., and interest are given, how do you find the time? Explain the principles of the rule.

ber of dollars on which it has accrued; the quotient, which is the percentage of $1, or 100 cents, will express the rate per cent.

EXAMPLES FOR PRACTICE.

2. A merchant purchases goods to the amount of $ 550; what per cent. profit must he make to gain $ 66? *Ans.* 12 per cent.

3. —— What per cent. profit must he make on the same purchase, to gain $ 38'50? —— to gain $ 24'75? —— to gain $ 2'75? *Ans.* to the last, '005, or ½ per cent.

4. Bought a hogshead of rum, containing 114 gallons, at 96 cents per gallon, and sold it again at $ 1'0032 per gallon; what was the whole gain, and what was the gain per cent.?

Ans. { $ 4'924, the whole gain.
4½, gain per cent. }

5. Bought 30 hogsheads of molasses, for $ 600; paid in duties $ 20'66; for freight, $ 40'78; for porterage, $ 6'05, and for insurance, $ 30'84; if I sell them at $ 26 per hogshead, how much shall I gain per cent.? *Ans.* 11'695 + per cent.

6. A spendthrift, who received an inheritance of $ 3000, spent $ 960 the first week in gambling; what per cent. of his money is gone? *Ans.* 32 per cent.

7. A farmer paid $ 2'50 for insuring buildings worth $ 1000; what was the rate per cent.? *Ans.* ¼ per cent.

8. A commission merchant receives $ 37'50 for selling goods to the amount of $ 1250; what was the rate per cent.? *Ans.* 3 per cent.

9. A broker receives $ 270 for selling $ 18000 worth of stocks; what is the per cent. for brokerage? *Ans.* 1½ per cent.

Bankruptcy.

¶ 170. An individual who fails in business sometimes makes an assignment of his property, which is divided among his creditors according to their respective debts.

In making calculations in bankruptcy, we find what can be paid on each dollar owed, and multiply this by the number of dollars which each man claims, to find his share.

EXAMPLES FOR PRACTICE.

1. An extensive banking house in New York fails for $ 800,000, the property of the house is found to be $ 300,000; what is paid on a dollar? *Ans.* $'37½, or 37½ per cent.

2. How much will a man receive on the above, whose dues are $ 16500? *Ans.* $ 6187'50.

Questions. - **¶ 169.** How is Ex. 1 explained? Rule.

3. A merchant fails in business, owing to A, $ 250 ; to B, $ 320 ; to C, $ 500, to D, $ 180 ; to E, $ 700 ; to F, $ 390 ; to G, $ 65‘50 ; to H, $ 1300 ; to I, $ 2200 ; to J, $ 850 ; his property is found to be $ 4653 ; how much does each receive?

General Average.

¶ 171. When a ship is in distress, the expense incurred, or the damage suffered by the ship, or any part of the cargo, is averaged upon the value of the ship ; upon the cargo, estimated at what the goods will bring at the destined port ; and upon the freight, deducting one third, on account of the seamen's wages.

To estimate general average, we find the proportion of the loss on each dollar, and then the loss on the number of dollars of each contributary interest.

The ship Silas Richards, in her voyage from New York to Charleston, became stranded on the coast of North Carolina, when it was found necessary to throw overboard 506 barrels of flour, belonging to Goodrich & Co., worth $ 6‘87 per barrel. The expense of getting the vessel off, was $ 197 ; of supplying new rigging, $ 240, of which one third is to be deducted, as the new is supposed to be better than the old. The ship is worth $ 10232 ; the freight is $ 4800, of which one third is to be deducted. Goodrich & Co. had on board 1000 barrels of flour ; M. H. Newman & Co., goods worth $ 4000 ; D. Appleton, goods worth $ 5236 ; Hyde & Duren, goods worth $ 9000 ; how much do Goodrich & Co. realize for all their flour ; what does each interest contribute towards the loss, and what is the rate per cent. on the contributary interests?

Goodrich & Co. realize	$6186‘669.
The ship's portion of the loss is	$1017‘735 +.
Portion of the freight,	$318‘291 +.
" M. H. Newman & Co.,	$397‘863 +.
" D. Appleton,	$520‘803 +.
" Hyde & Duren,	$895‘193 +.
" Goodrich & Co.,	$683‘331 +.
Rate per cent.,	$9\frac{6080}{6423}$.

Questions.—¶ **170.** What do you understand by bankruptcy? How are calculations in bankruptcy made?

¶ **171.** What do you understand by general average? What expenses and losses may it embrace? On what different interests are the expenses averaged?

Partnership.

¶ **172.** When two or more persons unite a part, or the whole, of their capital for the prosecution of business, they form a *Company* or *Firm*, and their business is called *Partnership business*. Each member of a firm is called a *partner*.

Capital or *Stock* is the money or other property employed in trade; *Joint Stock* is stock of a company or firm, *Dividend* is the gain, and *assessment* the loss, to be shared among the partners.

1. Two persons have a joint stock in trade; A put in \$250, and B \$350; they gain \$150; what is each man's share?

OPERATION.

6|00) \$1|50'00

\$'25

A's gain, '25 × 250 = \$62'50, } *Ans.*
B's gain, '25 × 350 = \$87'50, }

SOLUTION. — We divide the whole gain, \$150, by 600, which will give us the gain on \$1, = \$'25. Then 250 times \$'25 is A's gain, and 350 times \$'25 is B's gain. By the first operation, we get the rate per cent. of gain or loss, according to ¶ 169. The second may be performed by the ordinary rule for percentage.

Or the operation may be performed as follows:

A's gain will be $\frac{250}{600} = \frac{5}{12}$ of \$150 = \$62'50.
B's gain will be $\frac{350}{600} = \frac{7}{12}$ of \$150 = \$87'50. Hence,

RULE.

Take such a part of the whole gain or loss as each man's stock is part of the whole stock; the part thus taken will be his share of the gain or loss.

NOTE. — This rule may be applied to the operations in several following paragraphs.

EXAMPLES FOR PRACTICE.

2. A, B, and C trade in company; A's capital was \$175, B's \$200, and C's \$500; by misfortune they lose 250; what loss must each sustain?

Ans. { \$ 50, A's loss. / \$ 57'142$\frac{6}{7}$, B's loss. / \$142'857$\frac{1}{7}$, C's loss. }

3. Divide \$600 among 3 persons, so that their shares may be to each other as 1, 2, 3, respectively. *Ans.* \$100, \$200, and \$300.

Questions. — ¶ **172.** What is a company or firm? — partnership business? — a partner? — capital or stock? — joint stock? — dividend? Give the first solution of Ex. 1; — the second. Rule.

4. Two merchants A and B, loaded a ship with 500. hhds. of rum; A loaded 350 hhds., and B the rest; in a storm, the seamen were obliged to throw overboard 100 hhds.; how much must each sustain of the loss? *Ans.* A 70, and B 30 hhds.

5. A and B companied; A put in $ 45, and took out $\frac{3}{5}$ of the gain; how much did B put in? *Ans.* $ 30

NOTE. — They took out in the same ratio that they put in; if 3 fifths of the stock is $ 45, how much is 2 fifths of it?

6. A and B companied, and traded with a joint capital of $ 400; A received, for his share of the gain, $\frac{1}{2}$ as much as B; what was the stock of each? *Ans.* { $ 133'333$\frac{1}{3}$, A's stock; $ 266'666$\frac{2}{3}$, B's stock.

7. A and B ventured equal stocks in trade, and cleared $ 164; by agreement, A, because he managed the concerns, was to have $ 5 of the profits, as often as B had $ 2; what was each one's gain? and how much did A receive for his trouble?

Ans. A's gain was $117'142$\frac{6}{7}$, and B's $46'857$\frac{1}{7}$, and A received $70'285$\frac{5}{7}$ for his trouble.

8. A cotton factory, valued at $ 12000, is divided into 100 shares; if the profits amount to 15 per cent. yearly, what will be the profit accruing to 1 share? —— to 2 shares? —— to 5 shares? —— to 25 shares? *Ans.* to the last, $ 450.

9. In the above-mentioned factory, repairs are to be made which will cost $ 340; what will be the tax on each share, necessary to raise the sum? —— on 2 shares? —— on 3 shares? —— on 9 shares? *Ans.* to the last, $ 34.

10. Two men paid 10 dollars for the use of a pasture 1 month; A kept in 24 cows, and B 16 cows; how much should each pay?

PARTNERSHIP ON TIME.

¶ 173. 1. Two men hired a pasture for $10; A put in 8 cows 3 months, and B put in 4 cows 4 months; how much should each pay?

SOLUTION. — The pasturage of 8 cows for 3 months is the same as of 24 cows for 1 month, and the pasturage of 4 cows for 4 months is the same as of 16 cows for 1 month. The shares of A and B. therefore, are 24 to 16, as in the former question. Hence,

When *time* is regarded in partnership, *multiply each one's stock by the time he continues it in trade, and use the product for his share.*

Ans. A 6 dollars, and B 4 dollars.

2. A and B enter into partnership; A puts in $ 100 6 months, and then puts in $ 50 more; B puts in $ 200 4 months, and then takes out $ 80; at the close of the year they find that they have gained $ 95; what is the profit of each? *Ans.* { $ 43'711, A's share; $ 51'288, B's share.

Questions. — ¶ 173. What are we to understand by partnership on time? How do we proceed?

3. A, with a capital of $ 500, began trade Jan. 1, 1846, and, meeting with success, took in B as a partner, with a capital of $ 600, on the first of March following; four months after, they admit C as a partner, who brought $ 800 stock; at the close of the year they find the gain to be $ 700; how must it be divided among the partners?

Ans. A's share, $ 250; B's, $ 250; C's, $ 200.

NOTE. — Partnership is sometimes called Fellowship, or Single Fellowship; and partnership on time is called Compound Fellowship.

Banking.

¶ 174. A bank is an incorporated institution which traffics in money. Bank notes, or bank bills are promissory notes, payable to bearer.

Banks loan their money on notes, the interest always being paid in advance.

For example, B holds A's note for $100, payable in 90 days, without interest. But B is in immediate want of money. He carries A's note to a bank, and if their credit be undoubted, the bank will receive A's note and pay the face of it, *minus* the interest on it, for 3 days more than the given time, (90 + 3 = 93 days.) The note is then said to be discounted at the bank. These 3 days are called *days of grace.*

But the bank will require B to write his name on the back of the note. This is called *endorsing the note.* It subjects B to pay the note when the 90 days and the 3 days of grace shall expire, provided A, who gave the note, should fail so to do.

The money received from the bank for the note, is called *the avails of the note.* The note is said to be *mature* when the time that it should be paid shall arrive.

Again: B as principal, with C and D as sureties, may give their note jointly and severally to the bank, for the sum wanted, payable at a specified time, without interest. Then if B fails to pay the note, his sureties, C and D, either or both, will be holden to pay it.

$100'00 *Principal.*

$1'00 *Int.* 60 *days.*
'50 " 30 *days.*
'05 " 3 *days of grace.*

$1'55 *Discount for* 90 *days and grace.*

NOTE. — Bank interest, when the rate is 6 per cent., may be cast by inspection, as follows: Let the sum on which it is to be cast, be $100. The principal itself, removing the point two places to the left, is made to express the interest for 60 days, ($1'00) half of which ($'50) is the interest for 30 days, and $\frac{1}{10}$ of the interest for 30 days, $'50 ÷ 10 = $'05, is the interest for 3 days.

The sum of these results is the bank interest on $100, at 6 per cent., for 90 days, which sum $1'55, deducted from the face of the note, makes its avails to B, $98'45.

If the discount be other than 6 per cent., take such fractional part of the discount at 6 per cent. as the *required* rate is less or more than 6 per cent., which added to or subtracted from the discount at 6 per cent., as the case may require, will give the discount sought.

EXAMPLES FOR PRACTICE.

1. What will be received on a bank note of $500, due in 90 days, at 7 per cent.? *Ans.* $490'95$\frac{5}{8}$.

2. What is the discount of a bank note for $300, due in 90 days, at 5 per cent.? *Ans.* $3'875.

3. What is the discount of a note for $600, due in 90 days, at 8 per cent.? *Ans.* $12'40.

4. What is received on a note for $740, due in 90 days, at 6 per cent.? *Ans.* $728'53.

5. A man gets a note of $1000 discounted for 90 days, at 6 per cent. per annum, and lends the money immediately, till the time when he is obliged to pay: what does he lose? *Ans.* $'24 +.

Taxes.

¶ 175. A tax is a sum imposed on an individual for a public purpose.

A Poll tax is a specific sum assessed on male citizens above 21 years of age; each person so assessed is called a poll.

Taxes are usually assessed either on the person or property of the citizens, and sometimes on both.

Property is of two kinds, personal property and real estate.

Personal is *movable* property, such as money, notes, cattle, furniture, &c.

Real estate is *immovable* property, such as lands, houses, stores, &c.

An *Inventory* is a list of articles.

In assessing taxes, it is necessary to have an inventory of all the taxable property, both personal and real, of those on whom the tax is to be levied, and also, (if a poll tax is to be raised,) of the whole number of polls; and as the polls are

---

Questions.—¶ 174. What is a bank? When is bank interest paid? Illustrate by the example. What is meant by days of grace? How long is the interest cast on a note due in 90 days? What security does the bank require? How is bank interest found by inspection, when the rate is 6 per cent.? How, when it is any other rate?

rated as a certain sum each, we must first deduct from the whole tax the amount of the poll tax, and the *remainder* is to be assessed on the property.

The tax on $1 is found by dividing the amount to be assessed on the property by the *value* of the property taxed.

The tax on any *amount* of property is found by multiplying the *value* of the property by the tax on $1.

1. A tax of $917 is to be assessed on a town in which are 320 polls, assessed at 40 cents each; the inventoried value of the personal and real property of the town is $52600; what is the amount of the poll taxes? How much remains to be assessed on the property? What is the tax on $1?

SOLUTION. $‘40 × 320 = $128, amount of poll tax; then $917 — $128 = $789, amount to be assessed on property, and $789 ÷ $52600 = $‘015, the tax on $1.

NOTE. — In making out a tax list, form a table containing the taxes on 1, 2, 3, &c., to 10 dollars: then on 20, 30, &c., to 100 dollars; and then on 100, 200, &c., to 1000 dollars. Then, knowing the inventory of any individual, it is easy to find the tax upon his property.

Let us apply this method in assessing a tax on a town.

2. A certain town, valued at $64530, raises a tax of $2259‘90; there are 540 polls, which are taxed $‘60 each; what is the tax on a dollar, and what will be A's tax, whose *real estate* is valued at $1340, his personal property at $874, and who pays for 2 polls?

SOLUTION. 540 × $‘60 = $324, amount of the poll taxes, and $2259‘90 — $324 = 1935‘90, to be assessed on property, and $1935‘90 ÷ $64530 = $‘03, tax on $1.

TABLE.

dolls.	dolls.	dolls.	dolls.	dolls.	dolls.
Tax on 1 is	‘03	Tax on 10 is	‘30	Tax on 100 is	3‘
“ 2 “	‘06	“ 20 “	‘60	“ 200 “	6‘
“ 3 “	‘09	“ 30 “	‘90	“ 300 “	9‘
“ 4 “	‘12	“ 40 “	1‘20	“ 400 “	12‘
“ 5 “	‘15	“ 50 “	1‘50	“ 500 “	15‘
“ 6 “	‘18	“ 60 “	1‘80	“ 600 “	18‘
“ 7 “	‘21	“ 70 “	2‘10	“ 700 “	21‘
“ 8 “	‘24	“ 80 “	2‘40	“ 800 “	24‘
“ 9 “	‘27	“ 90 “	2‘70	“ 900 “	27‘
				“ 1000 “	30‘

Questions. — ¶ **175.** What is a tax? — a poll tax? How are taxes usually assessed? Property is of what kinds? What is personal property? — real estate? What is an inventory? In assessing taxes, what is to be done? When a poll tax is to be raised, what must first be done? How do you find the amount to be assessed on the property? How do you find the tax on $1? How on any amount of property? What course is usually pursued by assessors in making out a tax list? Explain the manner in which any individual's tax is made out from the assessor's tax table. How may a tax list be proved?

Now to find A's tax, his real estate being $134C we find, by the table, that

The tax on	$1000	is	$30'
The tax on	300	"	9'
The tax on	40	"	1'20
Tax on his real estate,			$40'20
In like manner, we find he tax on his personal prcperty to be			26'22
2 polls at '60 each, are			1'20
		Amount	$67'62

3. What will be the amount of B's tax, of the same town, whose inventory is 874 dollars *real*, and 210 dollars *personal* property, and who pays for 3 polls? *Ans.* $34'32.

4. —— of C's paying for 2 polls, whose property is valued at $3482? —— of D's, paying for 1 poll, whose property is valued at $4675? *Ans.* to the last, $140'85.

Proof. — After a tax list is made out, add together the taxes of all the individuals assessed; if the amount is equal to the whole tax assessed the work is right.

Duties.

¶ 176. Duties or customs are taxes on imported goods.

A custom-house is a house or an office where customs are paid.

Government has established a custom-house at every port in the United States into which foreign goods are imported.

Besides duties on merchandise, every vessel employed in commerce is required to pay a certain sum for entering the ports. This sum is governed by the size or tonnage of the vessel.

The income to the government, from duties and tonnage, is called Revenue.

All duties are imposed and regulated by the general government, and must be the same in all parts of the Union

Note. — A table of the duties imposed by government is called a Tariff.

The law requires that the cargoes of all vessels engaged in

Questions. — **¶ 176.** What are duties? — custom-houses, where, and by whom established? What tax is named besides duties, and how proportioned? What is revenue? How are duties imposed? What is a tariff? How is the value of the goods in a vessel ascertained? How many kinds of duties? What are specific duties? — ad valorem duties? Define ad valorem.

foreign commerce, shall be weighed, measured, or gauged, by the custom-house officers, for the purpose of ascertaining the amount or value of the goods on which duties are to be paid.

Duties on imported goods are of two kinds, Specific and Ad Valorem.

A Specific duty is a certain sum per ton, hundred weight, pound, hogshead, gallon, square yard, foot, &c., without regard to its value.

Ad Valorem signifies upon the value.

An Ad Valorem duty is a certain per cent. on the sum paid for the goods in the country from which they are brought.

Specific Duties.

¶ 177. In the custom-house weight and gauge of goods, certain deductions are made for the box, bag, cask, &c., containing the goods, and also for leakage, breakage, &c. These deductions must be made before the specific duties are imposed.

Gross weight is the weight of the goods together with the box, bale, bag, cask, &c., which contains them.

Draft is an allowance made for waste, which is to be deducted from the gross weight, and is as follows:

On	112 lbs.			1 lb.
Above	112 lbs.,	and not exceeding	224 lbs.,	2 lbs.
"	224 lbs.,	" "	336 lbs.,	3 lbs.
"	336 lbs.,	" "	1120 lbs.,	4 lbs.
"	1120 lbs.,	" "	2016 lbs.,	7 lbs.
"	2016 lbs.			9 lbs.

Tare is an allowance for the weight of the box, bag, cask, &c. It is to be deducted from the remainder of any weight or measure, after the draft or tret has been allowed.

Leakage is an allowance of 2 per cent. on all liquors in casks, paying duty by the gallon.

Breakage is an allowance of 10 per cent. on ale, beer, and porter in bottles, and of 5 per cent. on all other liquors in bottles; or the importer may have the bottles counted, and pay duties on the number remaining unbroken.

Questions. — ¶ 177. What deductions are made, if goods pay specific duties? What is gross weight? — draft, or tret? How much is it, and when deducted? What is tare, and when deducted? What is leakage? — breakage, and what privilege has the importer? — net weight? Rule for calculating specific duties.

Net weight is the weight of the goods after deducting the weight of the box, bale, &c., and making all other allowances.

1. What is the specific duty on 5 hogsheads of molasses, each containing 120 gallons, at $12\frac{1}{2}$ cents per gallon, the customary allowance being made for leakage?

SOLUTION. — Since there are 120 gallons in 1 hogshead, in 5 hogsheads there are 5 times 120 gallons = 600 gallons. 2 per cent. of 600 gallons is 600 × ‘02 = 12 gallons, and 600 gallons — 12 gallons = 588 gallons. Since the duty on 1 gallon is $12\frac{1}{2}$ cents, the duty on 588 galons is 588 times $‘125 = $73‘50. Hence,

To find the specific duty on any given quantity of goods,

RULE.

I. Deduct from the given quantity of goods the legal allowance for draft, tare, leakage or breakage.

II. Multiply the remainder by the duty on a unit of the weight or measure of the goods, and the product will be the duty required.

EXAMPLES FOR PRACTICE.

2. What is the specific duty on 75 barrels of figs, each weighing 83 pounds gross, tare in the whole 597 pounds, at 4 cents per pound? *Ans.* $225‘12.

3. What is the duty on 420 dozen bottles of porter, at $5\frac{1}{2}$ cents per bottle, the customary allowance being made for breakage? *Ans.* $249‘48.

4. What is the duty on 8 hogsheads of sugar, each weighing 10 cwt. 2 qrs. gross, tare 14 lbs. per cwt., at $2\frac{1}{4}$ cents per pound? *Ans.* $185‘22.

5. What is the duty on 4 barrels of Spanish tobacco, the first weighing 171 pounds gross, the second 125 pounds gross, the third 109 pounds gross, and the fourth 99 pounds gross, at $6\frac{1}{4}$ cents per pound, the customary allowance being made for draft, and 16 pounds per barrel for tare? *Ans.* $27·25.

AD VALOREM DUTIES.

¶ 178. Since ad valorem duties are estimated upon the actual cost of the goods, it is plain that they are found by simply multiplying the cost of the goods by the given per cent.

Questions. — ¶ 178. Upon what are ad valorem duties estimated? How are they found? In estimating ad valorem duties, are any allowances ever made?

Note. — In estimating ad valorem duties, no deductions of any kind are to be made.

1. What is the ad valorem duty, at 25 per cent, on 32 yds of English broadcloth, which cost $4'75 per yard?

Solution. — 32 yds. at $4'75, cost 4'75 × 32 = $152, and 25 per cent of 152, is $38, *Ans.*

EXAMPLES FOR PRACTICE.

2. What is the ad valorem duty, at 18 per cent., on 40 bags of Java coffee, each weighing 115 pounds, and which cost 11¼ cents per pound? *Ans.* $93'15.
3. What is the ad valorem duty, at 33⅓ per cent., on 1 gross of Sheffield cutlery, which cost $256'80? *Ans.* $85'60.
4. What is the duty, at 20 per cent., on a piece of Turkey carpeting, containing 140 yards, and which cost $1'92 per yard? What is the duty on 1 yard? For how much must it be sold per yard, to gain 25 per cent. on the cost and duty? *Ans.* to the last, $2'88.
5. What is the duty, at 35 per cent., on a case of Italian silks which cost $4821'50? *Ans.* $1687'52½.
6. What is the duty, at 15 per cent., on 3 dozen gold watches which cost $68 each? *Ans.* $367'20.
7. What is the duty, at 22 per cent., on 75 chests of tea, the net weight of each chest being 92 pounds, and the tea costing 41 cents per pound? *Ans.* $622'38

¶ 179. Review of Percentage.

Questions. — What is meant by percentage? — rate per cent.? — general rule? What are calculated by percentage? What is insurance? — mutual insurance? What is meant by stocks? — brokerage? — profit and loss? What is interest, and how calculated? How is 6 per cent. interest calculated by inspection? What is meant by partial payments? What difference between the U. S. and Conn. rules? To what case does the Vt. rule for partial payments apply? What is done when notes with partial payments are paid within a year? How does compound differ from annual interest? Like what case in interest are discount and commission? What do you say of bankruptcy? — of general average? — partnership? How does the calculation of bank interest differ from that of other interest? How are taxes assessed? How do duties differ from other taxes? What two kinds of duties, and how is each computed?

EXERCISES.

1. What is the interest of $273'51, for year and 10 days, at 7 per cent.? *Ans.* $19'677 +
2. What is the interest of $486, for 1 year 3 months 19 days, at 8 per cent.? *Ans.* $50'652.

3. What is the interest of $1600, for 1 year and 3 months, at 6 per cent.? *Ans.* $120.

4. What is the interest of $5‘811, for 1 year 11 months, at 6 per cent.? *Ans.* $‘668.

5. What is the interest of $2‘29, for 1 month 19 days, at 3 per cent.? *Ans.* ‘009.

6. What is the interest of $18, for 2 years 14 days, at 7 per cent.? *Ans.* $2‘569.

7. What is the interest of $17‘68, for 11 months 28 days, at 6 per cent.? *Ans.* $1‘054.

8. What is the interest of $200, for 1 day, at 6 per cent.? —— 2 days? —— 3 days? —— 4 days? —— 5 days?
Ans. for 5 days, $0‘166.

9. What is the interest of half a mill, for 567 years, at 6 per cent.? *Ans.* $0‘017.

10. What is the interest of $81, for 2 years 14 days, at ½ per cent.? —— ¾ per cent.? —— ⅝ per cent.? —— 2 per cent.? —— 3 per cent? —— 4½ per cent.? —— 5 per cent.? —— 6 per cent.? —— 7 per cent.? —— 7½ per cent.? —— 8 per cent.? —— 9 per cent.? —— 10 per cent.? —— 12 per cent.? —— 12½ per cent.?
Ans. to the last, $20‘643.

11. What is the interest of 9 cents, for 45 years 7 months 11 days, at 6 per cent.? *Ans.* $0‘246

12. A's note of $175 was given Dec. 6, 1838, on which was endorsed one year's interest; what was there due Jan. 1, 1843, interest at 7 per cent.?

13. B's note of $56‘75 was given June 6, 1841, on interest at 6 per cent., after 90 days; what was there due Feb. 9th, 1842?
Ans. $58‘197.

14. C's note of $365‘37 was given Dec. 3, 1837; June 7, 1840, he paid $97‘16; what was there due Sept. 11, 1840, interest at 5 per cent.? *Ans.* $318‘184.

15. D's note of $203‘17 was given Oct. 5, 1838, on interest at 6 per cent. after 3 months; Jan. 5, 1839, he paid $50; what was there due May 2, 1841? *Ans.* $174‘537.

16. E's note of $870‘05 was given Nov. 17, 1840, on interest at 6 per cent. after 90 days; Feb. 11, 1845, he paid $186‘06; what was there due Dec. 23, 1847? *Ans.* $1041‘58.

17. Supposing a note of $317‘92, dated July 5, 1837, on which were endorsed the following payments, viz., Sept. 13, 1839, $208‘04; March 10, 1840, $76; what was there due Jan. 1, 1841, interest at 7 per cent.? *Ans.* $93‘032.

18. What will be the annual insurance, at ⅝ per cent. on a house valued at $1600? *Ans.* $10.

19. What will be the insurance of a ship and cargo, valued at $5643, at 1½ per cent.? —— at ⅘ per cent.? —— at $\frac{7}{16}$ per cent? —— at $\frac{11}{12}$ per cent.? —— at ¾ per cent.?
Ans. at ¾ per cent. $42‘322.

20. A man having compromised with his creditors at 62½ cents on a dollar, what must he pay on a debt of $137'46? *Ans.* $85'912.

21. What is the value of $800 stock in the Utica and Schenectady railroad, at 112½ per cent.? *Ans.* $900.

22. What is the value of $560'75 of stock, at 93 per cent.? *Ans.* $521'497.

23. What principal, at 7 per cent., will, in 9 months 18 days, amount to $422'40? *Ans.* $400.

24. What is the present worth of $426, payable in 4 years and 12 days, discounting at the rate of 5 per cent.?

In large sums, to bring out the cents correctly, it will sometimes be necessary to extend the decimal in the divisor to five places. *Ans.* $354'507+.

25. A merchant purchased goods for $250 ready money, and sold them again for $300, payable in 9 months; what did he gain, discounting at 6 per cent.? *Ans.* $37'081.

26. Sold goods for $3120, to be paid, one half in 3 months, and the other half in 6 months; what must be discounted for present payment, at 6 per cent.? *Ans.* $68'491+.

27. The interest on a certain note, for 1 year 9 months, at 6 per cent., was $49'875; what was the principal? *Ans.* $475.

28. What principal, at 5 per cent., in 16 months 24 days, will gain $35? *Ans.* $500.

29. If I pay $15'50 interest for the use of $500, 9 months and 9 days, what is the rate per cent.?

30. If I buy candles at $'167 per lb., and sell them at 20 cents, what shall I gain in laying out $100? *Ans.* $19'76.

31. Bought 37 gallons of brandy, at $1'10 per gallon, and sold it for $40; what was gained or lost per cent.?

32. Bought cloth at $4'48 per yard; how must I sell it to gain 12½ per cent.? *Ans.* $5'04.

33. Bought 50 gallons of brandy, at 92 cents per gallon, but by accident 10 gallons leaked out; for what must I sell the remainder per gallon, to gain upon the whole cost at the rate of 10 per cent.? *Ans.* $1'265 per gallon.

34. A merchant bought 10 tons of iron for $950; the freight and duties were $145, and his own charges $25; how must he sell it per lb. to gain 20 per cent.? *Ans.* 6 cents per lb.

35. A note is given for $2000, at 6 per cent. annual interest, payable in 6 years; the date of the note is Dec. 1, 1841; there are endorsements upon it as follows: June 1, 1842, $163; Feb. 1, 1843, $12; Jan. 1, 1844, $300; April 1, 1845, $20; June 1, 1845, $20; Aug. 1, 1845, $400; Jan. 1, 1846, $100; Aug. 1, 1847, $150; Oct. 1, 1847, $75. What remained due, Dec. 1, 1847, calculated by the U. S. Court rule, by the Connecticut rule, and by the Vermont rule, and how do the results compare?

Ans U. S., $1368'81; Ct., $1372'56; Vt., 1274'78.

EQUATION OF PAYMENTS.

¶ 180. 1. A country merchant owes in Boston, $200 due in 2 months, and $200 due in 6 months, each without interest; at what time could he pay both debts, that neither party may lose?

SOLUTION. — He may keep the first as long after it is due as he pays the last before it is due. *Ans.* 4 months.

The method of finding the time when several debts, due at different times without interest, should be paid at once, is called *Equation of Payments.*

The time of payment thus found, is called the *mean time.*

2. A man owes $106, due in one year, and $106, due in 3 years, without interest; in what time shall he pay both at once?

SOLUTION. — At the end of 2 years. But this, which is the common method, is a gain, in this example, of $'36 to the debtor. He keeps the first debt a year after it is due, and thus gains (at 6 per cent.) the interest on $106 for a year, or $6'36. He pays the whole of the second debt a year before due, when he should pay only such a sum as would amount to $106 in a year, or $100. Thus he loses $6 on the second debt, while he has gained $6'36 on the first. The error, which results from considering the interest and discount on the same sum for the same time equal, is not usually regarded in business.

¶ 181. *To find a rule for the equation of payments.*

1. Borrowed of a friend $6'00, for 4 months; afterwards I lent him $1, to keep long enough to balance the use of the money borrowed; how long must this be?

SOLUTION. — He should keep $1 six times as long, or, *Ans.* 24 months.

2. In how long time will $8 be worth as much as $40 for 1 month?

SOLUTION. — Every $8 in the $40 will be worth as much in 1 month, as the first sum, $8, is worth in that time. Then as many times as $8 are contained in $40, so many months the $8 will require to be worth as much as the $40 for 1 month. *Ans.* 5 months.

3. I have 3 notes against a man: 1 of $12, due in 3

Questions. — ¶ 180. On what principle is the time of paying at once the two debts, Ex. 1, determined? What is understood by the mean time What is equation of payments? What error appears, Ex. 2, and why?

months; 1 of $9, due in 5 months; 1 of $6, due in 10 months, all without interest; when should he pay the whole at once?

SOLUTION. — $12 for 3 months, is the same as $36 for 1 month.
$9 " 5 " " $45 "
$6 " 10 " " $60 "

He has $27 long enough to balance $141 for 1 month. Every $27 in $141 will be worth as much in 1 month as the first $27. Then as many times as $27 is contained in $141, so many months he can keep the $27.
141 ÷ 27 = 5 mo. 6 + days, *Ans.*

Hence, *To find the mean time of several payments,*

RULE.

Multiply each sum by its time of payment, and divide the sum of the products by the sum of the payments.

EXAMPLES FOR PRACTICE.

4. A western merchant owes in New York city $200, due in 5 months; $325'50, due in 3 months, and $413'37, due in 2 months; but he finds that it will be more convenient to make payment at one time; in what time will this be?
Ans. 2'984 months = 2 months 29 + days.

5. I owe several debts, due in different times, without interest, namely, $309'50, in 8 months; $161, in 5 months and 18 days, and $63'25, in 10 months and 11 days; what shall I pay now to cancel the whole, the rate being 6 per cent.?

NOTE 1. First find the mean time, then the present worth, ¶ 164. The fraction of a day is never regarded in business operations.

Ans. $514'375 +.

6. A merchant has owing him $300, to be paid as follows: $50 in 2 months, $100 in 5 months, and the rest in 8 months: and it is agreed to make one payment of the whole; in what time ought that payment to be? *Ans.* 6 months.

7. A owes B $136, to be paid in 10 months; $96, to be paid in 7 months; and $260, to be paid in 4 months; what is the equated time for the payment of the whole? *Ans.* 6 months 7 days +.

8. A owes B $600, of which $200 is to be paid at the present time, 200 in 4 months, and 200 in 8 months; what is the equated time for the payment of the whole? *Ans.* 4 months.

9. A owes B $300, to be paid as follows: ⅓ in 3 months, ¼ in 4 months, and the rest in 6 months; what is the equated time?
Ans. 4½ months.

NOTE 2. Sometimes retailers sell on 6 months' credit, without interest. But as it would be difficult to settle each item of a long account just 6 months from the time of purchase, all the charges for a year are settled at its close. Presuming that the purchases each month are uniform, this gives 6 months as

---

Questions. — ¶ **181.** Give the solution of Ex. 2. Give the solution of the 3d example. Rule. Give the substance of the notes.

the mean time of a settlement. Should a settlement be made at the end of 8 months, the mean time would be 4 months; at the end of 6 months, the mean time would be 3 months.

RATIO.

¶ 182. How many times is 4 contained in 8?

Ans. 2 times.

To find how many times one number is contained in another, is to find the ratio between the numbers, which we do by dividing one of the numbers by the other. But without performing the division, we may express it,

First, by the sign of division: thus, $8 \div 4$.

Second, by a line without dots, writing the dividend in place of the upper, and the divisor in place of the lower dot: thus, $\frac{8}{4}$.

Third, by dots without a line: thus, 8 : 4.

The last is the usual method of expressing ratio, when the quotient in division receives this name. Hence,

Ratio is the quotient expressing how many times one number is contained in another, or how many times one quantity is contained in another of the same kind or denominations.

NOTE. — Ratio can only exist between quantities *of the same kind*, since the dividend and divisor must be of the same kind. It would be absurd to inquire how many times 3 bushels of rye are contained in 12 lbs. of butter. (See ¶ 31.) But a ratio can exist between numbers of different denominations when they can be reduced to the same denomination; thus, we can determine how many times 8 quarts are contained in 6 gallons, when we reduce the quarts to gallons, or the gallons to quarts.

A ratio requires two numbers, each of which is called a term of the ratio, and together they are called a couplet. The first term, which is the dividend, is called the antecedent; the second, or divisor, is called the consequent.

Hence it follows that multiplying the antecedent or dividing the consequent multiplies the ratio, (¶ 56,) dividing the antecedent or multiplying the consequent, divides the ratio, (¶ 57,) and multiplying or dividing both antecedent and consequent by the same number does not alter the ratio, (¶ 58.)

Inverted and Direct Ratios.

¶ 183. In the ratio 8 : 4, the first term is divided by

Questions. — ¶ 182. What is it to find the ratio of numbers, and how done? Describe the three ways of expressing the division. What is said of the last way? Define ratio. How does the note limit ratio? Why? Illustrate. How many numbers are required, and how many different names do they receive? Apply ¶ 56; — ¶ 57; — ¶ 58.

the second, and the ratio is said to be direct. But sometimes the second is divided by the first, and the ratio is then said to be inverse, since it is equivalent to inverting the terms, and writing the expression 4 : 8. The latter is also called a reciprocal ratio, since $\frac{1}{2}$, the quotient of 4 : 8, is the reciprocal of 2, the quotient of 8 : 4, (¶ 55.) Hence,

Direct ratio is the quotient of the antecedent divided by the consequent; and

Inverse or reciprocal ratio is the quotient of the consequent divided by the antecedent.

Compound Ratio.

¶ 184. We have seen, ¶ 79, that a compound fraction consists of several simple fractions, to be multiplied together. Thus, the numerators of the compound fraction $\frac{8}{4}$ of $\frac{15}{5}$, are to be multiplied together for a new numerator, and the denominators for a new denominator. But since the terms of a ratio may be written fractionally, we may call the expression a compound ratio.

Hence, *A compound ratio consists of several simple ratios to be multiplied together*, which is done by multiplying together the antecedents, and also the consequents.

PROPORTION.

¶ 185. 1. If 12 yards of cloth cost $18, what will 4 yards cost?

Solution. — As 4 yards are $\frac{1}{3}$ of 12 yards, they will cost $\frac{1}{3}$ of $18 or $6. The 12 yards contain 4 yards as many times as $18 contain $6; that is, $12 \div 4 = 18 \div 6$; or fractionally, $\frac{12}{4} = \frac{18}{6}$.

We have here two ratios, which are equal. But as the sign of division is written to express ratio without a line, 4 dots may be written to express the equality of the ratios. The four dots are used instead of the lines usually employed. The expression then becomes

$$12 : 4 :: 18 : 6, \text{ equivalent to } \frac{12}{4} = \frac{18}{6}.$$

Such an expression is called a proportion, and is read, 12 divided by 4 equals 18 divided by 6; or more commonly, 12 is to 4 as 18 is to 6. Hence,

---

Questions. — ¶ 183. How does inverse differ from direct ratio? Define each. What other name has inverse ratio? Why?

¶ 184. Whence arises compound ratio? Define it. Give an example Write it in the common form. Reduce it to a simple ratio.

Proportion is the combination of two equal ratios. The first and last terms are called the extremes, the second and third, the means. The two antecedents are called corresponding terms, as are also the two consequents, since these terms have always a certain reference to each other. The third term, $18, of this proportion, expresses the value of 12 yards, the first term; and the fourth term, $6, expresses the value of 4 yards, the second term.

NOTE.—A proportion requires four terms, two antecedents and two consequents. Three numbers may form a proportion if one is used in both ratios; thus, with the numbers 12, 6, and 3, we have the proportion, 12 : 6 :: 6 : 3.

Rule of Three.

¶ 186. When, as in the last example, the first three terms are given to find a fourth, we may find it by taking such a part of the third term as the second is of the first; or by a method called the *Rule of Three*, on the following principles. Take the proportion,

12 : 4 :: 18 : 6, fractionally expressed, $\frac{12}{4} = \frac{18}{6}$.

As the fractions are equal, if we reduce them to the common denominator, 24, by the rule, ¶ 70, the numerators will be equal, and the fractions will become, $\frac{72}{24} = \frac{72}{24}$.

The first numerator, 72, it may be seen, is the product of the extremes, and the second numerator, 72, is the product of the means, of the proportion. Hence,

The product of the extremes of a proportion is equal to the product of the means.

Take the first three terms to find the fourth.

```
yds. yds.  doll.  doll.
 12 : 4 :: 18 : ....
             4
            ---
       12 ) 72
            ---
             6 dollars, Ans.
```

SOLUTION.—We multiply together 4 and 18, the means, which gives the product of the extremes, of which one extreme, 12, is given, and we divide the product by the given factor to get the other extreme, or fourth term.

Questions.—¶ 185. How is the price of the 4 yards, Ex. 1, found? What two ratios are formed? By what sign is their equality expressed, and instead of what is it used? Give the two ways of reading a proportion. Define proportion. What are its terms? What are extremes? What are means? What are corresponding terms, and why? How many terms, and how many numbers, are required in proportion? Illustrate

NOTE.—This operation is strictly analytic, for had the price of 1 yard been 18 dollars, we should have multiplied it by 4 to get the price of 4 yards. But as it is 12 yards, which cost 18 dollars, the product of 18 by 4 is 12 times as large, as it should be, and must be divided by 12.

¶ 187. *To write down the three given numbers.*

2. At $90 for 15 barrels of flour, how many barrels can be bought for $30?

```
doll. doll.  bar.  bar.
90    30 :: 15 : ....
             30
      9|0)45|0
         5 bar., Ans.
```

SOLUTION.—We have the terms of one ratio, $90 and $30, being of the same kind. For the same reason, 15 barrels and the required number of barrels will form another ratio. We place the antecedent of the second ratio, 15 barrels, for the third term, and its correspondent, $90 for the first, and $30 for the second, that its correspondent may be the fourth term. We then multiply and divide as above, and get the *Ans.*, 5 barrels.

¶ 188. *To invert both ratios.*

In the proportion, 90 : 30 :: 15 : 5, the quotient of 90 ÷ 30 is 3, and that of 15 ÷ 5 is 3. Inverting the terms of each couplet, we still have the proportion, 30 : 90 :: 5 : 15, for each ratio is now $\frac{1}{3}$, the reciprocal of the former ratio, (¶ 183,) and consequently the two ratios are equal. This inversion often virtually occurs in operations; thus,

3. At $30 for 5 barrels of flour, how many barrels can be bought for $90?

```
doll. doll.  bar. bar.
90 : 30 :: .... 5
  By inversion.
30 : 90 :: 5 :
           5
3|0)45|0
   15 bar., Ans.
```

SOLUTION.—Writing the first ratio, 90 : 30, as above, 5 barrels must be the fourth term, as it corresponds to $30, the second term. But it is convenient always to regard the fourth as the unknown term, and it will become so by inverting both ratios, when the operation is the same as before.

Ans. 15 barrels.

Questions.—¶ 186. What is the object of the *rule of three?* Compare the two methods of expressing a proportion, and show what results from reducing the fractions to a common denominator. How is the first numerator obtained? — the second? What important conclusion is derived? How applied to finding the fourth term? How explained analytically?

¶ 187. What must be made the third term? Why? What must be made the first, and why? What the second? Why?

¶ 188. Why can both ratios be inverted? What is the object in so doing?

¶ 189. *To invert one ratio.*

4. If 3 men will build a wall in 10 days, in how long time will 6 men build it?

men. men. days. days.
3 : 6 :: 10 : ....
Inverting the first ratio, we have,
6 : 3 :: 10
3
6)30
5 *days, Ans.*

SOLUTION. — Writing the corresponding terms as already described, we have 3 to 6 as 10 to the required number of days, since 10 days is the time required by 3 men, and the unknown term the time required by 6 men. But since twice the number of men will build the wall in half the time, 3 men are such a part of (are contained in) 6 men as the required number of days are a part of 10 days. The first, then, is an inverse ratio, the consequent being divided by the antecedent. But if we invert its terms, the division will be as usual, and the operation as already described.

NOTE. — In the third example, more money would buy more flour; in the second, less money would buy less flour. In such examples, each antecedent is divided by its consequent, or if one ratio is inverted, the other is also. The proportion is then called *direct*. But when, as in the fourth example, more requires less, (more days, less time,) or, as may be the case, less requires more, one ratio is inverse, while the other is direct. The proportion is then called *inverse*. Hence,

Direct proportion is the combination of two direct, or two inverse ratios.

Inverse proportion is the combination of a direct and an inverse ratio. But if one of the ratios is inverted, it may be treated as a direct proportion.

¶ 190. Hence, *to find the fourth term of a proportion when three terms are given, we have the following*

RULE.

I. Make that one of the three numbers the third term which is of the same kind as the answer sought, reducing the other two, if necessary, to the same denomination, that the terms of each ratio may be of the same kind.

II. Write that one of the remaining two for the first term which corresponds to the third, and the other for the second term.

Questions. — ¶ 189. How are the numbers, Ex. 4, to be written according to the rule for the corresponding terms? What happens from this manner of writing them, and why? What then is done? How do the second and third differ from the fourth example? What is direct proportion? What inverse?

III. If the question involves the inverse proportion, more requiring less, or less requiring more, invert the first ratio.

IV. Multiply together the second and third terms, or two means, to get the product of the extremes, which being divided by the first term, or known extreme, the quotient will be the other extreme, or fourth term sought, of the same kind as the third term.

NOTE. — If the third term is a compound, or a mixed number, it must be made of but one denomination.

EXAMPLES FOR PRACTICE.

¶ 191. 1. If 6 horses consume 21 bushels of oats in 3 weeks, how many bushels will serve 20 horses the same time?

horses. horses. oats. oats.

$$\overset{2}{\not{6}} : 20 :: \overset{7}{\not{2}\not{1}} : \cdots$$

$$\begin{array}{r} 20 \\ 7 \\ \hline 2)\,140 \\ \hline 70 \end{array}$$

70 *bush.*, *Ans.*

NOTE 1. — Since in the operation there are always two numbers to be multiplied together, and their product to be divided by a third number, the process may frequently be shortened by cancelation, as shown; the factor, 3, being canceled in 21 and 6, and the remaining factors, 7 and 2, being used as multiplier and divisor.

2. *The above question reversed.* If 20 horses consume 70 bushels of oats in 3 weeks, how many bushels will serve 6 horses the same time? *Ans.* 21 bushels.

3. If 365 men consume 75 barrels of provisions in 9 months, how much will 500 men consume in the same time?

$$\overset{73}{\not{3}\not{6}\not{5}} : 500 :: \overset{15}{\not{7}\not{5}} :$$

Ans. $102\frac{54}{73}$ barrels.

4. If 500 men consume $102\frac{54}{73}$ barrels of provisions in 9 months how much will 365 men consume in the same time?

$\overset{100}{\not{5}\not{0}\not{0}} : \overset{73}{\not{3}\not{6}\not{5}} :: 102\frac{54}{73} : \cdots$ And reducing the third term to an improper fraction, we have, $100 : \not{7}\not{3} :: \frac{7500}{\not{7}\not{3}} : \cdots$ Now, since the denominator is a divisor, (¶ 64,) we cancel 73, and have, 100 : 1 : : 7500 : ·· ·· *Ans.* 75 barrels.

5. If the moon move 13° 10′ 35″ in 1 day, in what time does it perform one revolution? *Ans.* 27 days 7 h. 43 m. 6 s. +.

Questions. — **¶ 190.** What number is made the third term, and why? How are the other two numbers arranged? When is the first ratio inverted? For what is the multiplication? — the division? When is a reduction needed? Repeat the whole rule.

6. If a person, whose rent is \$145, pay \$12'63 parish taxes, how much should a person pay whose rent is \$378? *Ans.* \$32'925.

7. If I buy 7 lbs. of sugar for 75 cents, how many pounds can I buy for \$6? *Ans* 56 lbs.

NOTE 2.—Every example in proportion may be performed on general principles, ¶ 134. Thus, in the last example, we may divide \$'75, the price of 7 lbs., by 7, and we shall have the price of 1 lb., and then divide \$6, the price of the required number of lbs., by the price of 1 lb. now found, and it will give us the number of lbs.

8. If I give \$6 for the use of \$100 for 12 months, what must I give for the use of \$357'82 the same time? *Ans.* \$21'469+.

On general principles. Divide \$6 by 100 to get the gain on \$1, which multiply by the number of dollars, to get the gain on the number of dollars.

NOTE 3.—Let the pupil be required to perform all the subsequent examples both by proportion and on general principles, or analysis.

9. If a staff 5 ft. 8 in. in length, cast a shadow of 6 feet, how high is that steeple whose shadow measures 153 feet? *Ans* 144½ feet.

10. If a family of 10 persons use 3 bushels of malt in a month how many bushels will serve them when there are 30 in the family? *Ans.* 9 bushels.

By Analysis. If 10 persons use 3 bushels, 1 person will use $\frac{1}{10}$ of 3 bushels, or $\frac{3}{10}$ of a bushel. And if 1 person use $\frac{3}{10}$, 30 persons will use 30 times $\frac{3}{10} = \frac{90}{10} = 9$ bushels.

NOTE. 4—The seven following examples involve the inverse proportion.

11. There was a certain building raised in 8 months by 120 workmen; but the same being demolished, it is required to be built in 2 months; I demand how many men must be employed about it. *Ans.* 480 men.

12. There is a cistern having a pipe which will empty it in 10 hours; how many pipes of the same capacity will empty it in 24 minutes? *Ans.* 25 pipes.

13. A garrison of 1200 men has provisions for 9 months, at the rate of 14 oz. per day; how long will the provisions last, at the same allowance, if the garrison be reinforced by 400 men? *Ans.* 6¾ months.

14. If a piece of land 40 rods in length and 4 in breadth, make an acre, how wide must it be when it is but 25 rods long? *Ans.* $6\frac{2}{5}$ rods.

15. If a man perform a journey in 15 days, when the days are 12 hours long, in how many will he do it when the days are but 10 hours long? *Ans.* 18 days.

16. If a field will feed 6 cows 91 days, how long will it feed 21 cows? *Ans.* 26 days.

Questions.—¶ **191.** How is the operation in proportion shortened? How may every example in proportion be performed without stating it? What is the method described, ¶ 134, for examples requiring several operations?

17. Lent a friend 292 dollars for 6 months; some time after, he lent me 806 dollars; how long may I keep it to balance the favor? *Ans.* 2 months 5+ days.

18. If 7 lbs. of sugar cost $\frac{3}{4}$ of a dollar, what cost 12 lbs.? *Ans.* \$1$\frac{2}{7}$.

19. If 6$\frac{1}{2}$ yds. of cloth cost \$3, what cost 9$\frac{1}{4}$ yds? *Ans.* \$4'269+.

20. If 2 oz. of silver cost \$2'24, what costs $\frac{3}{4}$ oz.? *Ans.* \$0'84

21. If $\frac{5}{7}$ oz. cost \$$\frac{11}{12}$, what costs 1 oz.? *Ans.* \$1'283.

22. If $\frac{3}{5}$ yd. cost \$$\frac{7}{8}$, what will 40$\frac{1}{2}$ yds. cost? *Ans.* \$59'062+

23. A merchant, owning $\frac{4}{5}$ of a vessel, sold $\frac{2}{3}$ of his share for \$957; what was the vessel worth? *Ans.* \$1794'375.

24. If 12 acres 3 roods produce 78 quarters 3 pks of wheat, how much will 35 acres 1 rood 20 poles produce? *Ans.* 216 qrs. 5 bush. 1 pk. 4 qts.

25. A cistern has 4 pipes which will fill it, respectively, in 10, 20, 40, and 80 minutes; in what time will all four, running together, fill it? *Ans.* 5$\frac{1}{3}$ minutes.

26. At \$33 for 6 barrels of flour, what must be paid for 178 barrels? *Ans.* \$979.

27. If 2'5 lbs. of tobacco cost 75 cents, how much will 185 lbs cost? *Ans.* \$55'50.

28. What is the value of '15 of a hogshead of lime, at \$2'39 per hhd.? *Ans.* \$0'3585.

29. If '15 of a hhd. of lime cost \$0'3585, what is it per hhd.? *Ans.* \$2'39.

30. A bankrupt owes \$972, and his property, amounting to \$607'50, is distributed among his creditors; what does one receive whose demand is \$11$\frac{1}{3}$? *Ans.* \$7'083+.

31. When wheat is worth \$'93 per bushel, a 3 cent loaf weighs 12 oz.; what must it weigh when wheat is \$1'24 per bushel? *Ans.* 9 oz.

32. A company of 16 men are on an allowance of 6 oz. of bread a day; what will be their daily allowance for 28 days, if they receive an addition of 224 lbs.? *Ans.* 14 oz

---

Compound Proportion.

¶ 192. 1. If a man travel 240 miles in 8 days of 12 hours long, how far will he travel in 6 days of 10 hours long, traveling at the same rate?

SOLUTION. — The relation of the number of miles, 240, which must be made the third term, to the required distance, depends on two circumstances, the number, and the length of the days.

First, considering the number of days, without regard to their length, we shall have,

$$\begin{array}{cccc} \textit{days.} & \textit{days.} & \textit{miles.} & \textit{miles.} \\ 8 : & 6 :: & 240 : & \cdots \end{array}$$

We find that he will travel 180 miles in 6 days of the same length as the 8 days.

Second, considering the length of the days, without regard to their number, we shall have,

$$\begin{array}{cccc} \textit{hours.} & \textit{hours.} & \textit{miles.} & \textit{miles.} \\ 12 : & 10 :: & 180 : & \cdots \end{array}$$

He will travel 150 miles in 6 days, 10 hours long, when he travels 180 in the same number of days, 12 hours long.

By the dependence of the distance on the ratio 8 : 6, we get $\frac{6}{8}$ of 240, and on the ratio 12 : 10, we get $\frac{10}{12}$ of this $\frac{6}{8}$. But $\frac{10}{12}$ of $\frac{6}{8}$ is a compound fraction, and consequently, 8 : 6 and 12 : 10, on which the distance depends, constitute a compound ratio, which may be united with the given distance, as follows:

$$\textit{Comp. ratio.} \left\{ \begin{array}{c} 8 : 6 \\ 12 : 10 \end{array} \right\} :: 240 : \cdots$$

$$\textit{Reduced.} \quad 96 : 60 :: 240 : \cdots$$

The compound may be reduced to a simple ratio, after which the operation is the same as in a simple proportion

Such a proportion is called a *compound proportion*. Hence,

A compound proportion is the combination of a compound and a simple ratio, and exists when the relation of the given quantity to the required quantity of the same kind depends on two or more conditions.

NOTE 1.—That the compound does not differ from the simple proportion, may be seen from the fact that, in the above reduced proportion, 96 is the number of hours in which 240 miles are traveled, and 60 the number of hours in which the required distance is traveled.

2. If 264 men, in 5 days of 12 hours each, can dig a trench 240 yards long, 3 wide, and 2 deep, in how many days, of 9 hours long, will 24 men dig a trench, 420 yards long, 5 wide, and 3 deep?

SOLUTION.—The number of days is required, and its relation to the given quantity, 5, depends on five circumstances, the number of men, the length of the days, the length, breadth, and depth of the trench. Placing 5 for the third term, we connect it with a compound ratio composed of five simple ratios; thus,

---

Questions.—¶ **192.** Why two operations to Ex. 1? Describe the first; — the second. How does it appear that the relation of the distances depends upon a compound ratio? What is a compound proportion? How is it seen that it does not differ in principle from the simple proportion? What reduction is made on the compound ratio? How is the operation of reducing shortened? Give the rule.

			11	
Inverse	~~24~~		~~264~~	
	3		~~4~~	
Inverse	~~9~~	:	~~12~~	*days.* *aays.* :: 5 : ....
	~~4~~		7	
Direct,	~~240~~	:	~~420~~	
Direct,	~~3~~	:	5	
Direct,	2	:	~~3~~	

$3 \times 2 = 6. \quad 11 \times 7 \times 5 = 385.$

Each simple ratio is from one circumstance, regarding the other cir cumstances uniform. The first two ratios, it may be seen, are inverse since the less the number of men employed, or the shorter the days, the greater will be the number of days.

Reducing this compound ratio to a simple one, shortening the process by cancelation, we have the simple proportion,—

6 : 385 :: 5 : *Ans.*

```
    5
 ------
6)1925
 ------
 320⅚ days, Ans.
```

By this proportion we get the answer in the same manner as by any simple proportion.

¶ 193. Hence, *questions involving a compound proportion, may be performed by the following*

RULE.

I. Having made that number the third term which is of the same kind as the answer sought, we form a ratio, either direct or inverse, as required, from any two remaining numbers which are of the same kind, and place it for one couplet of the compound ratio. Form each two like remaining numbers into a ratio, and so on, till all are used.

II. Having reduced the compound to a simple ratio, by multiplying together the antecedents and the consequents, shortening the process by cancelation, the operation will be the same as in simple proportion.

EXAMPLES FOR PRACTICE.

1. If 6 men build a wall 20 ft. long, 6 ft. high, and 4 ft. thick, in 16 days, in what time will 24 men build one 200 ft. long, 8 ft. high, and 6 ft. thick?

2. If the freight of 9 hhds. of sugar, each weighing 1200 lb., 20 miles, cost $ 16, what must be paid for the freight of 50 tierces, each weighing 250 lbs., 100 miles.

Questions. — ¶ **193.** What is the rule for compound proportion? How many, and what circumstances in Ex. 2? How else may the examples be performed?

NOTE 1.—The price of freight depends on three circumstances, the number of casks, the size of the casks, and the distance.

Ans. \$92'59 +.

3. If 56 lbs. of bread be sufficient for 7 men 14 days, how much bread will serve 21 men 3 days? *Ans.* 36 lbs.

The same by analysis. If 7 men consume 56 lbs., 1 man will consume $\frac{1}{7}$ of $56 = 8$ lbs. in 14 days, and $\frac{1}{14}$ of 8 lbs., $= \frac{8}{14}$ of a lb., in 1 day. 21 men will consume 21 times what 1 man will consume. that is, 21 times $\frac{8}{14} = \frac{168}{14} = 12$ lbs. in 1 day, and 3 times 12 lbs $= 36$ lbs. in 3 days.

NOTE 2.—Having wrought the following examples by proportion, let the pupil be required to do the same by analysis.

4. If 4 reapers receive \$11'04 for 3 days' work, how many men may be hired 16 days for \$103'04? *Ans.* 7 men.

5. If 7 oz. 8 drs. of bread be bought for \$'06 when wheat is \$'76 per bushel, what weight of it may be bought for \$'18 when wheat is \$'90 per bushel? *Ans.* 1 lb. 3 oz.

6. If \$100 gain \$6 in 1 year, what will \$400 gain in 9 months?

NOTE 3.—This and the three following examples reciprocally prove each other.

7. If \$100 gain \$6 in 1 year, in what time will \$400 gain \$18?

8. If \$400 gain \$18 in 9 months, what is the rate per cent. per annum?

9. What principal, at 6 per cent. per annum, will gain \$18 in 9 months?

10. A usurer put out \$75 at interest, and at the end of 8 months, received, for principal and interest, \$79; I demand at what rate per cent. he received interest. *Ans.* 8 per cent.

¶ 194. Review of Proportion.

Questions.—What is ratio? Between what quantities does it exist? What is inverse ratio?—compound ratio? What is proportion?—rule of three? How is it shown that the product of the extremes equals the product of the means? What inversions of the ratios take place? How is it shown that the operation of the rule of three depends on analysis? What is inverse proportion?—direct proportion?—compound proportion?

EXERCISES.

1. If I buy 76 yds. of cloth for \$113'17, what does it cost per ell English? *Ans.* \$1'861 +.

2. Bought 4 pieces of Holland, each containing 24 ells English, for \$96; how much was that per yard? *Ans.* \$0'80.

3. A garrison had provision for 8 months, at the rate of 15 ounces

to each person per day; how much must be allowed per day, in order that the provision may last $9\frac{1}{2}$ months? *Ans.* $12\frac{12}{19}$ oz.

4. How much land, at \$2'50 per acre, must be given in exchange for 360 acres, at \$3'75 per acre? *Ans.* 540 acres.

5. Borrowed 185 quarters of corn when the price was 19s.; how much must I pay when the price is 17s. 4d.? *Ans.* $202\frac{41}{52}$ qrs.

6. A person, owning $\frac{3}{5}$ of a coal mine, sells $\frac{3}{4}$ of his share for £171; what is the whole mine worth? *Ans.* £380.

7. If $\frac{5}{8}$ of a gallon cost $\frac{5}{8}$ of a dollar, what costs $\frac{5}{9}$ of a tun? *Ans.* \$140

8. At £$1\frac{1}{2}$ per cwt., what cost $3\frac{1}{3}$ lbs.? *Ans.* $10\frac{5}{7}$d.

9. If $4\frac{1}{2}$ cwt. can be carried 36 miles for 35 shillings, how many pounds can be carried 20 miles for the same money? *Ans.* $907\frac{1}{5}$ lbs.

10 If the sun appears to move from east to west 360 degrees in 24 hours, how much is that in each hour? —— in each minute? —— in each second? *Ans.* to the last, 15″ of a deg.

11. If a family of 9 persons spend \$450 in 5 months, how much would be sufficient to maintain them 8 months, if 5 persons more were added to the family? *Ans.* \$1120.

ALLIGATION—Medial.

¶ 195. 1. A farmer mixed 8 bushels of corn, worth 60 cents per bushel, 4 bushels of rye, worth 80 cents per bushel, and 4 bushels of oats, worth 40 cents per bushel; what was a bushel of the mixture worth?

When several simples of different values are to be mixed the process of finding the average price, is called *Alligation Medial.* The average price is called the *mean price.*

OPERATION.

8 *bushels, at* \$'60; $60 \times 8 =$ \$4'80.
4 *bushels, at* \$'80; $80 \times 4 =$ \$3'20.
4 *bushels, at* \$'40; $40 \times 4 =$ \$1'60.

16 *bushels are worth* \$9'60.
1 *bushel is worth* \$'60.

SOLUTION. — Multiply ing \$'60, the price of 1 bushel of corn, by 8, the product is the price of 8 bushels. In like manner we find the price of the rye, and the oats in the mixture, and adding together the prices of the corn, the rye, and the oats, we have \$9'60, the price of $8+4+4=16$ bushels, which are contained in the whole mixture, and dividing the price of 16 bushels by 16, we get the price of 1 bushel. *Ans.* 60 cents.

Hence, the

Questions. — ¶ **195.** What is alligation medial? — mean rate? What is the rule? To what does it apply?

RULE.

Find the prices of the several simples, and add them together for the price of the whole compound, which divide by the number of pounds, bushels, &c., to get the price of 1 pound or bushel.

NOTE. — The principles of the rule are applicable to many examples not embraced by the above definition of Alligation Medial.

EXAMPLES FOR PRACTICE.

2. A grocer mixed 5 lbs. of sugar worth 10 cents per lb., 8 lbs. worth 12 cents, 20 lbs. worth 14 cents; what is a pound of the mixture worth? *Ans.* \$‘$12\frac{10}{11}$.

3. A goldsmith melted together 3 ounces of gold 20 carats fine, and 5 ounces 22 carats fine; what is the fineness of the mixture? *Ans.* $21\frac{1}{4}$ carats.

4. A grocer puts 6 gallons of water into a cask containing 40 gallons of rum, worth 42 cents per gallon; what is a gallon of the mixture worth? *Ans.* $36\frac{12}{23}$ cents.

5. On a certain day the mercury was observed to stand in the thermometer as follows: 5 hours of the day it stood at 64 degrees; 4 hours, at 70 degrees; 2 hours, at 75 degrees; and 3 hours at 73 degrees; what was the *mean* temperature of that day? *Ans.* $69\frac{3}{14}$ degrees.

6. A farm contains 16 acres of land worth \$90 per acre, 22 acres worth \$75, 18 acres worth \$64, 10 acres worth \$55, 30 acres worth \$36, and 42 acres worth \$25 per acre; what is the average value of the farm per acre? *Ans.* \$50‘16 nearly per acre.

7. A dairyman has 20 cows, 3 of which are worth \$35 each, 4 are worth \$30, 6 are worth \$24, 4 are worth \$20, 2 are worth \$18 each, and 1 is worth \$13; what is the average value? *Ans.* \$24‘90.

Alligation Alternate.

¶ 196. 1. A farmer has 1 bushel of corn, worth 50 cents. How many bushels of oats, worth 40 cents, must be put with it, to make the mixture worth 42 cents?

SOLUTION. — The 1 bushel of corn is worth 8 cents more than the price of the mixture, and as each bushel of oats is worth 2 cents less, he must take 4 bushels of oats *Ans.*

When the price of several simples, (corn and oats,) and the price of a mixture to be formed from them are given, the method of finding the quantity of each simple is called *Alligation Alternate.*

2. How many bushels of oats must the farmer take to mix with 2 bushels of corn, prices the same as above?

SOLUTION. — Evidently, twice as many as were required for 1 bushel, or, *Ans.* 8 bushels

NOTE 1. — We see that 2, the number of bushels of corn, equals the number of cents that the oats are worth *less* than the mixture, and 8, the number of bushels of oats, equals the number of cents that the corn is worth *more* than the mixture.

Then, if the three prices had been given, we might have taken 2, the difference between the price of the oats and of the mixture for the number of bushels of corn, and 8, the difference between the price of the corn and of the mixture, for the number of bushels of oats, and we should have such a mixture.

That is, *we take the difference between the price of each simple and of the mixture for the number of bushels of the other simple.*

NOTE 2. — By this process, the sum of the excesses, found by multiplying 8 cents, the excess of 1 bushel of corn, by 2, the number of bushels, equals the sum of the deficiencies, found by multiplying 2, the deficiency of 1 bushel of oats, by 8, the number of bushels. Or, $8 \times 2 = 2 \times 8$, since the factors are the same in each.

3. A merchant has two kinds of sugar, worth 6 cents and 17 cents per pound, of which he would make a mixture worth 10 cents; how much of each must he take?

OPERATION.

Price of mixture, 10 cents. { 6 — 7 lbs.
17 — 4 lbs.

SOLUTION. — Take the difference between 10 and 17 for the number of lbs. at 6 cents, and the difference between 6 and 10 for the number of lbs. at 17 cents. The sum of the deficiencies, 7 times 4 cents, equals the sum of the excesses, 4 times 7 cents.

NOTE 3. — This gives a mixture of 11 pounds, and if 3 times, 5 times, one half, or any other proportion of the mixture be required, the same proportion of each simple must evidently be taken. The same would be done if 3 times, 5 times, &c., one simple were given.

4. A merchant has two kinds of sugar, worth 8 cents and 13 cents; how much of each must he take for a mixture worth 10 cents per pound?

OPERATION.

Price of mixture, 10 cents. { 8 — 3 lbs. at 8 cents.
13 — 2 lbs. at 13 cents.

NOTE 4. — The price of the mixture in the last two examples is the same.

5. A merchant has four kinds of sugar, worth 6, 8, 13, and 17 cents per pound; how much of each must he take for a mixture worth 10 cents?

OPERATION.

Price of mixture, 10 cents. $\begin{cases} 6 & 7. \\ 8 & 3. \\ 13 & 2. \\ 17 & 4. \end{cases}$

SOLUTION.—Connecting 6 with 17, to show that sugar at those prices are to be mixed, we have 7 lbs. at 6 cents and 4 lbs. at 17 cents, forming a mixture of 11 lbs. worth 10 cents. In like manner, we have a mixture of 5 lbs. worth 10 cents, by connecting 8 and 13. And $11+5=16$ lbs. in all.

NOTE 5.—We may see in this and all examples, that the sum of the deficiencies equals the sum of the excesses.

6. A merchant has 3 kinds of sugar, worth 6, 8, and 15 cents; how much of each must he take for a mixture worth 11 cents per pound?

OPERATION.

Price of mixture, 11 cents. $\begin{cases} 6 & 4. \\ 8 & 4. \\ 15 & 5+3=8. \end{cases}$

SOLUTION.—We have not two pairs, as in the former example, but 4 lbs. at 6 cents, may be mixed with 5 lbs. at 15 cents, and 4 lbs. at 8 cents, with 3 lbs. at 15 cents. The 15 is connected with both, because we take two portions at this price, 5 lbs. to mix with 4 lbs. at 6 cents, and 3 lbs. to mix with 4 lbs. at 8 cents. We have, then, 8 lbs. at 15 cents.

¶ 197. From the examples explained, we have the following

RULE.

I. Write the prices of the simples directly under each other, beginning with the least, and the price of the mixture at the left hand.

II. Connect each price less than the mean price with one or more greater, and each price greater with one or more that is less.

III. Write the difference between the price of the mixture and of each simple opposite to the price or prices with which it is connected; the number or numbers opposite to the price of each simple will express its quantity in the mixture.

IV. If any measure or multiple, as one third, or three times one of the simples is to be taken, the same measure or multi-

---

Questions.—¶ **196.** Why 4 bushels of oats to 1 of corn, Ex. 1 and 2? What equalities appear from Ex. 2? How could we have made such a mixture as we have, if the prices only had been known? Give the general method of forming such mixtures. What are equal, as explained in note 2? Why? What must be done if a greater or less quantity of the mixture be required?—of one simple? Why are numbers connected? What is done, in Ex. 6, when there are not two pairs? What is alligation alternate

¶ **197.** What is the rule?

ple of each of the other simples will be required. Or, if any measure or multiple of the whole compound be required, the same measure or multiple of each simple must be taken.

EXAMPLES FOR PRACTICE.

1. What proportions of sugar, at 8 cents, 10 cents, and 14 cents per pound, will compose a mixture worth 12 cents per pound?

Ans. In the proportion of 2 lbs. at 8 and 10 cents to 6 lbs. at 14 cents.

2. A grocer has sugars, worth 7 cents, 9 cents, and 12 cents per lb., which he would mix so as to form a compound worth 10 cents per pound; what must be the *proportions* of each kind?

Ans. 2 lbs. of the first and second, to 4 lbs. of the third kind.

3. If he use 1 lb. of the first kind, how much must he take of the others? —— if 4 lbs., what? —— if 6 lbs., what? —— if 10 lbs., what? —— if 20 lbs., what?

Ans. to the last, 20 lbs. of the second, and 40 of the third.

4. A merchant has spices at 16d., 20d., and 32d. per pound; he would mix 5 pounds of the first sort with the others, so as to form a compound worth 24d. per pound; how much of each sort must he use? *Ans.* 5 lbs. of the second, and 7½ lbs. of the third.

5. How many gallons of water, of no value, must be mixed with 50 gallons of rum, worth 80 cents per gallon, to reduce its value to 70 cents per gallon? *Ans.* 8$\frac{4}{7}$ gallons.

6. A man would mix 4 bushels of wheat, at $1'50 per bushel, with rye at $1'16, corn at $'75, and barley at $'50, so as to sell the mixture at $'84 per bushel; how much of each must he use?

7. A goldsmith would mix gold 17 carats fine with some 19, 21, and 24 carats fine, so that the compound may be 22 carats fine; what proportions of each must he use?

Ans. 2 of the first 3 sorts to 9 of the last.

8. If he use 1 oz. of the first kind, how much must he use of the others? What would be the quantity of the compound?

Ans. to the last, 7½ ounces.

9. If he would have the whole compound consist of 15 oz., how much must he use of each kind? —— if of 30 oz., how much of each kind? —— if of 37½ oz., how much?

Ans. to the last, 5 oz. of the first 3, and 22½ oz. of the last.

10. A man would mix 100 pounds of sugar, some at 8 cents, some at 10 cents, and some at 14 cents per pound, so that the compound may be worth 12 cents per pound; how much of each kind must he use?

20 lbs. at 8 cts.
20 lbs. at 10 cts. } *Ans.*
60 lbs. at 14 cts.

11. A grocer has currants at 4d., 6d., 9d., and 11d. per lb., and he would make a mixture of 240 lbs., so that the mixture may be sold at 8d. per lb.; how many pounds of each sort may he take?

Ans. 72, 24, 48, and 96 lbs., or 48, 48, 72, 72, &c.

Note. — This question may have five different answers.

EXCHANGE.

¶ 198. If a farmer, A, has corn, and a manufacturer, B, has cloth, each more than he needs himself, while he wants some of the article possessed by the other, an exchange will be made for mutual accommodation. But if B does not want A's corn, while A still wants the cloth, the latter must find a third person, if possible, who wants his corn, and can give him something for it which B *may* want. This might be difficult, unless some article was settled on, which every one would take; then A might exchange his corn for it, as B would part with his cloth for such an article, since, if every one would take it, he could procure with it whatever he might desire, though he did not want it himself.

Such an article is called money. Gold and silver, containing great value in little space, are employed for money among civilized nations, and sometimes copper, to represent small values. This exchange between individuals is called trade, or commerce.

NOTE 1. — Since gold and silver, in their pure state, are too flexible for the purposes of a circulating medium, nine parts of pure gold and one part of silver and copper, in equal quantities, are used by the U. S. government, for gold coins, nine parts of silver and one of copper, for silver coins. The baser metal, in each instance, is called *alloy*. The English government uses only one part of alloy to eleven of the gold, and a little less alloy in silver coins. Hence, English coins are more valuable than ours of the same weight. Again, as coins are troublesome, and sometimes expensive to transport, and also suffer loss by wear, bank bills are much used for circulation, which, though valueless themselves, are readily taken as money, being payable in specie, on demand at the banks which issued them. The coins are called specie, in distinction from paper money, and together they form what is called the circulating medium, or currency.

NOTE 2. — By the *fineness* of gold, is meant its purity, a twenty-fourth part of any quantity being called a carat. When, for example, there are two parts of alloy to twenty-two of pure gold, it is said to be twenty-two carats fine.

¶ 199. Bank bills can be used in trade by individuals of the same country, but are not convenient in trade between those of different countries, since they would be removed too far from the place where payable. But the cost and risk of

---

Questions. — ¶ **198.** What trade are A and B supposed to make? What will A do, if B does not want his corn? What is exchange? What is money? What are used for money, and why? How much do coins want of being pure gold and silver? What is the value of English coins compared with ours, and why? Why are bank bills used? What do you understand to be the difference between a bank bill now described, and a bank note, ¶ 174? What is specie? What is currency? What is meant by fineness of gold? — by carat? — by carats fine? Illustrate..

transporting specie is considerable. If, for instance, Boston merchants purchase goods in Hamburgh to the amount of $2,050,000 a year, and the expense of transporting specie was 3 per cent., which is only a moderate allowance, this expense would be $61,500 to make payments for one year. If, on the other hand, Hamburgh merchants should purchase $2,000,000 worth of goods, it would cost them $60,000 to make the payments in specie.

To reduce this expense, the following method has been devised. A B, a Boston merchant, has $10,000 due him in Hamburgh, from C D. He writes an order for the sum, and finds some one who owes $10,000 to another merchant in Hamburgh. He sells to him this order, and the purchaser sends it to his creditor, who goes to C D, A B's debtor, and receives the money. In this way, $2,000,000 of the Boston purchase might be balanced by the $2,000,000 purchased in Boston by Hamburgh merchants, and the Boston merchants would have to send only $50,000, the balance, thus reducing the expense of making payments between the two ports from $121,500 to $1500!

Such an order is called *a Bill of Exchange.*

Note 1. — Lest a bill of exchange may be lost, or delayed, three copies are sent, by different conveyances, and when one is paid, the others are canceled. The form of one bill, to which the others agree, except in the numbers of the bills, is here given.

Exchange for $10,000. Boston, Jan. 1, 1848.

Three months after date, pay this, my first of exchange, (second and third of the same tenor and date not paid,) to the order of Flemming and Johnson, ten thousand dollars, value received, with, or without further advice from me. A B.

C D,

Merchant at Hamburgh.

Note 2. If A B had to be at the expense of bringing from Hamburgh the specie on his due, he could afford to sell his bill for less than $10,000, on account of freight. This he would have to do when more was to be paid from Hamburgh to Boston than from Boston to Hamburgh, that is, if the balance of trade were in favor of Boston, as there would not be demand for all the

---

Questions. — ¶ **199.** Why are not bank bills convenient in exchange between different countries? What difficulty in paying with specie? Illustrate by the example of trade between Boston and Hamburgh. How is A B supposed to get his pay from Hamburgh, on a debt of $10,000? How much is saved by this means, in the supposed trade between the two ports? Why must $50,000 of specie be sent to Hamburgh? What is a bill of exchange? What care is taken in sending bills? Give a form. What would be the form of the second bill? — of the third? When, and why, is exchange below par? — above par? Why do brokers deal in exchange?

orders on Hamburgh, and specie would have to be brought on some. Exchange is then said to be *below par*. But if the purchaser had to pay for freighting the money he owes to Hamburgh, he could afford to pay more than \$10,000 for A B's bill, which he would have to do when the balance of trade was against Boston, that is, when Boston owed more to Hamburgh than it had owing, for then there would not be orders on Hamburgh sufficient to pay the debts, and some must send specie. Exchange is then said to be *above par*, or at a premium.

NOTE 3. — The broker is the medium between the seller and purchaser of bills, since the former might not know to whom he could sell, or the latter of whom he could buy. Brokers purchase bills, or take them to sell on commission.

The illustrations of the principles of exchange will be applied to exchange with England and France.

Exchange with England.

¶ 200. The nominal value of the pound sterling is \$4'44$\frac{4}{9}$, consequently, a bill of exchange for £1000 is said to be worth \$4444'44$\frac{4}{9}$. But by comparing the materials of the English sovereign, a gold piece representing a pound, and the eagle of our currency, the former is worth about \$4'86$\frac{6}{10}$. Sovereigns, however, are more or less worn by use, those dated as far back as 1821 being worth no more than \$4'80. It is presumed that a quantity will average \$4'84 each, and at this value they are taken in payment of duties. If, now, the nominal pound, \$4'44$\frac{4}{9}$, be multiplied by '09, \$4'44$\frac{4}{9}$ × '09 = '40, and the product be added to it, \$4'44$\frac{4}{9}$ + '40 = \$4'84$\frac{4}{9}$ it will be changed to about its value, in the custom-house estimations. If '09$\frac{1}{2}$ be added to the nominal pound, it will become \$4'86$\frac{2}{3}$, nearly its commercial value. When, then, sterling exchange is quoted at 9 or 9$\frac{1}{2}$ per cent. advance, we must understand that bills sell for their par value. When above these rates, they are at a premium; when below, at a discount.

NOTE 1. — In the following examples we shall consider 9$\frac{1}{2}$ per cent. above the nominal, the par value. The pupil must not suppose, however, that 10 per cent. above the nominal, would be $\frac{1}{2}$ per cent. above the real value of a bill.

1. A merchant sells a bill of exchange for £5000 at its par value; what does he receive? *Ans.* \$24333'33$\frac{1}{3}$.

2. A merchant sold a bill of exchange for £7000 sterling, at 11 per cent. advance; what did he receive more than its real value? *Ans.* \$466'66$\frac{2}{3}$.

3. A merchant sells a bill on London for £4000 at 8 per cent. above its nominal value, instead of importing specie at an expense of 2 per cent.; what does he save? *Ans.* \$122'66$\frac{2}{3}$

4. A broker sold a bill of exchange for £2000, on commission, at 10 per cent. above its nominal value, receiving a commission of $\frac{1}{10}$ per cent. on the real value, and 5 per cent. on what he obtained for the bill above its real value ; what was his commission?

Ans. \$11'95$\frac{5}{9}$.

NOTE 2.—Though dollars and cents are the denominations of U. S. money, shillings and pence are much used in common calculations. But the dollar has different values in different states, as expressed in shillings; thus, in New York, Ohio, and Michigan, it is 8 shillings; in North Carolina, 10 shillings; in New Jersey, Pennsylvania, Delaware, and Maryland, it is 7 shillings 6 pence; in Georgia and South Carolina it is 4 shillings 8 pence, while in the other states it is 6 shillings. The change of money from one of these currencies to the other is not now worthy of a formal discussion, as a method will readily suggest itself to the practised arithmetician, and the custom of using these denominations, it is hoped, will be speedily given up for the simpler system of our federal currency. Thus, as 6 shillings in New England equal 8 shillings in New York, add one third of any number of shillings N. E currency to the number, and we have the value expressed in shillings N. Y currency.

Exchange with France.

¶ 201. The unit of French money is the franc, the value of which is \$'18$\frac{3}{5}$. In the quotations of French exchange, we have the number of francs that the dollar is rated at. As \$1'00 is equal to $5\frac{7}{18\frac{6}{10}}$ francs = 5'376 +, hence, when a dollar is worth 5'37$\frac{6}{10}$ francs, it is at par.

1. A New York merchant sold a bill of exchange for \$2500 on Havre, at 5'4 francs per dollar; what did he obtain for it more than its value? *Ans.* \$11.

2. A merchant bought a bill on Havre of \$2800 at 5'31 francs per dollar; what did he give less than its value? *Ans.* \$34'552.

Questions.—¶ **200.** What is the nominal value of the English pound? —the real value? What are sovereigns of 1821 worth? Why no more? What is the average value of sovereigns supposed to be, and where are they taken at this value? How is the pound changed from its nominal to its real value? What is added to the nominal value in the examples? What is said of 10 per cent.? What denominations are still used in common calculations? What are the different values of the dollar in different states?

¶ **201.** What is the unit of French money? —its value? What is the par value of the dollar, as expressed in francs?

¶ 202. Value of Gold Coins

According to the Laws passed by Congress, May and June, 1834.]

Names of Coins.	d. c. m.
UNITED STATES.	
Eagle, coined before July 31, 1834,	10 66 5
Shares in proportion.	
FOREIGN GOLD.	
AUSTRIAN DOMINIONS.	
Souverein,	3 37 7
Double Ducat,	4 58 9
Hungarian, do.,	2 29 6
BAVARIA.	
Carolin,	4 95 7
Max d'or, or Maximilian,	3 31 8
Ducat,	2 27 5
BERNE.	
Ducat, double in proportion,	1 98 6
Pistole,	4 54 2
BRAZIL.	
Johannes, ½ in proportion,	17 6 4
Dobraon,	32 70 6
Dobra,	17 30 1
Moidore, ½ in proportion,	6 55 7
Crusade,	63 5
BRUNSWICK.	
Pistole, double in proportion,	4 54 8
Ducat,	2 23 0
COLOGNE.	
Ducat,	2 26 7
COLOMBIA.	
Doubloons,	15 53 5
DENMARK.	
Ducat, Current,	1 81 2
Ducat, Specie,	2 26 7
Christian d'or,	4 02 1
EAST INDIES.	
Rupee, Bombay, 1818,	7 09 6
Rupee, Madras, 1818,	7 11 0
Pagoda, Star,	1 79 8
ENGLAND.	
Guinea, half in proportion,	5 07 5
Sovereign, do.,	4 84 6
Seven Shilling Piece,	1 69 8
FRANCE.	
Double Louis, coined before 1786,	9 69 7
Louis, do.,	4 84 6
Double Louis, coined since 1786,	9 15 3
Louis, do. do.,	4 57 6
Double Napoleon, or 40 francs,	7 70 2
Napoleon, or 20 do.,	3 85 1
FRANKFORT ON THE MAIN.	
Ducat,	2 27 9
GENEVA.	
Pistole, old,	3 98 5
Pistole, new,	3 44 4
GENOA.	
Sequin,	2 30 2
HAMBURG.	
Ducat, double in proportion,	2 27 9
HANOVER.	
Double George d'or, single in proportion,	7 87 9
Ducat,	2 29 6
Gold Florin, double in proportion,	1 67 0
HOLLAND.	
Double Ryder,	12 20 5
Ryder,	6 04 3
Ducat,	2 27 5
Ten Guilder piece, 5 do. in proportion,	4 03 4
MALTA.	
Double Louis,	9 27 8
Louis,	4 85 2
Demi Louis,	2 33 6
MEXICO.	
Doubloons, shares in proportion,	15 53 5
MILAN.	
Sequin,	2 29 0
Doppia, or Pistole,	3 80 7
Forty Livre Piece, 1808,	7 74 2
NAPLES.	
Six Ducat Piece, 1783,	5 24 9
Two do., or Sequin, 1762,	1 59 1
Three do., or Oncetta, 1818,	2 49 0
NETHERLANDS.	
Gold Lion, or Fourteen Florin Piece,	5 04 6
Ten Florin Piece, 1820,	4 01 9
PARMA.	
Quadruple Pistole, double in proportion,	16 62 8
Pistole or Doppia, 1787,	4 19 4
do. do., 1796,	4 13 5
Maria Theresa, 1818,	3 86 1
PIEDMONT.	
Pistole, coined since 1785, half in proportion,	5 41 1
Sequin, half in proportion,	2 28 0
Carlino, coined since 1785, half in proportion,	27 34 0
Piece of 20 francs, called Marengo,	3 56 4
POLAND.	
Ducat,	2 27 5
PORTUGAL.	
Dobraon,	32 70 6
Dobra,	17 30 1
Johannes,	17 06 4
Moidore, half in proportion,	6 55 7
Piece of 16 Testoons, or 1600 Rees,	2 12 1
Old Crusado, of 400 Rees,	58 5
New do., 480 do.,	63 5
Milree, coined in 1775,	78 0
PRUSSIA.	
Ducat, 1748,	2 27 9
do., 1787,	2 26 7

Names of Coins.	d. c. m.
Frederick, double, 1769,	7 95 5
do. do., 1800,	7 95 1
do. single, 1778,	3 99 7
do. do., 1800,	3 97 5
ROME.	
Sequin, coined since 1760,	2 25 1
Scudo of Republic,	15 81 1
RUSSIA.	
Ducat, 1796,	2 29 7
do., 1763,	2 26 7
Gold Ruble, 1756,	96 7
do., 1799,	73 7
do. Poltin 1777,	35 5
Imperial, 1801,	7 82 9
Half do., 1801,	3 93 3
SARDINIA.	
Carlino, half in proportion,	9 47 2
SAXONY.	
Ducat, 1784,	2 26 7
do., 1797,	2 27 9
Augustus, 1754,	3 92 5
do., 1784,	3 97 4
SICILY.	
Ounce, 1751,	2 50 4
Double do., 1758,	5 04 4
SPAIN.	
Doubloon, 1772, double and single, and shares in proportion,	16 02 8
Doubloon,	15 53 5
Pistole,	3 88 4
Coronilla, Gold Dollar, or Vintern, 1801,	98 3
SWEDEN.	
Ducat,	2 23 5
SWITZERLAND.	
Pistole of Helvetic Republic, 1800,	56
TREVES.	
Ducat,	2 26
TURKEY.	
Sequin Fonducli, of Constantinople, 1773,	1 86
do., 1789,	1 84
Half Misseir, 1818,	52
Sequin Fonducli,	1 83 [illegible]
Yeermeeblekblek,	3 02 8
TUSCANY.	
Zechino, or Sequin,	2 31 8
Ruspone of the kingdom of Etruria,	6 93 8
VENICE.	
Zechino, or Sequin, shares in proportion,	2 31 0
WIRTEMBURG.	
Carolin,	4 89 8
Ducat,	2 23 5
ZURICH.	
Ducat, double and half in proportion,	2 26 7

DUODECIMALS

¶ 203. Duodecimals are fractions of a foot. The word is derived from the Latin word *duodecim*, which signifies *twelve*. A foot, instead of being divided *decimally* into *ten* equal parts, is divided *duodecimally* into *twelve* equal parts, called *primes*, marked thus ($'$). Again, each of these parts is conceived to be divided into twelve other equal parts, called *seconds*, ($''$). In like manner, each second is conceived to be divided into twelve equal parts, called *thirds* ($'''$); each third into twelve equal parts, called *fourths* ($''''$); and so on to any extent.

In this way of dividing a foot, it is obvious, that

1 prime is . . . . $\frac{1}{12}$ of a foot.
second is $\frac{1}{12}$ of $\frac{1}{12}$, ... $=$ $\frac{1}{144}$ of a foot.
1 third is $\frac{1}{12}$ of $\frac{1}{12}$ of $\frac{1}{12}$, . . $=$ $\frac{1}{1728}$ of a foot.
1 fourth is $\frac{1}{12}$ of $\frac{1}{12}$ of $\frac{1}{12}$ of $\frac{1}{12}$, $=$ $\frac{1}{20736}$ of a foot.
1 fifth is $\frac{1}{12}$ of $\frac{1}{12}$ of $\frac{1}{12}$ of $\frac{1}{12}$ of $\frac{1}{12}$, $=$ $\frac{1}{248832}$ of a foot, &c.

TABLE.

12''''	fourths make	1'''	third,
12'''	thirds . . .	1''	second,
12''	seconds . . .	1'	prime,
12'	primes, . . .	1	foot.

NOTE 1. — The marks, ', '', ''', '''', &c., which distinguish the different parts, are called the *indices* of the *parts* or denominations.

NOTE 2. — The divisions of a unit in duodecimals are uniform, just as in decimal fractions, with this difference: they decrease in a *twelve-fold* proportion, 12 of a lower denomination making 1 of a higher. Operations in them are consequently the same as in whole numbers or decimals, except that 12 is the carrying number instead of 10.

Multiplication of Duodecimals.

¶ 204. Duodecimals are used in measuring *surfaces* and *solids*.

1. How many square feet in a board 16 feet 7 inches long and 1 foot 3 inches wide?

NOTE. — Length × breadth = superficial contents, (¶ 48.)

OPERATION.

	ft.		
Length,	16	7'	
Breadth,	1	3'	
	4	1'	9''
	16	7'	
Ans.	20	8'	9''

SOLUTION. — 7 inches, or primes, $=\frac{7}{12}$ of a foot, and 3 inches $=\frac{3}{12}$ of a foot; consequently, the product of $7' \times 3' = \frac{21}{144}$ of a foot, that is, $21'' = 1'$ and $9''$; wherefore we set down the 9'', and reserve the 1' to be carried forward to its proper place. To multiply 16 feet by 3', is to take $\frac{3}{12}$ of $\frac{16}{1}=\frac{48}{12}$, that is, 48'; and the 1' which we reserved makes 49', = 4 feet 1'; we therefore set down the 1', and carry forward the 4 feet to its proper place. Then, multiplying the multiplicand by the 1 foot in the multiplier, and adding the two products together, we obtain the *Answer,* 20 feet 8' and 9''.

NOTE 1. — In all cases the *product of any two denominations will always be of the denomination denoted by the sum of their* INDICES. Thus, in the above example, the sum of the indices of 7' × 3' is ''; consequently, the product is 21''; and thus *primes* multiplied by *primes*, produce *seconds; primes* multiplied by *seconds*, produce *thirds; fourths* multiplied by *fifths*, produce *ninths*, &c.

Questions. — ¶ **203.** What are duodecimals? Explain the duodecimal divisions and subdivisions of a foot. Repeat the table. What are indices? What part of a foot is 1'? — 1''? — 1'''? — 1''''? — 1'''''? What difference between the decimal and duodecimal divisions of a unit? How are operations on duodecimals performed?

2. How many solid feet in a block 15 ft. 8′ long 1 ft. 5′ wide, and 1 ft. 4′ thick?

OPERATION.

	ft.			
Length,	15	8′		
Breadth,	1	5′		
	6	6′	4″	
	15	8′		
	22	2′	4″	
Thickness,	1	4′		
	7	4′	9″	4‴
	22	2′	4″	
Ans.	29	7′	1″	4‴

The length multiplied by the breadth, and that product by the thickness, gives the *solid contents*, (¶ 51.)

Hence, To multiply duodecimals,

RULE.

I. Write the multiplier under the multiplicand, like denominations under like, and in multiplying, remember that the product of any two denominations will be of that denomination denoted by the *sum of their indices.*

II. Add the several products together, and their sum will be the product required.

EXAMPLES FOR PRACTICE.

3. How many square feet in a stock of 15 boards, each of which is 2 ft. 8′ in length, and 13′ wide? *Ans.* 205 ft. 10′.

4. What is the product of 371 ft. 2′ 6″ multiplied by 181 ft. 1′ 9″? *Ans.* 67242 ft. 10′ 1″ 4‴ 6⁗.

5. There is a room plastered, the compass of which is 47 ft. 3′, and the hight 7 ft. 6′; what are the contents? *Ans.* 39 yds. 3 ft. 4′ 6″.

6. What will it cost to pave a court yard, 26 ft. 8′ long by 24 ft. 9′ wide, at $'90 per square yard? *Ans.* $66

7. There is a house containing two rooms, each 16 ft. by 15 ft. 4′; a hall 24 ft. by 10 ft. 6′; three bed-rooms, each 11 ft. 4 by 8 ft.; a pantry 7 ft. by 9 ft. 6′; a kitchen 14 ft. 2′ by 18 ft., and two cham-

---

Questions.—¶ 204. For what are duodecimals used? Of what denomination is the product of any two denominations? Repeat the rule for the multiplication of duodecimals. How do you carry from one denomination to another? How is masons' work estimated? What is understood by girt, and for what used?

bers, each 16 ft. by 20 ft. 8′; what did the work of flooring cost, at \$‘02 per square foot? *Ans.* \$39‘95.

NOTE 2. — Masons' work is estimated by the perch of $16\frac{1}{2}$ feet in length, $1\frac{1}{2}$ feet in width, and 1 foot in hight. A perch contains 24‘75 cubic feet. If any wall be $1\frac{1}{2}$ feet thick, its contents in perches may be found by dividing its superficial contents by $16\frac{1}{2}$; but if it be any other thickness than $1\frac{1}{2}$ feet, its cubic contents must be divided by 24‘75, ($=24\frac{3}{4}$,) to reduce it to perches.

Joiners, painters, plasterers, brick-layers, and masons, make no allowance for windows, doors, &c. Brick-layers and masons make no allowance for corners to the walls of houses, cellars, &c., but estimate their work by the girt, that is, the length of the wall on the outside.

8. The side walls of a cellar are each 32 ft. 6′ long, the end walls 24 ft. 6′, and the whole are 7 ft. high, and $1\frac{1}{2}$ ft. thick; how many perches of stone are required, allowing nothing for waste, and for how many must the mason be paid?

Ans. { $45\frac{9}{11}$ perches in the wall.
The mason must be paid for $48\frac{4}{11}$ perches. }

9. How many cord feet of wood in a load 7 feet long, 3 feet wide, and 3 feet 4 inches high, and what will it cost at \$‘40 per cord foot? *Ans.* $4\frac{3}{8}$ cord feet, and it will cost \$1‘75.

10. How much wood in a load 10 ft. in length, 3 ft. 9′ in width and 4 ft. 8′ in hight? and what will it cost at \$1‘92 per cord? *Ans.* 1 cord and $2\frac{1}{16}$ cord feet, and it will cost \$2‘62½.

¶ 205. By some surveyors of wood, dimensions are taken in feet and *decimals* of a foot. For this purpose, make a rule or scale 4 feet long, and divide it into feet, and each foot into ten equal parts. Such a rule will be found very convenient for surveyors of wood and of lumber, for painters, joiners, &c.; for the dimensions taken by it being in feet and decimals of a foot, the casts will be no other than so many operations in decimal fractions.

1. How many square feet in a hearth stone, which, by a rule, as above described, measures 4‘5 feet in length, and 2‘6 feet in width? and what will be its cost, at 75 cents per square foot? *Ans.* 11‘7 feet; and it will cost \$8‘775.

2. How many cords in a load of wood, 7‘5 feet in length, 3‘6 feet in width, and 4‘8 in hight? *Ans.* 1 cord $1\frac{6}{10}$ cu. ft.

3. How many cord feet in a load of wood 10 feet long, 3‘4 feet wide, and 3‘5 high? *Ans.* $7\frac{7}{16}$.

Questions. — ¶ **205.** How do some surveyors of wood take dimensions? Explain the rule used in measuring. How are dimensions taken by it estimated?

INVOLUTION.

First power.

Second power.

Third power.

Fourth power.

¶ 206. Three feet in length (¶ 111) are a yard, linear measure; 3 in length and 3 in width, $3 \times 3 = 9$ square feet, are a yard, square measure; 3 in length, 3 in width, and 3 in hight, $3 \times 3 \times 3 = 27$ solid feet, are a yard, cubic measure, (¶113.)

When a number, as 3, is multiplied into itself, and the product by the original number, and so on, the several numbers produced are called *powers*, and the process of producing them is called *Involution.*

The first number, represented by a line, is called the first power, or *root;* the second, represented by a square, is called the square, or 2d power; the third, represented by a cube, is called the cube, or 3d power.

The 4th power of 3 is 3 times the 3d power, 3 blocks like that employed to represent the 3d power, and may be represented by a figure 3 times as large, that is, 3 feet wide, 3 feet high, and 9 feet long.

Fifth power.

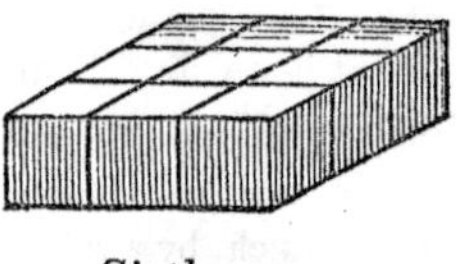

The 5th power, by 3 times such a figure, or one 3 high, 9 wide, and 9 long.

The 6th power, 3 times this, by a figure 9 long, 9 wide, and 9 high, or a cube.

Sixth power

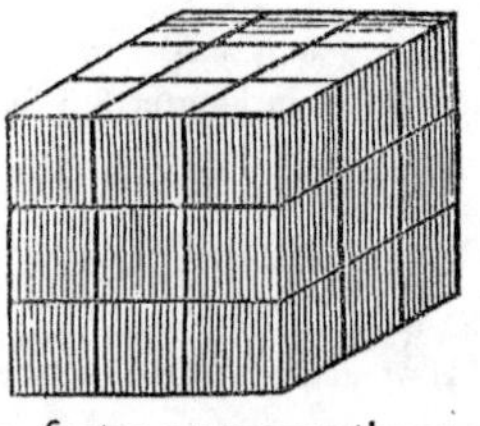

Thus it may be shown that the 9th, 12th, 15th, 18th, &c., powers, may be represented by cubes; the 7th, 10th, 13th, 16th, &c., by figures having greater length than width and hight; the 8th, 11th, 14th, 17th, &c., by figures having greater length and width than hight.

To involve a number, take it as a factor as many times as is indicated by the required power.

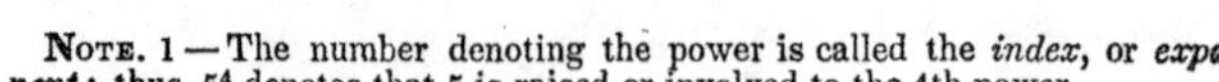

NOTE. 1—The number denoting the power is called the *index*, or *exponent;* thus, 5^4 denotes that 5 is raised or involved to the 4th power.

EXAMPLES FOR PRACTICE.

1. What is the square, or 2d power, of 7? *Ans.* 49
2. What is the square of 30? *Ans.* 900.
3. What is the square of 4000? *Ans.* 16000000.
4. What is the cube, or 3d power, of 4? *Ans.* 64.
5. What is the cube of 800? *Ans.* 512000000.
6. What is the 4th power of 60? *Ans.* 12960000.
7. What is the square of 1? —— of 2? —— of 3? —— of 4? *Ans.* 1, 4, 9, and 16.
8. What is the cube of 1? —— of 2? —— of 3? —— of 4? *Ans.* 1, 8, 27, and 64.
9. What is the square of $\frac{2}{3}$? —— of $\frac{4}{5}$? —— of $\frac{7}{8}$? *Ans.* $\frac{4}{9}$, $\frac{16}{25}$, and $\frac{49}{64}$.
10. What is the cube of $\frac{2}{3}$? —— of $\frac{4}{5}$? —— of $\frac{7}{8}$? *Ans.* $\frac{8}{27}$, $\frac{64}{125}$, and $\frac{343}{512}$.
11. What is the square of $\frac{1}{2}$? —— the 5th power of $\frac{1}{2}$? *Ans.* $\frac{1}{4}$ and $\frac{1}{32}$
12. What is the square of 1‘5? —— the cube? *Ans.* 2‘25, and 3‘375.
13. What is the 6th power of 1‘2? *Ans.* 2‘985984.
14. Involve $2\frac{1}{4}$ to the 4th power.

NOTE 2. — A mixed number, like the above, may be reduced to an im proper fraction before involving; thus, $2\frac{1}{4} = \frac{9}{4}$; or it may be reduced to a decimal; thus, $2\frac{1}{4} = 2‘25$.

Ans. $\frac{6561}{256} = 25\frac{161}{256}$.

15. What is the square of $4\frac{7}{8}$? *Ans.* $\frac{1521}{64} = 23\frac{49}{64}$.
16. What is the value of 7^4, that is, the 4th power of 7? *Ans.* 2401.
17. How much is 9^3? —— 6^5? —— 10^4? *Ans.* 729, 7776, 10000.
18. How much is 2^7? —— 3^6? —— 4^5? —— 5^3? —— 6^5? —— 10^8? *Ans.* to the last, 100000000.

NOTE 3. — The powers of the nine digits, from the first power to the fifth may be seen in the following

TABLE.

Roots . .	or 1st Powers	1	2	3	4	5	6	7	8	9
Squares .	or 2d Powers	1	4	9	16	25	36	49	64	81
Cubes . .	or 3d Powers	1	8	27	64	125	216	343	512	729
Biquadrates	or 4th Powers	1	16	81	256	625	1296	2401	4096	6561
Sursolids .	or 5th Powers	1	32	243	1024	3125	7776	16807	32768	59049

Questions. — ¶ **206.** What are powers? How is the first power represented? Why is the second power called the square? Why the third called the cube? How is the fourth power represented? — the fifth? — the sixth? — thirtenth? — the fourteenth? — the twenty-first? — the twenty-third? — the twenty-fifth? What is involution? How is a number involved to any power? What is the index, and how written? How is a mixed number involved?

EVOLUTION.

¶ 207. Evolution, or the *extracting* of roots, is the method of finding the *root* of any power or number.

The *root*, as we have seen, is that number, which, by a continual multiplication into itself, produces the given power, and to find the square root of a number (one side of a square when the contents are given) is to find a number, which, being squared, will produce the given number; to find the cube root of a number (the length of one side of a cubic body when the solid contents are given) is to find a number, which, being cubed or involved to the 3d power, will produce the given number: thus, the *square root* of 144 is 12, because $12^2 = 144$; and the *cube root* of 343 is 7, because 7^3, that is, $7 \times 7 \times 7 = 343$; and so of other numbers.

NOTE. — Although there is no number which will not produce a perfect power by involution, yet there are many numbers of which *precise roots* can never be obtained. But, by the help of *decimals*, we can approximate, or approach, towards the root to any assigned degree of exactness. Numbers, whose precise roots cannot be obtained, are called *surd* numbers, and those whose roots can be exactly obtained are called *rational* numbers.

The square root is indicated by this character $\sqrt{}$ placed before the number; the other roots by the same character, with the index of the root placed over it. Thus, the square root of 16 is expressed $\sqrt{16}$; and the cube root of 27 is expressed $\sqrt[3]{27}$; and the 5th root of 7776, $\sqrt[5]{7767}$.

When the power is expressed by several numbers, with the sign + or — between them, a line, or *vinculum*, is drawn from the top of the sign over all the parts of it; thus, the square root of 21 — 5 is $\sqrt{21-5}$.

Extraction of the Square Root.

¶ 208. 1. Supposing a man has 625 yards of carpeting, a yard wide, what is the length of one side of a square room, the floor of which the carpeting will cover? that is, what is one side of a square, which contains 625 square yards?

SOLUTION. — We may find one side of a square containing 625 square yards, that is, the square root of 625, by a sort of trial; and,

---

Questions. — ¶ 207. What is evolution? What is a root? — the square root, and how found? — the cube root, and how found? Give examples. What do you say of perfect powers and perfect roots? Give the distinction between surd and rational numbers. How is the square root indicated? — the cube root? Describe the manner of using the vinculum.

1st. We will endeavor to ascertain how many figures there will be in the root. This we can easily do, by pointing off the number, from units, into periods of two figures each; for the square of and root always contains just *twice* as many, or one figure *less* than twice as many figures, as are in the root. The square of 3 ($3 \times 3 = 9$) contains 1 figure, the square of 4 ($4 \times 4 = 16$) contains 2 figures; the square of 9 ($9 \times 9 = 81$) contains 2 figures; the square of 10 ($10 \times 10 = 100$) contains 3 figures; the square of 32 ($32 \times 32 = 1024$) contains 4 figures; the square of 99 ($99 \times 99 = 9801$) contains 4 figures; the square of 100 ($100 \times 100 = 10000$) contains 5 figures, and so of any number. Pointing off the number, we find that the root will consist of *two* figures, a ten and a unit.

OPERATION

$\dot{6}2\dot{5}$ (2
4

225

Fig. I.

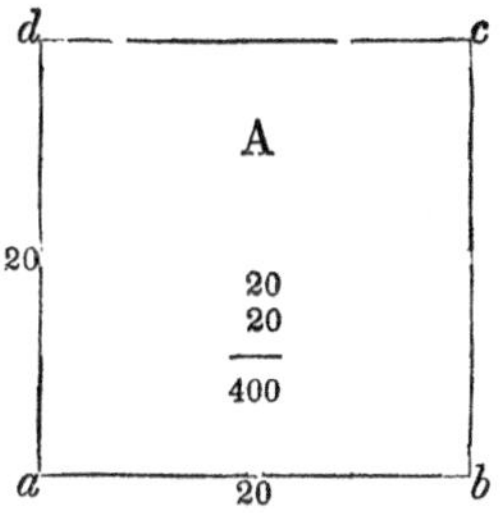

d c
A
20
20
20
400
a 20 b

2d. We will now seek for the first figure, that is, for the *tens* of the root, which we must extract from the left hand period, 6, (hundreds.) The greatest square in 6 (hundreds) we find to be 4, (hundreds,) the root of which is 2, (tens, = 20; therefore, we set 2 (tens) in the root. Since the *root* is *one side* of a square, let us form a square, (A, Fig. I.,) each side of which shall be regarded 2 tens, = 20 yards long.

The contents of this square are $20 \times 20 = 400$ yards, now disposed of, and which, consequently, are to be deducted from the whole number of yards, (625,) leaving 225 yards. This deduction is most readily performed by subtracting the square number, 4, (hundreds,) or the square of 2, (tens,) from the period 6, (hundreds,) and bringing down the next period to the remainder, making 225.

3d. The square A is now to be enlarged by the addition of the 225 remaining yards; and in order that the figure may retain its *square form*, the addition must be made on *two* sides. Now, if the 225 yards be divided by the *length* of the *two* sides, ($20 + 20 = 40$,) the quotient will be the *breadth* of this new addition of 225 yards to the sides *c d* and *b c* of the square A.

But our root already found, = 2 tens, is the length of *one* side of the figure A; we therefore take *double* this root, = 4 tens, for a divisor.

The divisor, 4, (tens,) is in reality 40, and we are to seek how many times 40 is contained in 225, or, which is the same thing, we may seek how many times 4 (tens) is contained in 22, (tens,) rejecting the right hand figure of the dividend, because we have rejected the cipher in the

Questions. — ¶ **208.** How may one side of a square, when the contents are given, be found? Why, in the trial, must we point off the number into periods of two figures each? Illustrate. Why is the first root figure 2 tens? Of what number is 2 tens the root? What may now be formed? How large? How must it be increased? What will be the divisor? — the dividend? — the quotient? Why is the divisor too small? What is the entire divisor? How are the contents of the addition found? What does Fig. II. represent? What is the first method of proof? — the second method?

OPERATION—CONTINUED.

```
 . .
625(25
4
——
45)225
   225
   ———
```

FIG II

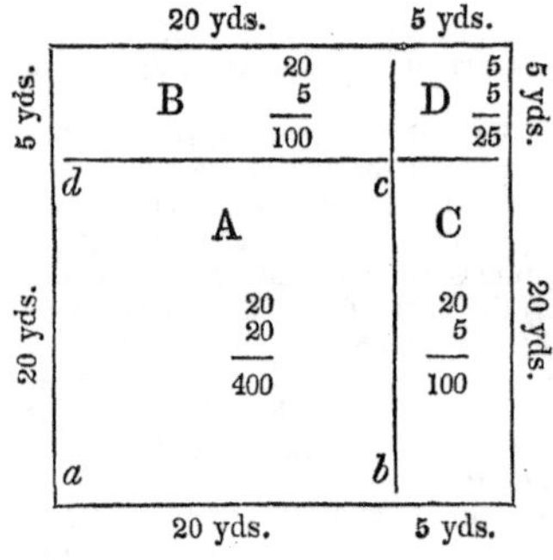

divisor. We find our quotient *that is,* the *breadth* of the addi tion to be 5 yards; but, if we look at Fig II., we shall perceive that this addition of 5 yards to the *two* sides does not complete the square; for there is still wanting, in the corner D, a small square, each side of which is equal to this last quotient, 5; we must, therefore, add this quotient, 5, to the divisor, 40, that is, place it at the right hand of the 4, (tens,) making the length of the whole addition formed by 225 square yards, 45 yards; and then the whole length of the addition, 45 yards, multiplied by 5, the number of yards in the width, will give the contents of the whole addition around the sides of the figure A, which, in this case, being 225 yards, the same as our dividend, we have no remainder, and the work is done. Consequently, Fig. II. represents the floor of a square room, 25 yards on a side, which 625 square yards of carpeting will exactly cover.

The proof may be seen by adding together the several parts of the figure, thus:—

The square A contains 400 yards.
The figure B " 100 "
" " C " 100 "
" " D " 25 "
Proof, 625 "

Or we may prove it by involution, thus:—$25 \times 25 = 625$, as before.

¶ 209. *From this example and illustration we derive the following general*

RULE

FOR THE EXTRACTION OF THE SQUARE ROOT.

I. Point off the given number into periods of two figures each, by putting a dot over the units, another over the hundreds, and so on. These dots show the number of figures of which the root will consist.

II. Find the greatest square number in the left hand period, and write its root as a quotient in division. Subtract the square number from the left hand period, and to the remainder bring down the next period for a dividend.

III. Double the root already found for a divisor; seek how many times the divisor is contained in the dividend, excepting the right hand figure, and place the result in the root, and *also* at the right hand of the divisor; the divisor thus increased will be the length of the whole addition now made to two sides of the square; multiply the divisor, or length of the addition, by the last figure of the root, (the breadth of the addition,) and subtract the contents of the addition thus obtained from the dividend, and to the remainder bring down the next period for a new dividend.

IV. Double the root already found for a new divisor, and continue the operation as before, until all the periods are brought down.

NOTE 1. — As the value of figures, whether integers or decimals, is determined by their distance from the place of units, we must always begin at units' place to point off the given number, and if it be a mixed number, we must point it off *both* ways from units, and if there be but one figure in any period of decimals, a cipher must be added to it. And as the *root* must always consist of as many integers and decimals as there are periods belonging to each in the given number, when it is necessary to carry the operation to a greater degree of exactness by decimals in the root, after all the periods are brought down, two ciphers, a whole period, must be annexed for every decimal figure which we would obtain in the root.

EXAMPLES FOR PRACTICE.

2. What is the square root of 61504?

ILLUSTRATION.

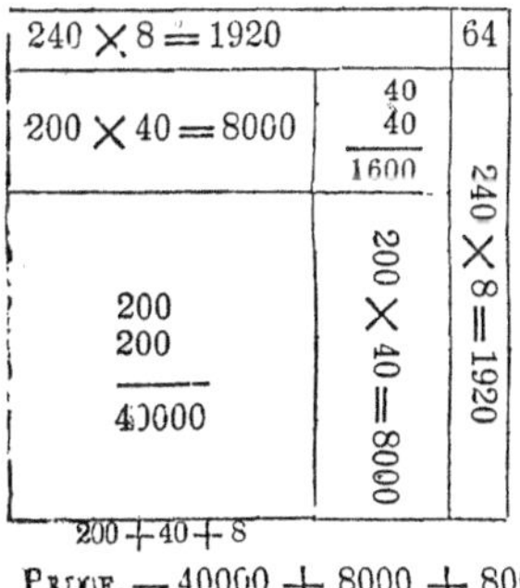

240 × 8 = 1920		64
200 × 40 = 8000	40 40 1600	240 × 8 = 1920
200 200 40000	200 × 40 = 8000	

200 + 40 + 8

OPERATION.

```
     . . .
     61504(248 Ans
     4
    ----
 44)215
    176
   -----
488)3904
    3904
    ----
    0000
```

The pupil will easily illustrate the operation by the annexed diagram.

PROOF. — 40000 + 8000 + 8000 + 1600 + 1920 + 1920 + 64 = 61504, or 248 × 248 = 61504.

Questions. — ¶ 209. What is the general rule for extracting the square root? Where must we begin to point off? What is done when one decimal place is wanting? How is the operation continued, when all the periods are brought down? Why cannot the precise root be found when there is a remainder? How is the root of a fraction obtained? Why? What is done when the terms of the fraction are not exact squares?

3. What is the square root of 43264? *Ans.* 208.
4. What is the square root of 998001? *Ans.* 999.
5. What is the square root of 234‘09? *Ans.* 15‘3.
6. What is the square root of 964‘5192360241? *Ans.* 31‘05671.
7. What is the square root of ‘001296? *Ans.* ‘036.
8. What is the square root of ‘2916? *Ans.* ‘54.
9. What is the square root of 36372961? *Ans.* 6031.
10. What is the square root of 164? *Ans.* 12‘8+

NOTE 2.—In the last example, there was a remainder, after all the figures were brought down. In such cases, the precise root can never be obtained. For, as the operation is continued by annexing ciphers, the last figure of every dividend must be a cipher. But the root figure obtained from this dividend, is also placed at the right hand of the divisor, and consequently is multiplied into itself, and the last figure of the product placed under the cipher, which is the last figure of the dividend, to be subtracted from it. And as the product of no one of the significant figures ends in a cipher, there will always be a remainder.

11. What is the square root of 3? *Ans.* 1‘73+.
12. What is the square root of 10? *Ans.* 3‘16+.
13. What is the square root of 184‘2? *Ans.* 13‘57+.
14. What is the square root of $\frac{4}{9}$?

NOTE 3.—Since, from the rule for multiplying one fraction by another, a fraction is involved by involving its numerator and its denominator, the root of a fraction is obtained by finding the root of its numerator, and of its denominator. *Ans.* $\frac{2}{3}$.

15. What is the square root of $\frac{4}{25}$? *Ans.* $\frac{2}{5}$.
16. What is the square root of $\frac{16}{100}$? *Ans.* $\frac{4}{10}$.
17. What is the square root of $\frac{81}{144}$? *Ans.* $\frac{9}{12}=\frac{3}{4}$.
18. What is the square root of $20\frac{1}{4}$? *Ans.* $4\frac{1}{2}$.

NOTE 4.—When the numerator and denominator are not *exact squares*, the fraction may be reduced to a decimal, and the *approximate* root found.

19. What is the square root of $\frac{3}{4}$ = ‘75? *Ans.* ‘866+.
20. What is the square root of $\frac{35}{42}$. *Ans.* ‘912+.

PRACTICAL EXERCISES IN THE EXTRACTION OF THE SQUARE ROOT.

¶ 210. 1. A general has 4096 men; how many must he place in rank and file to form them into a square? *Ans.* 64.

2. If a square field contains 2025 square rods, how many rods does it measure on each side? *Ans.* 45 rods.

3. How many trees in each row of a square orchard containing 5625 trees? *Ans.* 75.

4. There is a circle whose *area*, or superficial contents, is 5184 feet; what will be the length of the side of a square of equal area? $\sqrt{5184} = 72$ feet, *Ans.*

5. A has two fields, one containing 40 acres, and the other containing 50 acres, for which B offers him a square field containing the

same number of acres as both of these; how many rods must each side of this field measure? *Ans.* 120 rods.

6. If a certain square field measure 20 rods on each side, how much will the side of a square field measure, containing 4 times as much? $\sqrt{20 \times 20 \times 4} = 40$ rods, *Ans.*

7. If the side of a square be 5 feet, what will be the side of one 4 times as large? —— 9 times as large? —— 16 times as large? —— 25 times as large? —— 36 times as large?

Answers, 10 ft.; 15 ft.; 20 ft; 25 ft., and 30 ft.

8. It is required to lay out 288 rods of land in the form of a parallelogram which shall be twice as many rods in length as it is in width.

NOTE 1.—If the field be divided in the middle, it will form two *equal* squares.

Ans. 24 rods long, and 12 rods wide.

9. I would set out, at equal distances, 784 apple trees, so that my orchard may be 4 times as long as it is broad; how many rows of trees must I have, and how many trees in each row?

Ans. 14 rows, and 56 trees in each row.

10. There is an oblong piece of land, containing 192 square rods, of which the width is $\frac{3}{4}$ as much as the length; required its dimensions. *Ans.* 16 by 12.

11. There is a circle, whose diameter is 4 inches; what is the diameter of a circle 9 times as large?

NOTE 2.—A square 4 inches on one side, contains 16 square inches one twice as long, or 8 inches on each side, contains 64 square inches, 4 times 16; one 3 times as long, or 12 inches on each side, contains 144 = 9 times 16 square inches. It may also be shown by geometry, that if the diameter of a circle be doubled, its contents will be increased 4 times; if the diameter be trebled, the contents will be increased 9 times. That is, the contents of squares are to each other as the squares of their sides, and the contents of circles are to each other as the squares of their diameters. Hence, to perform the above example, square the diameter, multiply the square by 9, and extract the square root of the product.

Ans. 1[illegible] inches.

12. There are two circular ponds in a gentleman's pleasure ground; the diameter of the less is 100 feet, and the greater is 3 times as large; what is its diameter? *Ans.* 173·2 + feet.

13. If the diameter of a circle be 12 inches, what is the diameter of one $\frac{1}{4}$ as large? *Ans.* 6 inches.

14. A carpenter has a large *wooden square;* one part of it is 4 feet long, and the other part 3 feet long; what is the length of a pole, which will just reach from one end to the other?

Fig. 1.

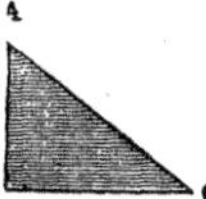

NOTE 3.—A figure of 3 sides is called a triangle, and if one of the corners be a *square corner*, or *right angle*, like the angle at B in the annexed figure, it is called a *right angled triangle*. It is proved by a geometrical demonstration that the square contents of a square formed on the longest side, A C, are equal to the square contents of the two squares, one formed on each of the other two sides, A B, and C B. Thus, Fig. 2, a square formed on A B, the shortest side, will contain 9 square feet, the square on C B

Fig. 2.

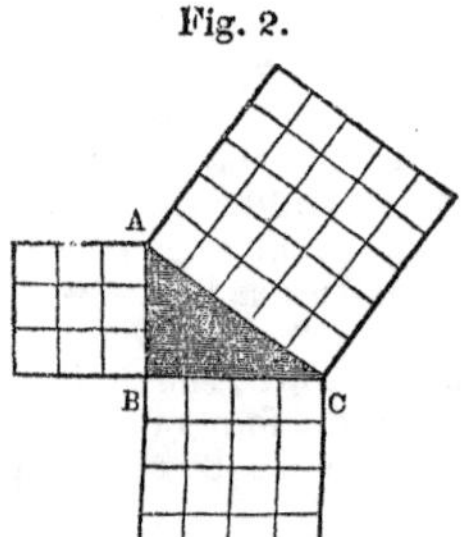

will contain 16 square feet, $9+16=25$ square feet, in both squares. The square on A C contains 25 small squares of the same size as the squares on the other two sides are divided into, or 25 square feet and the square root of 25 will be the length of the longest side, or, *Ans.*, 5 feet.

Hence, if the length of the two short sides are given, *square each, add the squares together, and extract the square square root of the sum; the root will be the length of the long side.*

If the long side, and one of the short sides are given, *square each, subtract the square of the short side from the square of the long side; the square root of the remainder will be the other short side.*

EXAMPLES.

15. If, from the corner of a square room, 6 feet be measured off one way, and 8 feet the other way, along the sides of the room, what will be the length of a pole reaching from point to point?
Ans. 10 feet.

16. A wall is 32 feet high, and a ditch before it is 24 feet wide; what is the length of a ladder that will reach from the top of the wall to the opposite side of the ditch? *Ans.* 40 feet.

17. If the ladder be 40 feet, and the wall 32 feet, what is the width of the ditch? *Ans.* 24 feet.

18. The ladder and ditch given, required the wall.
Ans. 32 feet.

19. The distance between the lower ends of two equal rafters is 32 feet, and the hight of the ridge, above the beam on which they stand, is 12 feet; required the length of each rafter. *Ans.* 20 feet.

20. There is a building 30 feet in length and 22 feet in width, and the eaves project beyond the wall 1 foot on every side; the roof terminates in a point at the centre of the building, and is there supported by a post, the top of which is 10 feet above the beams on which the rafters rest; what is the distance from the foot of the post to the corners of the eaves? and what is the length of a rafter, reaching to the middle of one *side?* —— a rafter reaching to the middle of one *end?* and a rafter reaching to the *corners* of the eaves?

Answers, in order, 20 ft.; 15'62+ ft.; 18'86+ ft.; and 22'36+ feet.

21. There is a field 800 rods long and 600 rods wide; what is the distance between two opposite corners? *Ans.* 1000 rods.

22. There is a square field containing 90 acres; how many rods

---

Questions. — ¶ **210.** How does it affect the contents of a square to double its length? — to treble its length? How does it affect the contents of circles to double or treble their diameters? How will you find the diameter of a circle nine times as large as one of a given diameter? What is a right angled triangle? What is said of the squares on its sides? How shown by Fig. 2? When both short sides are given, how do you find the long side? When the long side, and one short side are given, how do you find the other?

in length is each side of the field? and how many rods apart are the opposite corners? *Answers*, 120 rods, and 169'7 + rods.

23. There is a square field containing 10 acres; what distance is the centre from each corner? *Ans.* 28'28 + rods.

Extraction of the Cube Root.

¶ 211. 1. How many feet in length is each side of a cubic block, containing 125 solid feet?

Solution. — As the solid contents of a cubical body are found, when one side is known, by involving the side to the third power, or cube, (¶ 206,) so when the solid contents are known, we find the length of one side by extracting the cube root, a number, which, taken as a factor 3 times, will produce the given number, (¶ 207.) The cube root of 125 we find by inspection, or by the table, ¶ 206, to be 5. *Ans.* 5 feet.

2. What is the side of a cubic block, containing 64 solid feet? — 27 solid feet? —— 216 solid feet? —— 512 solid feet?
Answers, 4 ft., 3 ft., 6 ft., and 8 ft.

3. Supposing a man has 13824 feet of timber, in separate blocks of 1 cubic foot each; he wishes to pile them up in a cubic pile; what will be the length of each side of such a pile?

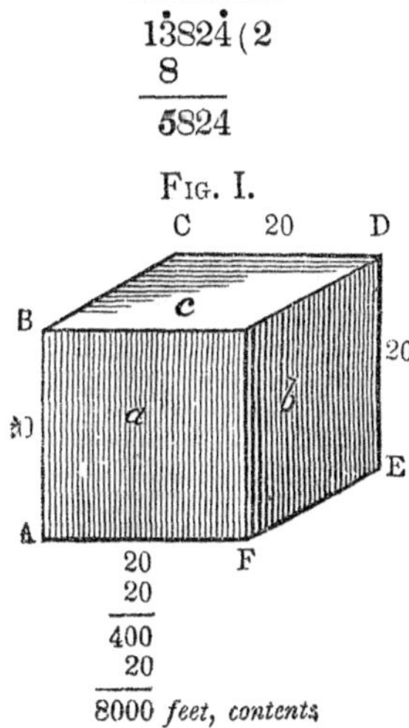

Solution. — It is evident that, as in the former examples, we must find the length of one side of a cubical pile which 13824 such blocks will make by extracting the cube root of 13824. But this number is so large, that we cannot so easily find the root as in the former examples; — we will endeavor, however, to do it by a *sort of trial;* and,

1st. We will try to ascertain the number of figures, of which the root will consist. This we may do by pointing the number off into periods of *three* figures each. For the cube of any figure will contain 3 times as many, or 1 or 2 less than 3 times as many, figures as the number itself. The cube of 2 contains 1 figure; the cube of 5 contains 2 figures; the cube of 9 contains 3 figures; the cube of 10 contains 4 figures, and so on.

Pointing off, we see that the root will consist of two figures, a *ten* and a *unit.* Let us, then, seek for the first figure, or tens of the root, which must be extracted from the left hand period, 13, (thousands.) The greatest cube in 13 (thousands) we find *by inspection*, or by the *table of powers*, to be 8, (thousands,) the root of which is 2, (tens;) therefore, we place 2 (tens) in the root. As the root is one side of a cube, let us form a cube, (Fig. I.,) each side of which shall be regarded 20 feet, expressed by the root now obtained. The contents of this cube are 20 × 20 × 20 = 8000 solid feet, which are now disposed of, and which, consequently, are to be deducted from the whole number of feet, 13824. 8000 taken from 13824 leave 5824 feet. This deduction is most readily performed by subtracting the cubic number, 8, or the cube of 2, (the figure of the root already found,) from the period 13, (thousands,) and bringing down the next period by the side of the remainder, making 5824, as before.

2d. The cubic pile A D is now to be enlarged by the addition of 5824 solid feet, and, in order to preserve the cubic form of the pile, the addition must be made on one half of its sides, that is, on 3 sides, *a*, *b*, and *c*. Now as each side is 20 feet square, its square contents are 400 square feet, and the square contents of the 3 sides are 1200 square feet. Hence, an addition of 1 foot thick would require 1200 solid feet, and dividing 5824 solid feet by 1200 solid feet, the contents of the addition 1 foot thick, and we get the thickness of the addition. It will be seen that the quotient figure must not always be as large as it can be. There might be enough, for instance, to make the three additions now under consideration 5 feet thick, when there would not then be enough remaining to complete the additions.

OPERATION—CONTINUED.

```
                  13824 (24 Root.
                   8
                  ----
Divisor, 1200 ) 5824 Dividend.
                  ----
                  4800
                   960
                    64
                  ----
                  5824
                  ----
                  0000
```

The divisor, 1200, is contained in the dividend 4 times; consequently, 4 feet is the thickness of the addition made to each of the three sides, *a*, *b*, *c*, and 4 × 1200 = 4800, is the solid feet contained in these additions; but there are still 1024 feet left, and if we look at Fig. II., we shall perceive that this addition to the 3 sides does not complete the cube; for there are deficiencies in the 3 corners, *n*, *n*, *n*. Now the *length* of each of these *deficiencies* is the same as the *length* of *each side*, that is, 2 (tens) = 20, and their *width* and *thickness* are each equal to the *last quotient* figure, (4;) their contents, therefore, or the number of feet required to *fill* these deficiencies, will be found by multiplying the *square* of the last quotient figure. (4^2,) = 16, by 20; 16 × 20 =

FIG. II.

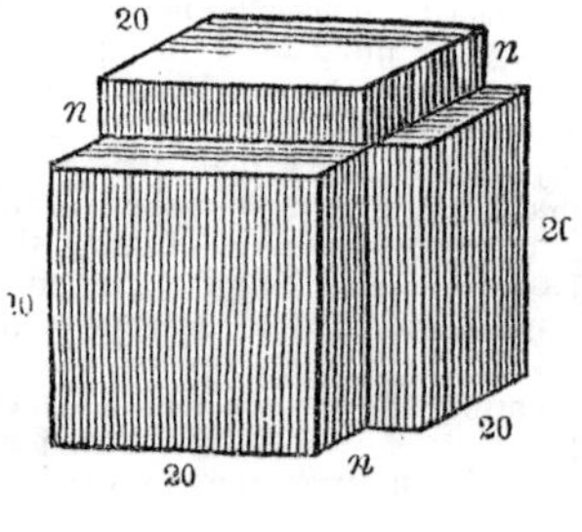

Fig. III.

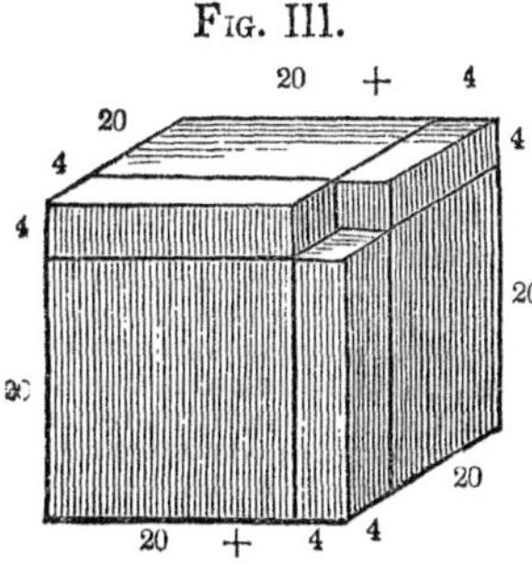

320 solid feet, required for one deficiency, and multiplying 320 by 3, $320 \times 3 = 960$ solid feet, required for the 3 deficiencies, *n*, *n*, *n*.

Looking at Fig. III., we perceive there is still a deficiency in the corner where the last blocks meet. This deficiency is a cube, each side of which is equal to the last quotient figure, 4. The cube of 4, therefore, ($4 \times 4 \times 4 = 64$,) will be the solid contents of this corner, which in Fig. IV. is seen filled.

Now, the sum of these several additions, viz., $4800 + 960 + 64 = 5824$, will make the subtrahend, which, subtracted from the dividend, leaves no remainder, and the work is done.

Fig. IV.

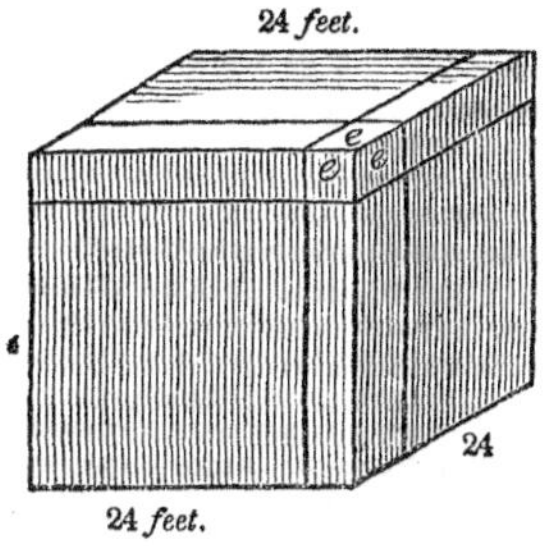

24 feet.

24 feet.

Fig. IV. shows the pile which 13824 solid blocks of one foot each would make, when laid together, and the root, 24, shows the length of one side of the pile. The correctness of the work may be ascertained by cubing the side now found, 24^3, thus, $24 \times 24 \times 24 = 13824$, the given number; or it may be proved by adding together the contents of all the several parts, thus,

Feet.
8000 = contents of Fig. I.
4800 = addition to the sides, *a*, *b*, and *c*, Fig. I.
960 = addition to fill the deficiencies *n*, *n*, *n*, Fig. II.
64 = addition to fill the corner, *e*, *e*, *e*, Fig. IV.

13824 = contents of the whole pile, Fig. IV., 24 feet on each side.

¶ 212. *From the foregoing example and illustration we derive the following*

Questions. — **¶ 211.** How is the length of one side of a cube found, when the contents are known? Why, Ex. 3, is the number pointed off as it is? How many figures in the cube of any number? Illustrate by cubing some numbers. What is 2, the first figure of the root? Of what is it the root? For what is the subtraction? What is to be done with the remainder? On how many sides is it to be added, and why? What is the divisor, 1200? What is the object in dividing? The quotient expresses what? Why should it not be made as large as it can be? What additions are next made, and what are the contents of each? How are the contents found? What deficiency yet remains, and how large? Of what parts of the last figure does the subtrahend consist? Describe Fig. I.; — Fig. II.; — Fig. III.; — Fig. IV. How is the work proved?

RULE

FOR EXTRACTING THE CUBE ROOT.

I. Place a point over the unit figure, and over every third figure at the left of the place of units, thereby separating the given number into as many periods as there will be figures in the root.

II. Find the greatest complete cube number in the left hand period, and place its cube root in the quotient.

III. Subtract the cube thus found from the period taken, and bring down to the remainder the next period for a dividend.

IV. Calling the quotient, or root figure now obtained, so many tens, multiply its square by 3, and use the product for a divisor.

V. Seek how many times the divisor is contained in the dividend, and diminishing the quotient, if necessary, so that the whole subtrahend, when found, may not be greater than the dividend, place the result in the root; then multiply the divisor by this root figure, and write the product under the dividend.

VI. Multiply the square of this root figure by the former figure or figures of the root, regarded as so many tens, and the resulting product by 3, add the product thus obtained, together with the cube of the last quotient, to the former product for a subtrahend.

VII. Subtract the subtrahend from the dividend, and to the remainder bring down the next period for a new dividend, with which proceed as before, till the work is finished.

NOTE 1.—If it happens that the divisor is not contained in the dividend a cipher must be put in the root, and the next period brought down for a dividend.

NOTE 2.—The same rule must be observed for continuing the operation, and pointing off for decimals, as in extracting the square root.

EXAMPLES FOR PRACTICE.

4. What is the cube root of 1860867?

Questions —¶ **212.** What is the general rule? —note 1? —note 2? —note 3?

OPERATION.

$$\dot{1}86\dot{0}86\dot{7}\,(\,123\ \textit{Ans.}$$
$$1$$

$$10^2 \times 3 = 300)\,860 \text{ first Dividend.}$$

$$600$$
$$2^2 \times 10 \times 3 = 120$$
$$2^3 = 8$$

$$728 \text{ first Subtrahend.}$$

$$120^2 \times 3 = 43200)\,132867 \text{ second Dividend.}$$

$$129600$$
$$3^2 \times 120 \times 3 = 3240$$
$$3^3 = 27$$

$$132867 \text{ second Subtrahend.}$$

$$000000$$

5. What is the cube root of 373248? *Ans.* 72.
6. What is the cube root of 21024576? *Ans.* 276.
7. What is the cube root of 84·604519? *Ans.* 4·39.
8. What is the cube root of ·000343? *Ans.* ·07.
9. What is the cube root of 2? *Ans.* 1·25+.
10. What is the cube root of $\frac{8}{27}$? *Ans.* $\frac{2}{3}$.

NOTE 3. — The cube root of a fraction is the cube root of the numerator divided by the cube root of the denominator. (¶ 209.)

11. What is the cube root of $\frac{125}{216}$? *Ans.* $\frac{5}{6}$.
12. What is the cube root of $\frac{343}{1728}$? *Ans.* $\frac{7}{12}$.
13. What is the cube root of $\frac{1}{500}$? *Ans.* ·125+.
14. What is the cube root of $\frac{1}{125}$? *Ans.* $\frac{1}{5}$

PRACTICAL EXERCISES IN EXTRACTING THE CUBE ROOT.

¶ 213. 1. What is the side of a cubical mound, equal to one 288 feet long, 216 feet broad, and 48 feet high? *Ans* 144 feet.

2. There is a cubic box, one side of which is 2 feet; how many solid feet does it contain? *Ans.* 8 feet.

3. How many cubic feet in one 8 times as large? and what would be the length of one side?

Ans. 64 solid feet, and one side is 4 feet.

4. There is a cubical box, one side of which is 5 feet, what would be the side of one containing 27 times as much? —— 64 times as much? —— 125 times as much? *Ans.* 15, 20, and 25 feet.

5. There is a cubical box, measuring 1 foot on each side; what is the side of a box 8 times as large? —— 27 times? —— 64 times?
Ans. 2, 3, and 4 feet.

NOTE. — It appears from the above examples that the sides of cubes are as the cube roots of their solid contents, and their solid contents as the cubes of their sides. It is also true that if a globe or ball have a certain contents, the contents of one whose diameter is double are 8 times as great, or having a treble diameter are 27 times as great, and so on; that is, the contents are proportional to the cubes of their diameters. The same proportion is true of the *similar sides*, or of the *diameters* of *all* solid figures of similar forms.

6. If a ball, weighing 4 pounds, be 3 inches in diameter, what will be the diameter of a ball of the same metal, weighing 32 pounds? $4 : 32 :: 3^3 : 6^3$. *Ans.* 6 inches.

7. If a ball 6 inches in diameter weigh 32 pounds, what will be the weight of a ball 3 inches in diameter? *Ans.* 4 lbs.

8. If a globe of silver, 1 inch in diameter, be worth $6, what is the value of a globe 1 foot in diameter? *Ans.* $10368.

9. There are two globes; one of them is 1 foot in diameter, and the other 40 feet in diameter; how many of the smaller globes would it take to make 1 of the larger? *Ans.* 64000.

10. If the diameter of the sun is 112 times as much as the diameter of the earth, how many globes like the earth would it take to make one as large as the sun? *Ans.* 1404928.

11. If the planet Saturn is 1000 times as large as the earth, and the earth is 7900 miles in diameter, what is the diameter of Saturn?
Ans. 79000 miles.

12. There are two planets of equal density; the diameter of the less is to that of the larger as 2 to 9; what is the ratio of their solidities? *Ans.* $\frac{8}{729}$; or, as 8 to 729.

¶ 214. Review of Involution and Evolution.

Questions. — What is involution? What are powers? How are the different powers represented? How is a number involved? What is evolution? What is a root? How do you find the square root, or the cube root of a number? What is a rational, and what a surd number? How is the square root indicated? — the cube root? Give briefly the solution of the example in the extraction of the square root; — rule. How are decimals pointed off? How is the operation continued, when there is a remainder? Why cannot the precise root be ascertained? How is the square root of a vulgar fraction found? What is said of the relation between the sides and contents of squares? — the diameters and

---

Questions. — ¶ 213. What proportion exists between the sides of cubes, and their solid contents? Illustrate. What between the diameters of globes and their contents? If you increase the diameter of a ball 5 times, how much are its contents increased?

contents of circles? — the squares on the sides of a right-angled triangle? Repeat briefly the solution of the example in cube root; — the rule. What is said of the relation of the sides of cubes to the contents? — of the diameters of globes to their contents?

EXERCISES.

1 What is the difference of the contents of 6 fields, each 20 rods square, and 1 field 50 rods square? *Ans.* 100 square rods.

2. What is the difference between 56 cubical stacks of hay, each 10 feet on a side, and 1 stack 40 feet on a side? *Ans.* 8000 solid feet.

3. How many times larger is a circular pond, 1 mile in diameter, than one that is 40 rods in diameter? *Ans.* 64 times.

4. What is one side of a cubical pile of wood which contains 4 cords? *Ans.* 8 feet.

5. What is one side of a cubical pile of bricks which will lay up the walls of a house 36 feet high and 16 inches thick for the first 12 feet, 12 inches the next 12, and 8 inches the upper 12, the house being 60 feet long and 34 wide on the outside, no allowances being made for windows, doors, &c.? What are the solid contents? *Ans.* to the last, 6013 cu. ft., 576 cu. in.

NOTE. — The principal object in evolution is to find one side of a square or of a cube, when the contents are known, or to extract the square and cube roots. There are methods of demonstrating these operations different from those here given, which are preferable in some respects, but they are deficient in one important particular — intelligibleness to those for whom they are designed. In a "higher arithmetic" they might be appropriate.

Other roots may be extracted arithmetically, but the methods of demonstrating the operations, even where any are given, are difficult of comprehension. The fourth root, however, may be found by taking the square root of the square root, the sixth root by taking the square root of the cube root, and so of many other roots. Any root is easily taken by what are called logarithms, used in the more advanced departments of mathematics.

ARITHMETICAL PROGRESSION.

¶ 215. 1. A teamster starts with 5 barrels of flour; he passes by 4 mills, at each of which he takes on 3 barrels; how many barrels has he then?

SOLUTION. — He has 8 barrels after the first addition, 11 after the second, 14 after the third, and 17 after the fourth. *Ans.* 17 barrels.

2. A peddler having 17 hats, sold 3 at each of 4 stores how many had he left?

SOLUTION. — He had 14 after the first sale, 11 after the second, 8 after the third, and 5 after the fourth. *Ans.* 5 hats.

A series of numbers increasing by a constant addition or

decreasing by a constant subtraction of the same number, is called an *Arithmetical Progression* or *series.*

The first of the above examples is called an ascending, the second a descending series.

NOTE 1.—The numbers which form the series are called the *terms* of the series. The *first* and *last* terms are the *extremes*, and the other terms are called the *means*.

There are five things in an arithmetical progression, any *three* of which being given, the other *two* may be found:—

1st. The *first* term.
2d. The *last* term.
3d. The *number* of terms.
4th. The *common difference.*
5th. The *sum* of all the terms.

NOTE 2.—The *common difference* is the number added or subtracted at one time.

¶ 216. *One of the extremes, the common difference, and the number of terms being given, to find the other extreme.*

1. A man bought 100 yards of cloth, giving 4 cents for the first yard, 7 cents for the second, 10 cents for the third, and so on, with a common difference of 3 cents; what was the cost of the last yard?

SOLUTION.—We add 3 to 4 cents, ($4+3=7$,) to get the price of the second yard, 3 to 7 to get the price of the third yard, and so on, thus making 99 additions to 4, of 3 cents each; or, we may take 3, 99 times, (the multiplication being a short way of performing the 99 additions,) and add the product to 4, for the price of the last yard, 3×99; or, since either factor may be the multiplier, $99\times 3=297$, and $4+297=301$ cents, the price of the last yard. *Ans.* 301 cents.

NOTE 1.—The prices, 4, 7, 10, 13 cents, &c., are an ascending series, which has as many terms as there are yards, namely, 100; 3 is the common difference, and 4 the first term, to which 99 times 3 must be added to find the price of the last yard, or the last term. It is added 1 time less than the number of terms, since 4 is the price of the first yard without any addition.

Hence, *To find the last term of an ascending series when the first term, common difference, and number of terms are given,*

RULE.

Multiply the common difference by the number of terms less one, to get the sum of the additions, and add this sum to the first term; the amount will be the last term.

Questions.—¶ **215.** How is Ex. 1 explained? — Ex. 2? What are the extremes? — the means? — the terms? How many, and what things are there, of which, if three are given, the others may be found? What is the common difference?

NOTE 2. — If the same things are given of a descending series, we must evidently take the sum of the subtractions from the first term to find the last. In the same manner we may find the first term of an ascending series when the last term and the other things named are given; but having these things given of a descending series, we find the first term by the rule above for finding the last term of an ascending series.

EXAMPLES FOR PRACTICE.

2. There are 23 pieces of land, the first containing 95 acres, the second 91, the third 87, and so on, decreasing by a common difference of 4; what is the number of acres in the last piece? *Ans.* 7.

3. The first term of a series is 6, the common difference is 3, and the number of terms is 57; what is the last term? *Ans.* 174.

4. The last term of a series is 117, the common difference is 8, and the number of terms is 15; what is the first term? *Ans.* 5.

5. The last term is 6, the number of terms 21, and the common difference 10; what is the first term? *Ans.* 206.

Simple Interest by Progression.

¶ 217. 1. A man puts out $10, at 6 per cent., simple interest; to what does it amount in 20 years?

SOLUTION. — The first sum is $10, the amount at the end of the first year is $10‘60, at the end of the second year $11‘20, increasing each year by the constant addition of $‘60. Hence, simple interest is a case of arithmetical progression, the principal being the first term, the interest for one year being the common difference, the number of terms one more than the number of years, since there is one term, the principal, at the commencement of the first year, and one term, the amount for a year, at its close, and the last term, which we wish to find, is the amount for the number of years. To find the last term, or this amount, *multiply the interest for 1 year by the number of years,* (one less than the number of terms,) *and add the product to the first term.* *Ans.* $22.

2. Two lads, at 14 years of age, commence labor for themselves; the one lays up nothing, but the other, by prudence, lays up $300 by the time he is 20 years old, which he puts out at 7 per cent. simple interest; afterwards, each earns his living, and no more; at the age of 70 the one is worth nothing, and comes upon public charity; what is the other worth at that age? *Ans.* $1350.

Questions. — ¶ **216.** What things are given, and what are required, in ¶ 216? How many cases may there be, and what are they? What are given in Ex. 1? How much is the first term increased to make the last? Why are only 99 times 3 added? Give the rule. To what case does it apply? What is done in each case, when other things are given?

¶ **217.** How does it appear that simple interest is a case of progression? What things, according to ¶ 216, are given, and what is required? Why is there one more term than the number of years? How is simple interest performed by progression?

3. What will a watch, purchased at 21 for $25, cost an individual by the time he is 75, reckoning nothing for repairs but simple interest at 6 per cent, on the purchase money? —— at 8 per cent.?

Ans. to the last, $133.

¶ 218. *The extremes and the number of terms given, to find the common difference.*

1. The prices of 100 yards are in arithmetical progression, the first being 4, the last being 301 cents; what is the common increase of price on each succeeding yard?

Solution. — As the first yard costs 4 cents, 297 cents have been added to 4 for the price of the last yard, at 99 times, and dividing the number added at 99 times by 99, we get the number added at 1 time. Hence,

RULE.

Divide the whole number added or subtracted, by the number of additions or subtractions, that is, the difference of the extremes by the number of terms less 1, and the quotient is the number added or subtracted at one time, or the common difference.

EXAMPLES FOR PRACTICE.

2. If the extremes be 5 and 605, and the number of terms 151, what is the common difference? *Ans.* 4.

3. A man had 8 sons, whose ages differed alike; the youngest was 10 years old, and the eldest 45; what was the common difference of their ages? *Ans.* 5 years.

Note. — If the extremes and common difference are given, we may find the number of terms by dividing the difference of the extremes by the common difference, and adding 1 to the quotient.

4. The extremes are 5 and 1205, and the common difference 8; what is the number of terms? *Ans.* 151.

¶ 219. *The extremes and the number of terms being given, to find the sum of all the terms.*

1. What is the amount of the ascending series, 3, 5, 7, 9, 11, 13, 15, 17, 19?

Solution. — The sum may be found by adding together the terms, but in an extended series this process would be tedious. We will therefore seek for a shorter method; and first, will write down the terms of

---

Questions. — ¶ 218. What are given and what required, ¶ 218? Explain how the common difference is found, Ex. 1. Give the rule. How is the number of terms found when the extremes and the common difference are given?

the series in order, and beginning with the last, write the terms of the same series under these, placing the last term under the first, the next to the last under the second, the third from the last under the third, and so on, thus: —

3	5	7	9	11	13	15	17	19
19	17	15	13	11	9	7	5	3
22	22	22	22	22	22	22	22	22

Adding together each pair, we see that the sums are alike, and the amount of the whole is as many times 22, the first sum, as there are terms in either series, which is 9. $22 \times 9 = 198$, the number in both series, and $198 \div 2 = 99$ must be the sum of the first series, which we wish to find. But 22 is the sum of the extremes of the series; hence, *when the extremes and the number of terms are given to find the sum of the terms,*

RULE.

Multiply the sum of the extremes by the number of terms, and half the product will be the sum of the terms.

EXAMPLES FOR PRACTICE.

2. If the extremes be 5 and 605, and the number of terms 151, what is the sum of the series? *Ans.* 46055.
3. What is the sum of the first 100 numbers, in their natural order, that is, 1, 2, 3, 4, &c.? *Ans.* 5050.
4. How many times does a common clock strike in 12 hours? *Ans.* 78.

Annuities by Arithmetical Progression.

¶ 220. An annuity, (from the Latin word annus, meaning a year,) is a uniform sum, due at the end of every year. When payment is not made at the end of the year, the annuity is said to be in arrears, and the sums of the annuities should draw interest just as any other debts not paid when due. When on simple interest, the several years' annuities, with the interest on each, form an arithmetical progression, and the calculation to ascertain the whole sum due, is — finding the sum of an arithmetical series; thus: —

Questions. — ¶ 219. What are given, and what is required, ¶ 219? How might the sum be found? What difficulty in this method? What is the process by which a shorter method is found? What fact do we discover from the additions, Ex. 1? What does the product express, and why is it divided by 2? What is the quotient? Why is 22 the sum of the extremes? Give the rule.

1. A man, whose salary is $100 a year, does not receive anything till the end of 8 years; what was then his due, simple interest on the sums in arrears at 6 per cent.?

Solution. — The first year's salary not being paid till 7 years after it is due, since it was due at the end of the first year, is on interest years. The interest of $100, 7 years, is $42, and $100 + 42 = $142 which he should receive on account of his first year's salary. His second year's salary, on interest 6 years, will amount to $136, his third year's salary will amount to $130, and so on, decreasing uniformly by $6, the interest of $100 for a year, till the last year, when the salary, being paid at the end of the year for which it has accrued, will not be on interest, but will yield him $100. The sums, $142, $136, $130 $124, $118, $112, $106, $100, form a descending arithmetical progression, and to find the sum due, multiply the sum of the extremes by the number of terms, and take half the product, thus: —

$142 + $100 = $242; and $242 × 8 = $1936, which ÷ 2 = $968, *Ans.*

EXAMPLES FOR PRACTICE.

2. A soldier of the revolution did not establish his claim to a pension of $96 a year till 10 years after it should have begun; what was then his due, simple interest on the sums in arrears at 6 per cent.? *Ans.* $1219'20.

3. A man uses tobacco at an expense of $5 a year from the age of 18 till the age of 79, when he dies, leaving to his heirs $300; what might he have left them if he had dispensed with the worse than useless article, and loaned the money, which it cost him, at the end of each year, for 7 per cent., simple interest? *Ans.* $1245'50.

4. A and B have the same income, but the expenses of A are $210 a year, and those of B are $250; at the end of 40 years B is worth $1500; what is A worth, having loaned what he saved more than B at 7 per cent. simple interest at the end of each year? *Ans.* $5284.

5. Refer to ¶ 164, Ex. 8: what would the merchant gain if he continued in trade 31 years to borrow money instead of purchasing on credit, loaning the money saved at the end of each year at 7 per cent. simple interest? *Ans.* $20151'70 nearly.

EXERCISES.

¶ 221. 1. If a triangular piece of land, 30 rods in length, be 20 rods wide at one end, and come to a point at the other, what number of square rods does it contain? *Ans.* 300.

Questions. — ¶ 220. What is an annuity? Why so called? When are annuities in arrears? Why should they then draw interest? When will they form an arithmetical progression? What is the calculation to find the whole sum due? Why, Ex. 1, was the first year's salary on interest 7 years? Show how long each year's salary was on interest. Why was not the last year's salary on interest? Why do the sums form a descending series? How may the sum be found? Why is not the number of terms one more than the number of years, as in ¶ 217?

2. A debt is to be discharged at 11 several payments, in arithmetical series, the first to be \$5, and the last \$75; what is the whole debt? —— the common difference between the several payments?

Ans. Whole debt, \$440; common difference, \$7.

3. What is the sum of the series 1 3, 5, 7, 9, &c., to 1001?

Ans. 251001.

NOTE. — The number of terms must first be found.

4. A man bought 100 yards of cloth in arithmetical series; he gave 4 cents for the *first* yard, and 301 cents for the *last* yard; what was the amount of the whole? *Ans.* \$152'50.

5. What annuity, in 20 years, at 6 per cent. simple interest, will amount to \$1570? *Ans.* \$50.

6. What is the sum of the arithmetical series, 2, 2½, 3, 3½, 4, 4½, &c., to the 50th term inclusive? *Ans.* 712½.

7. What is the sum of the decreasing series, 30, 29⅔, 29⅓, 29, 28⅔, &c., down to 0? *Ans.* 1365.

8. A laboring female was able to put \$30, at the end of each year, in the savings bank, at 5 per cent., simple interest, from the age of 18 till 77, when she died; how much had she become worth?

Ans. \$4336'50.

GEOMETRICAL PROGRESSION.

¶ 222. 1. A man, having 5 acres of land, doubled the quantity at the end of each year for 4 years; how many acres had he then?

SOLUTION. — Having 5 acres at first, he had 2 times 5, or 10, at the end of the first year, 2 times 10, or 20, at the end of the second year, 2 times 20, or 40, at the end of the third year, and 2 times 40, or 80, at the end of the fourth year. *Ans.* 80 acres.

2. A lady, having \$80, traded at 4 stores, expending one half at the first, half of what she had left at the second, and so on, expending half the money in her possession at each store till the last: what had she left?

SOLUTION. — She leaves the first store, to which she came with \$80 ÷ 2 = \$40, the second with \$40 ÷ 2 = \$20, the third with \$20 ÷ 2 = \$10, the fourth with \$10 ÷ 2 = \$5. *Ans.* \$5.

Any series of numbers like 5, 10, 20, 40, 80, increasing by the same multiplier, or like 80, 40, 20, 10, 5, decreasing by the same divisor, is called a geometrical progression.

The multiplier or divisor is called the ratio. The first and last terms are called the extremes.

24*

The first is called an increasing, the second is called a decreasing geometrical series.

NOTE. — As in arithmetical, so also in geometrical progression, there are five things, any *three* of which being given, the other *two* may be found. —

1st. The *first* term.
2d. The *last* term.
3d. The *number* of terms.
4th. The *ratio*.
5th The *sum* of all the terms.

¶ 223. *The first term, number of terms, and ratio of an increasing geometrical series being given, to find the last term.*

1. A man agreed to pay for 13 valuable houses, worth $5000 each, what the last would amount to, reckoning 7 cents for the first, 4 times 7 cents for the second, and so on, increasing the price 4 times on each to the last; did he gain or lose by the bargain, and how much?

SOLUTION. — We multiply 7 cents, the sum reckoned for the first house, by 4, to get the sum reckoned for the second house, and this by 4 to get the sum reckoned for the third house, and so on, multiplying 12 times by 4. But to multiply twice by 4 is the same as to multiply once by 16, which is the second power of 4, or 4 times 4. Thus, $7 \times 4 = 28$, and $28 \times 4 = 112$. So, also, $7 \times 16 = 112$. Hence, multiplying 7 by the twelfth power of 4 is the same as multiplying by 4 12 times. And 4^{12}, that is, the twelfth power of 4, is 16777216; and 7×16777216, (using, if we choose, the larger factor for the multiplicand, ¶ 21,) produces 117440512 cents, = $1174405'12; the houses were worth $5000 × 13 = $65000, and $1174405'12 — $65000 = $1109405'12, loss, *Ans.*

Hence, *when the first term, number of terms, and ratio of an increasing series are given, to find the last term,*

RULE.

Multiply the first term by the ratio raised to a power one less than the number of terms.

NOTE 1. — To get a high power of a number, it is convenient to write down a few of the lower powers, and multiply them together, thus: powers of 4.

1st power.	2d power.	3d power.	4th power.	5th power.	6th power.	7th power.
4	16	64	256	1024	4096	16384, &c.

Now the 7th power, multiplied by the 5th power, will produce the 12th power, as also the 6th by the 4th, and the product by the 2d; the 5th by the 4th, and the product by the 3d; the 7th by the 4th, and the product by the 1st, &c Multiplying all the powers now written, will produce the 28th; all but the 5th, will produce the 23d; all but the 2 1, will produce the 26th, &c.

Questions. — ¶ **222.** What is a geometrical progression? — an increasing series? — a decreasing series? — the extremes? — the ratio? What five things are there, of which, if three are given, the others may be found

EXAMPLES FOR PRACTICE.

2. A man plants 4 kernels of corn, which, at harvest, produce 32 kernels: these he plants the second year; now, supposing the annual increase to continue 8 fold, what would be the produce of the 15th year, allowing 1000 kernels to a pint?

NOTE 2. — The 4 kernels planted is the first term, and the 32 kernels harvested the second term, both within the first year.

Ans. 2199023255'552 bushels.

3. Suppose a man had put out one cent at compound interest in 1620, what would have been the amount in 1824, allowing it to double once in 12 years? $2^{17} = 131072$. *Ans.* $1310'72.

NOTE 3. — When the ratio, the number of terms, and the last term of a decreasing series is given, the first term is evidently found by the same rule. But when the same things are given of a decreasing series, as those of the ascending series required by the rule, that is, the first term, ratio, and number of terms, we must divide the first term by the ratio raised to a power which is one less than the number of terms. In the same way we may find the first term of an increasing series, when the ratio, number of terms, and last term are given.

4. If the last term of a decreasing series be 5, the ratio 3, and the number of terms 7, what is the first term? *Ans.* 3645.

5. If the first term of a decreasing series be 10935, the ratio 3 and the number of terms 8, what is the last term? *Ans.* 5.

6. If the last term of an increasing series be 196608, the number of terms 17, and the ratio 2, what is the first term? *Ans.* 3.

NOTE 4. — When the first and last terms, and the ratio are given, to find the number of terms, we may divide the greater term by the less, the quotient by the ratio, and so on, continually dividing by the ratio till nothing remains: the number of divisions will be equal to the number of terms.

7. The first term is 7, the ratio 10, and the last term 700000000; what is the number of terms? *Ans.* 9

Compound Interest by Progression.

¶ 224. 1. To what will $40 amount in 4 years, compound interest at 6 per cent.?

Questions. — **¶ 223.** What are given, and what is required, in ¶ 223? How is the last term found, as first described? What may be done instead of this? Why? How may a high power be obtained? How the 20th power? — the 16th power? — the 27th power? — the 11th power? What is the rule? Give the substance of note 3; — of note 4.

OPERATION.

$40‘, *prin., or 1st term*
1‘06
———
240
40
———
42‘40, *2d term*
1‘06
———
25440
42‘40
———
44‘9440, *3d term.*
1‘06
———
269664O
449440
———
47‘640640, *4th term.*
106
———
285843840
47640640
———
50‘49907840, *5th term.*

SOLUTION. — The amount of $40 for one year is once $40 + $\frac{6}{100}$ of $40, or 1‘06 ($\frac{106}{100}$) of $40, and to obtain it, we multiply $40 by 1‘06. This product, multiplied by 1‘06, gives the amount for 2 years. Hence, compound interest is a case of an increasing geometrical progression, of which there are given the first term, or principal, the ratio, or the amount of $1 for 1 year, and the number of terms, which is one more than the number of years, there being 2 terms the first year, the principal at the commencement, and the amount with one year's interest at the close; and we are to find the last term — the amount for the time — by the rule in the last ¶. The several terms, it will be seen, increase by the common multiplier, 1‘06.

Ans. $50‘499 +.

NOTE 1. — The powers of the ratio may usually be found in the table, ¶ 161, since they are the same with the amounts of $1 for the number of years which indicates the power of the ratio that we wish. It appears also that the only difference in finding the amount of a sum at compound interest by the table, and by progression, is the order in which we take the factors. By ¶ 161, we multiply the amount of $1 for the time by the number of dollars; by progression, we multiply the number of dollars by the ratio raised to a power denoted by the number of years, or the amount of $1 for the time.

EXAMPLES FOR PRACTICE.

2. What is the amount of 40 dollars, for 11 years, at 5 per cent. compound interest? *Ans.* $68‘413 +.

3. What is the amount of $6, for 4 years, at 10 per cent., compound interest? *Ans.* $8‘784$\frac{6}{10}$.

4. In what time will $1000 amount to $1191‘016, at 6 per cent compound interest?

NOTE 2. — The case is evidently one of finding the number of terms, (one more than the number of years,) when the ratio and the first and last terms are given, ¶ 223, note 4. *Ans.* 3 years.

Questions. — **¶ 224.** How does it appear from the illustration that compound interest is a case of progression? What are the things given? — to find what? What is the ratio, Ex. 1? Why? How may the powers of the ratio usually be found? Why? What does it appear are given, and what is required, Ex. 4?

Compound Discount.

¶ 225. 1. What is the present worth of $304‘899, due 4 years hence without interest, money being worth 6 per cent. compound interest?

Solution. — We find, by the table, ¶ 161, that $1, in 4 years, amounts to $ 1‘26247, and as many times as this amount of $1 is contained in the given sum, so many dollars it will be worth; for it is worth a sum, which, put at compound interest 4 years, would amount to it, and dividing the amount of the number of dollars by the amount of one dollar, — we have the number of dollars, or, *Ans.* 241‘509 +.

Note. — The case is evidently one of a geometrical progression, in which the ratio, (1‘06,) the number of terms, (5,) and the greater term are given, to find the less, as in ¶ 223, note 3.

TABLE

Showing the present worth of $1, or £1, from 1 year to 40, allowing compound discount, at 5 and 6 per cent.

Years.	5 per cent.	6 per cent.	Years.	5 per cent.	6 per cent.
1	‘952381	‘943396	21	‘358942	‘294155
2	‘907029	‘889996	22	‘341850	‘277505
3	‘863838	‘839619	23	‘325571	‘261797
4	‘822702	‘792094	24	‘310068	‘246979
5	‘783526	‘747258	25	‘295303	‘232999
6	‘746215	‘704961	26	‘281241	‘219810
7	‘710681	‘665057	27	‘267848	‘207368
8	‘676839	‘627412	28	‘255094	‘195630
9	‘644609	‘591898	29	‘242946	‘184557
10	‘613913	‘558395	30	‘231377	‘174110
11	‘584679	‘526788	31	‘220359	‘164255
12	‘556837	‘496969	32	‘209866	‘154957
13	530321	‘468839	33	‘199873	‘146186
14	‘505068	‘442301	34	‘190355	‘137912
15	‘481017	‘417265	35	‘181290	‘130105
16	‘458112	‘393646	36	‘172657	‘122741
17	‘436297	‘371364	37	‘164436	‘115793
18	‘415521	‘350344	38	‘156605	‘109239
19	‘395734	‘330513	39	‘149148	‘103056
20	‘376889	‘311805	40	‘142046	‘097222

¶ 226. *The extremes and the ratio given to find the sum of the series.*

Questions. — ¶ 225. What is compound discount? How is the present worth found? Like what case in a geometrical progression is it?

1. A man bought 4 yards of cloth, giving 2 cents for the first yard, 6 for the second, 18 for the third, and 54 for the fourth; what does he pay for all?

SOLUTION. — We may add together the prices of the several yards thus:

$$2+6+18+54=80.$$

But in a lengthy series, this process would be tedious; we will therefore seek for a shorter method. Writing down the terms of the series, we multiply the first term by the ratio, and place the product over the second, to which it will be equal, since the second term is the product of the first into the ratio. Multiply, also, the second term, placing the product over its equal, the third; multiply the third, placing the product over the fourth; multiply the fourth, and place the product at the right of the last product thus:

Second series,		6	18	54	162
First series,	2	6	18	54	
		0	0	0	

162
2
———
2) 160, *twice the first series.*
———
80, *sum of the first series.*

The second series is three times the first series, and subtracting the first from it, there will remain twice the first series. But the terms balance each other, except the first term of the first series, the sum of which we wish to find, and the last term of the second, which is 3 times the last term of the series whose sum we wish. Subtracting the former from the latter, we have left 160, twice the sum of the first, which dividing by 2, the quotient is 80, sum of the series required. Hence,

RULE.

Multiply the larger term by the ratio, and subtract the less term from the product, divide the remainder by the ratio less 1; the quotient will be the sum of the series.

EXAMPLES FOR PRACTICE.

2. If the extremes be 4 and 131072, and the ratio 8, what is the sum of the series? *Ans.* 149796.

3. What is the sum of the decreasing series, 3, 1, $\frac{1}{3}$, $\frac{1}{9}$, $\frac{1}{27}$, &c *extended to infinity?*

Questions. — ¶ **226.** What are given, to find what, ¶ 226? How might the sum be found? What difficulty in this? What is the manner of proceeding to find a shorter method? Give the rule. Explain the reason of the rule What is an infinite series, and what its last term?

NOTE. — Such a series is called an infinite series, the last term of which is so near nothing that we regard it 0; hence when the extremes are 3 and 0, and the ratio 3, what is the sum of the series?

Ans. $4\frac{1}{2}$.

4. What is the value of the infinite series, $1 + \frac{1}{4}, + \frac{1}{16}, + \frac{1}{64}$, &c.? *Ans.* $1\frac{1}{3}$.

5. What is the value of the infinite series, $\frac{1}{10} + \frac{1}{100} + \frac{1}{1000} + \frac{1}{10000}$, &c., or, what is the same, the decimal '11111, &c., continually repeated? *Ans.* $\frac{1}{9}$.

6. What is the value of the infinite series, $\frac{2}{100} + \frac{2}{10000}$, &c decreasing by the ratio 100, or, which is the same, the repeating decimal '020202, &c.? *Ans.* $\frac{2}{99}$.

¶ 227. *The first term, ratio, and number of terms given to find the sum of the series.*

1. A lady bought 6 yards of silk, agreeing to pay 5 cents for the first yard, 15 for the second, and so on, increasing in a three fold proportion; what did the whole cost?

SOLUTION. — We may find the prices of the several yards, and add them together, or, having found the last term by ¶ 223, we can find the sum by the last ¶. But our object is to find a still more expeditious method. Let us find the several terms and write them down as a first series, and below it write a series which we will call the second, having 1 for the first term, and the same number of terms, thus:

First series,	5	15	45	135	405	1215	
							6th power of ratio.
Third series,		3	9	27	81	243	729
Second series,	1	3	9	27	81	243	

729 — 1 = 728, which ÷ 2 = 364, and 364 × 5 = 1820 cents.

Now multiplying the second series by the ratio, 3, and writing the products as directed in the last ¶, we have a series three times the second The last term of the third series, it must be carefully noticed, is the 6th power of 3, the ratio, the power denoted by the number of terms. Subtracting the second series from the third, which is done by taking 1 from the last term of the third, the other terms balancing, 729 — 1 = 728, we have twice the second series, and dividing 728 by 2, 728 ÷ 2 = 364, we have once the second series. Now the first series, the sum of which is required, is 5 times the second, since, as the first term is 5 times greater, each term is 5 times greater than the corresponding term of the second series; and multiplying 364, the sum of the second, by 5, we have the required sum, or 1820 cents = $18'20, *Ans.*

Hence, *the first term, ratio, and number of terms being given to find the sum of the series,*

RULE.

Raise the ratio to a power whose index is equal to the

number of terms, from which subtract 1, and divide the remainder by the ratio less 1; the quotient is the sum of a series with 1 for the first term; then multiply this quotient by the first term of any required series; the product will be its amount.

EXAMPLES FOR PRACTICE.

2. A gentleman, whose daughter was married on a new year's day, gave her a dollar, promising to triple it on the first day of each month in the year; to how much did her portion amount?

Applying this rule to the example, $3^{12} = 531441$, and $\frac{531441-1}{3-1} \times$ $= 265720$. *Ans.* \$265,720.

3. A man agrees to serve a farmer 40 years without any other reward than 1 kernel of corn for the first year, 10 for the second year, and so on, in tenfold ratio, till the end of the time; what will be the amount of his wages, allowing 1000 kernels to a pint, and supposing he sells his corn at 50 cents per bushel?

$$\frac{10^{40}-1}{10-1} \times 1 = \left\{ \begin{array}{c} 1,111,111,111,111,111,111,111,111, \\ 111,111,111,111,111 \text{ kernels.} \end{array} \right.$$

Ans. \$8,680,555,555,555,555,555,555,555,555,555,555‘555$\frac{555}{1000}$.

4. A gentleman, dying, left his estate to his 5 sons; to the youngest \$1000, to the second \$1500, and ordered that each son should exceed the younger by the ratio of $1\frac{1}{2}$; what was the amount of the estate?

NOTE.—Before finding the power of the ratio $1\frac{1}{2}$, it may be reduced to an improper fraction $= \frac{3}{2}$, or to a decimal, 1‘5.

$$\frac{\frac{3}{2}^5-1}{\frac{3}{2}-1} \times 1000 = \$13187\tfrac{1}{2}; \text{ or, } \frac{1‘5^5-1}{1‘5-1} \times 1000 = \$13187‘50,$$

Answer.

Annuities at Compound Interest.

¶ 228. 1. A man rented a dwelling-house for \$100 a year, but did not receive anything till the end of 4 years, when the whole was paid, with compound interest at 6 per cent., on the sums not paid when due; what did he receive?

Questions.—**¶ 227.** What is the first method of finding the sum, when the things are given, named in ¶ 227? — the second? Describe the process for finding a third method. What terms constitute the first series? — the second? How is the third found? What do you say of its last term? How does it appear to be so? How much is the remainder, after subtracting the second from the third series? How, then, is the sum of the second series found? How the sum of the first, and why? Give the rule. Why raise the ratio, &c.?

SOLUTION. — As annuities in arrears at simple interest form an arithmetical series, so the several years' rents with compound interest on those in arrears, are so many terms of a geometrical series. The last year's rent is $100 only, since it is paid when due, at the end of the year; the third year's rent is on interest 1 year, and is found by multiplying $100 by 1'06, producing $106', and this product multiplied by 1'06, will give the second year's rent, paid 2 years after it is due, and so on. The first term, $100, the number of terms, 4, and the ratio, 1'06, are given to find the sum, as in the last ¶, and we may apply the same rule, thus: —

$$\frac{1'06^4-1}{'06} \times 100 = 437'45. \qquad \textit{Ans.}\ \$437'45.$$

NOTE. — The powers of the ratio, see ¶ 224, may be found in the table ¶ 161.

EXAMPLES FOR PRACTICE.

2. What is the amount of an annuity of $50, it being *in arrears* 20 years, allowing 5 per cent. compound interest? *Ans.* $1653'29.

3. If the annual rent of a house, which is $150, be in arrears 4 years, what is the amount, allowing 10 per cent. compound interest? *Ans.* $696'15.

4. To how much would a salary of $500 per annum amount in 14 years, the money being improved at 6 per cent. compound interest? —— in 10 years? —— in 20 years? —— in 22 years? —— in 24 years? *Ans.* to the last, $25407'75.

5. Two men commence life together; the one pays cash down, $200 a year to mechanics and merchants; the second gets precisely the same value of articles, but on credit, and proving a negligent paymaster, is charged 20 per cent. more than the other; what is the difference in 40 years, compound interest being calculated at 6 per cent.? *Ans.* $6190'478+.

6. A family removes once a year for 30 years, at an expense and loss of $100 each time; what is the amount, 6 per cent. compound interest being calculated? *Ans.* $7905'818+.

Present Worth of Annuities at Compound Interest.

¶ 229. 1. A man, dying, left to his nephew, 21 years old, the use of a house, which would rent at $300 a year for 10 years, after which it was to come in the possession of his own children; the young man, wishing ready money to commence business in a small shop, rented the house for 10

Questions. — ¶ 228. How does it appear that an annuity is an example of geometrical progression? Why is the number of terms only equal to the number of years? What is to be found, and by what rule?

years, receiving in advance such a sum as was equivalent to $300 paid at the end of each year, reckoning compound dis count at 6 per cent.; what did he receive?

Solution. — First, we find what he would receive at the end of 10 years, if nothing had been paid before, by the last ¶. Now what he should receive at the commencement of the 10 years, is a sum, which, on compound interest at the rate given, would amount to this in 10 years, and we divide it by the amount of $1, found as above, for the present worth. *Ans.* $2208'024.

EXAMPLES FOR PRACTICE.

2. What is the present worth of an annual pension of $100, to continue 4 years, allowing 6 per cent. compound interest? *Ans.* $346'503+.

3. What is the present worth of an annual salary of $100, to continue 20 years, allowing 5 per cent.? *Ans.* $1246'218+.

¶ 230. The operations under this rule being somewhat tedious, we subjoin a

TABLE,

Showing the present worth of $1 or £1 annuity, at 5 and 6 per cent. compound interest, for any number of years from 1 to 40.

Years.	5 per cent.	6 per cent.	Years.	5 per cent.	6 per cent.
1	0'95238	0'94339	21	12'82115	11'76407
2	1'85941	1'83339	22	13'163	12'04158
3	2'72325	2'67301	23	13'48807	12'30338
4	3'54595	3'4651	24	13'79864	12'55035
5	4'32948	4'21236	25	14'09394	12'78335
6	5'07569	4'91732	26	14'37518	13'00316
7	5'78637	5'58238	27	14'64303	13'21053
8	6'46321	6'20979	28	14'89813	13'40616
9	7'10782	6'80169	29	15'14107	13'59072
10	7'72173	7'36008	30	15'37245	13'76483
11	8'30641	7'88687	31	15'59281	13'92908
12	8'86325	8'38384	32	15'80268	14'08398
13	9'39357	8'85268	33	16'00255	14'22917
14	9'89864	9'29498	34	16'1929	14'36613
15	10'37966	9'71225	35	16'37419	14'49824
16	10'83777	10'10589	36	16'54685	14'62098
17	11'27407	10'47726	37	16'71128	14'73678
18	11'68958	10'8276	38	16'86789	14'84601
19	12'08532	11'15811	39	17'01704	14'94907
20	12'46221	11'46992	40	17'15908	15'04629

NOTE 1.—From the table it appears that, instead of $1 a year for 30 years, paid at the end of each, which would be $30, one would receive at the commencement, $15'37245, at 5 per cent., or $13'76483, at 6 per cent compound discount, and for $50 a year 50 times as much. Hence, for finding the present worth at compound discount by the table,

RULE.

Multiply the present worth of $1 by the number of dollars.

EXAMPLES FOR PRACTICE.

1. What ready money will purchase an annuity of $150, to continue 30 years, at 5 per cent. compound interest?

Ans. $2305'8675

2. What is the present worth of a yearly pension of $40, to continue 10 years, at 6 per cent. compound interest? —— at 5 per cent.? —— to continue 15 years? —— 20 years? —— 25 years? —— 34 years? *Ans.* to last, $647'716.

NOTE 2.—The practised arithmetician will have no difficulty in calculating the present worth of annuities at simple interest, from principles heretofore presented.

Annuities at Compound Interest in Reversion.

¶ 231. NOTE.—An annuity is said to be in reversion when it does not commence immediately.

1. In Ex. 1, ¶ 229, supposing the uncle had reserved the use of the house to his sister for 2 years after the young man was 21, and given it to him for 10 years after this time should have expired, how much could he have obtained with which to commence business?

SOLUTION.—If he should wait till he is 23 years old, he could obtain $2208'024, as already found, and he can, at 21, obtain a sum which, at compound interest, would amount, in two years, to $2208'024, or the present worth of this sum paid two years before due, found by ¶ 225 to be *Ans.* $1965'13+

Hence, to find the present worth of an annuity in reversion,

RULE.

Find the present worth, were it to commence now, and the present worth of this sum for the time in reversion.

Questions.—¶ **229.** Give the first example. What sum should he receive now? How is it found?

¶ **230.** What appears from the table? How is the present worth of $50 found? Rule.

EXAMPLES FOR PRACTICE.

2. What ready money will purchase the reversion of a lease of $60 *per annum*, to continue 6 years, but not to commence till the end of 3 years, allowing 6 per cent. compound interest to the purchaser?

The present worth, to commence immediately, we find to be $295'039, and $\frac{295'039}{1'06^3} = 247'72$. *Ans.* $247'72.

3. What is the present worth of $100 annuity, to be continued 4 years, but not to commence till 2 years hence, allowing 6 per cent compound interest? *Ans.* $308'392+

4. What is the present worth of a lease of $100, to continue 20 years, but not to commence till the end of 4 years, allowing 5 per cent.? —— what, if it be 6 years in reversion? —— 8 years? —— 10 years? —— 14 years? *Ans.* to last, $629'426.

5. The revolutionary war closed in 1783; one of the soldiers commenced receiving, in 1817, a pension of $96 a year, which continued till 1840; what was the pension worth to him at the close of the war, the rate being 6 per cent. compound interest? *Ans.* $162'89+.

Perpetual Annuities.

¶ 232. 1. A farm rents for $60 a year, at 6 per cent.; what is its value?

Solution. — This is a perpetual annuity, since the owner is supposed to receive $60 a year forever. On every dollar which the farm is worth he receives 6 cents, and consequently the farm is worth as many dollars as the number of times 6 cents are contained in $60. $60 ÷ $'06 = $1000, *Ans.*

Hence, to find the worth of a perpetual annuity,

RULE.

Divide the annuity by the rate per cent.; the quotient will be the perpetual annuity.

2. A city lot is rented 999 years, at $800 a year; what is it worth, the rate being 7 per cent.?

Note 1. — This is the same as a perpetual annuity. *Ans.* $11428'57+

3. What is the worth of $100 annuity, to continue forever, allowing to the purchaser 4 per cent.? —— allowing 5 per cent.? —— 6 per cent.? —— 10 per cent.? —— 15 per cent.? —— 20 per cent.? *Ans.* to last, $500.

4. A farm is left me which will rent for $60 a year, but is

Questions. — ¶ 231. What do you understand by annuities in reversion? How is the worth of an annuity in reversion found?

not to come into my possession till the end of 2 years; what is it worth to me, the rate being 6 per cent. compound interest?

SOLUTION. — The farm will be worth $1000 to me 2 years hence, and it is now worth a sum which, put at compound interest 2 years, will amount to $1000. $\frac{\$1000}{1{`}06^2} = \$889{`}996$, *Ans.*

5. What is the present worth of a perpetual annuity of $100, to commence 6 years hence, allowing the purchaser 5 per cent. compound interest? —— what, if 8 years in reversion? —— 10 years? —— 4 years? —— 15 years? —— 30 years?

Ans. to last, $462'755.

NOTE 2. — The foregoing examples, in compound interest, have been confined to *yearly* payments; if the payments are *half* yearly, we may take *half* the *principal or annuity*, *half* the *rate per cent.*, and *twice* the *number of years* and work as before, and so for *any* other part of a *year*.

PERMUTATION.

¶ 233. Permutation is the method of finding how many different ways the order of any number of things may be varied or changed.

1. Four gentlemen agreed to dine together so long as they could sit, every day, in a different order or position; how many days did they dine together?

SOLUTION. — Had there been but *two* of them, *a* and *b*, they could sit only in 2 times 1 ($1 \times 2 = 2$) different positions, thus, *a b*, and *b a*. Had there been *three*, *a*, *b*, and *c*, they could sit in $1 \times 2 \times 3 = 6$ different positions; for, beginning the order with *a*, there will be 2 positions, viz., *a b c*, and *a c b*; next, beginning with *b*, there will be 2 positions, *b a c*, and *b c a*; lastly, beginning with *c*, we have *c a b*, and *c b a*, that is, in all, $1 \times 2 \times 3 = 6$ different positions. In the same manner, if there be *four*, the different positions will be $1 \times 2 \times 3 \times 4 = 24$.

Ans. 24 days.

Hence *to find the number of different changes or permutations, of which any number of different things is capable*,— Multiply continually together all the terms of the natural

---

Questions. — ¶ 232. What do you understand by a perpetual annuity? How is its value found? Rule. When it does not begin immediately, how is its worth calculated? What do you say of other than yearly payments?

¶ 233. What is permutation? Illustrate by the first example. What is the rule?

series of numbers, from 1 up to the given number, and the last product will be the answer.

2. How many variations may there be in the position of the nine digits? *Ans.* 362880.

3. A man bought 25 cows, agreeing to pay for them 1 cent for every different order in which they could all be placed; how much did the cows cost him? *Ans.* $15511210043330985984000000.

4. Christ Church, in Boston, has 8 bells; how many changes may be rung upon them? *Ans.* 40320.

MISCELLANEOUS EXAMPLES.

¶ 234. 1. $\overline{7+4-2+3+40} \times 5 =$ how many? *Ans.* 230.

NOTE. — A line drawn over several numbers, signifies that the whole are to be taken as one number.

2. The sum of two numbers is 990, and their difference is 90; what are the numbers?

3. There are 4 sizes of chests, holding respectively 48, 76, 87 and 90 lbs.; what is the least number of pounds of tea that will exactly fill some number of chests of either of the 4 sizes? *Ans.* 396720 lbs.

4. How many bushels of wheat, at $1'50 per bushel, must be given for 15 yards of cloth worth 2s. 3d. sterling per yard?

Ans. $5\frac{89}{200}$ bushels.

5. If oats, worth $'30 per bushel, are sold for $'35 on account, for what ought cloth to be sold on account, worth $3'75 per yard cash? *Ans.* $4'37½.

6. Bought a book, marked $4'50, at 33⅓ per cent. discount for cash, what did I pay? *Ans.* $3'00.

7. Bought 120 gallons of molasses for $42; how must I sell it per gallon to gain 15 per cent.? *Ans.* $'40¼.

8. What sum, at 6 per cent. interest, will amount to $150 in 2 years and 6 months? *Ans.* $130'434 +.

9. What is the present worth of $1000, payable in 4 years and 2 months, discounting at the rate of 6 per cent.? *Ans.* $800.

10. Bought cloth at $3'50 per yard, and sold it for $4'25 per yard; what did I gain per cent.? *Ans.* $21\frac{3}{7}$ per cent.

11. If 20 men can build a bridge in 60 days, how many would be required to build it in 50 days? *Ans.* 24 men.

12. How much Silesia, 1⅛ yards wide, will line 12 yards of plaid ⅝ yd. wide? *Ans.* 5 yards.

13. A cistern, holding 400 gallons, is supplied by a pipe at the rate of 7 gallons in 5 minutes, but 2 gallons leak out in 6 minutes; in what time will it be filled? *Ans.* 6 hours 15 minutes.

14. A ship has a leak which would cause it to sink in 10 hours, but it could be cleared by a pump in 15 hours; in what time would it sink? *Ans.* 30 hours.

15. How long must I keep $300, to balance the use of $500, which I lent a friend 4 months? *Ans.* 6⅔ months

6. If 800 men have provisions for 2 months, how many must leave that the remainder may subsist 5 months on the same? *Ans.* 480.

17. Bought 45 barrels of beef, at $3'50 per barrel, except 16 barrels, for 4 of which I pay no more than for 3 of the others; what do the whole cost? *Ans.* $143'50.

18. A hare, running 36 rods a minute, has 57 rods the start of a dog; how far must the dog run to overtake him, running 40 rods per minute? *Ans.* 570 rods.

19. The hour and minute hands of a watch are together at 12 o'clock; when are they next together? *Ans.* 1 h. 5 m. $27\frac{3}{11}$ s. P. M.

20. Three men start together to travel the same way around an island 20 miles in circumference, at the rate of 2, 4, and 6 miles per hour; in what time will they be together again? *Ans.* 10 hours

21. Two boats, propelled by steam engines 8 miles an hour, start at the same time, the one up, the other down a river, from places 300 miles apart; at what distance from the place where each started will they meet, if the one is retarded, and the other accelerated 2 miles an hour by the current?

Ans. $112\frac{1}{2}$ miles from the lower, $187\frac{1}{2}$ from the upper place.

22. The third part of an army were killed, the fourth part taken prisoners, and 1000 fled; how many in the army? *Ans.* 2400.

23. A farmer has his sheep in 5 fields: $\frac{1}{4}$ in the first, $\frac{1}{6}$ in the second, $\frac{1}{8}$ in the third, $\frac{1}{12}$ in the fourth, 450 in the fifth; how many sheep has he? *Ans.* 1200.

24. If a pole be $\frac{1}{8}$ in the mud, $\frac{3}{5}$ in the water, and 6 feet out of the water, what is its length? *Ans.* 90 feet.

25. If $\frac{1}{16}$ of a school study grammar, $\frac{3}{8}$ geography, $\frac{3}{10}$ arithmetic, $\frac{3}{20}$ learn to write, and 9 read, what number in the school? *Ans.* 80.

26. A man being asked how many geese he had, replied, if I had $\frac{1}{2}$ as many as I now have, and $2\frac{1}{2}$ geese more, added to my present number, I should have 100; how many had he? *Ans.* 65.

27. In a fruit orchard, $\frac{1}{2}$ the trees bear apples, $\frac{1}{4}$ pears, $\frac{1}{6}$ plums, 100 peaches and cherries; how many in all? *Ans.* 1200.

28. The difference between $\frac{1}{8}$ and $\frac{1}{5}$ of a number is 6; required the number. *Ans.* 80.

29. What number is that, to which, if $\frac{1}{2}$ and $\frac{1}{4}$ of itself be added, the sum will be 84? *Ans.* 48.

30. B's age is $1\frac{1}{2}$ times the age of A, C's age $2\frac{1}{10}$ times the age of both, and the sum of their ages is 93 years; required the age of each. *Ans.* A 12 years, B 18 years, C 63 years.

31. If a farmer had as many more sheep as he now has, $\frac{3}{4}$, $\frac{1}{2}$, $\frac{1}{4}$, and $\frac{1}{8}$ as many, he would have 435; how many has he? *Ans.* 120.

32. Required the number, which, being increased by $\frac{2}{3}$ and $\frac{2}{5}$ of itself, and by 22, will be three times as great as it now is. *Ans.* 30.

33. A and B commence trade with equal sums; A gained a sum equal to $\frac{1}{5}$ of his stock, B lost $200 when A's money was twice B's; what stock had each? *Ans.* $500.

34. A man was hired 50 days, receiving $'75 for every day he worked, and forfeiting $'25 for every day he was idle; he received $27'50; how many days did he work? *Ans.* 40.

35. A and B have the same income; A saves $\frac{1}{5}$ of his; B, spending

$30 a year more than A, is $40 in debt at the end of 8 years; what did B spend each year? *Ans.* $205.

36. A man left to A ½ his property, wanting $20, to B ⅓, to C the rest, which was $10 less than A's share; what did each receive?

Ans. A received $80, B $50, C $70.

37. The head of a fish is 4 feet long, the tail as long as the head and ½ the length of the body, the body as long as the head and tail; what is the length of the fish? *Ans.* 32 feet.

38. A can build a wall in 4 days, B in 3 days; in what time can both together build it? *Ans.* $1\frac{5}{7}$ days.

39. A and B can build a wall in 4 days, B and C in 6 days, A and C in 5 days; required the time if they work together. *Ans.* $3\frac{9}{37}$ days.

40. A and B can build a wall in 5 days; A can build it in 7 days; in how many days can B build it? *Ans.* 17½ days.

41. A man left his two sons, one 14, the other 18 years old, $1000, so divided, that their shares, being put at 6 per cent. interest, should be equal when each should be 21 years old; what was the share of each?

Ans. $546'153 +; $453'846 +.

42. What is paid for the rent of a house 5 years, at $60 a year, in arrears for the whole time at 6 per cent. simple interest?

Ans. $336.

43. If 3 dozen pairs of gloves be equal in value to 40 yards of calico, and 100 yards of calico to 90 yards of satinet, worth $'50 a yard, how many pairs of gloves will $4 buy? *Ans.* 8 pairs.

44. A, B, and C divide $100 among themselves, B taking $3 more than A, C $4 more than B; what is C's share? *Ans.* $37.

45. A man would put 30 gallons of mead into an equal number of 1 pint and 2 pint bottles; how many of each? *Ans.* 80.

46. A merchant puts 12 cwt. 3 qrs. 12 lbs. of tea into an equal number of 5 lb., 7 lb., and 12 lb. canisters; how many of each? *Ans.* 60.

47. If 18 grs. of silver make a thimble, and 12 pwts make a tea-spoon, how many of each can be made from 15 oz. 6 pwts.?

Ans. 24.

48. If 60 cents be divided among 3 boys so that the first has 3 cents as often as the second has 5 and the third 7, what does each receive?

Ans. 12, 20, and 28 cents.

49. A gentleman paid $18'90 among his laborers, to each boy $'06, to each woman $'08, to each man $'16; there were three women for each boy, and 2 men for each woman; how many men were there?

Ans. 90.

50. A man paid $82'50 for a sheep, a cow, and a yoke of oxen; for the cow 8 times, for the oxen 24 times as much as for the sheep; what did he pay for each? *Ans.* $2'50, $20, and $60.

51. Three merchants accompanied; A furnished ⅖ of the capital, B ⅜, and C the rest; what is C's share of $1250 gain? *Ans.* $281'25.

52. A puts in $500, B $350, and C 120 yards of cloth; they gain $332'50, of which C's share is $120; what is C's cloth worth per yard, and what is A's and B's share of the gain?

Ans. C's cloth $4 per yd., A's share $125, B's do. $87'50.

53. A, B, and C bought a farm, of which the profits were $580'80 a year; A paid towards the purchase $5 as often as B paid $7, and B $4 as often as C paid $6; what is each one's share of the gain?

Ans. A's share $129'066⅔, B's $180'693⅓, C's $271'04.

54. A gentleman divided his fortune among his sons, giving A \$9 as often as B \$5, and C \$3 as often as B \$7; C received \$7442'10½; what was the whole estate? *Ans.* \$56063'857⅗.

55. A and B accompany; A put in \$1200 Jan. 1st, B put in such a sum, April 1st, that he had half the profits at the end of the year; how much did B put in? *Ans.* \$1600.

56. Three horses, belonging to 3 men, do work to the amount of \$26'45; A and B's horses are supposed to do ¾ of the work, A and C's $\frac{9}{10}$, B and C's $\frac{13}{20}$, on which supposition the owners are paid proportionally; what does each receive? *Ans.* A \$11'50, B \$5'75, C \$9'20.

57. A gay fellow spent $\frac{2}{7}$ of his fortune, after which he gave \$7260 for a commission, and continued his profusion till he had only \$2178 left, which was $\frac{3}{8}$ of what he had after purchasing his commission; what was his fortune? *Ans.* \$18295'20.

58. A younger brother received £1560, which was $\frac{7}{12}$ of his elder brother's fortune, and 5⅜ times the elder brother's fortune was ⅝ of twice as much as the father was worth; what was he worth?

Ans. £19165 14s. 3$\frac{3}{7}$d.

59. A gentleman left his son a fortune, $\frac{5}{16}$ of which he spent in 3 months; ¾ of ⅚ of the remainder lasted him 9 months longer, when he had only £537 left; what was the sum bequeathed him by his father?

Ans. £2082 18s. 2$\frac{2}{11}$d.

60. A general, placing his army in a square, had 231 men left, which number was not enough by 44 to enable him to add another to each side; how many men in the army? *Ans.* 19000.

61. A military officer placed his men in a square; being reinforced by three times his number, he placed the whole again in a square; again being reinforced by three times his last number, he placed the whole a third time in a square, which had 40 men on each side; how many men had he at first? *Ans.* 100.

62. Suppose that a man stands 80 feet from a steeple, that a line to him from the top of the steeple is 100 feet long, and that the spire is three times as high as the steeple; what is the length of a line reaching from the top of the spire to the man? *Ans.* 197 feet nearly.

63. Two ships sail from the same port; one sails directly east at the rate of 10 miles, the other directly south at the rate of 7½ miles an hour; how far are they apart at the end of 3 days? *Ans.* 900 miles.

64. How many acres in a square field measuring 70'71 rods between the opposite corners? *Ans.* 15⅝ acres.

65. Supposing that the river Po is 1320 feet wide and 10 feet deep and runs 4 miles an hour; in what time will it discharge a cubic mile of water into the sea?

NOTE. — A linear mile is 5280 feet. *Ans.* 22 days.

66. If the country which supplies the river Po with water be 380 miles long and 120 broad, and the whole land upon the surface of the earth be 62,700,000 square miles, and if the quantity of water discharged by the rivers into the sea be everywhere proportional to the extent of land by which the rivers are supplied, how many times greater than the Po will the whole amount of the rivers be? *Ans.* 1375 times.

67. Upon the same supposition, what quantity of water, altogether will be discharged by all the rivers into the sea in a year of 365 days? *Ans.* 22812½ cubic miles.

68. If the ratio of sea to land be as $10\frac{1}{2}$ to 5, and the average depth of the sea be $1\frac{1}{2}$ miles, in how long time, if the sea were empty, would it be filled? *Ans.* 8657 years 275 days.

69. If a cubic foot of water weigh 1000 oz., and mercury be $13\frac{1}{2}$ times heavier than water, and the hight of the mercury in the barometer (which weighs the same as a column of air on the same base and extending to the top of the atmosphere) be 30 inches, what will the air weigh on a square foot? — on a square mile? What will the whole atmosphere weigh?

Ans., in order, 2109'375 lbs., 58806000000 lbs., 11430122220000000000 lbs.

70. A traveler who had set a perfectly accurate watch by the sun in Boston, 71° 4′ W. lon., being in Detroit, 82° 58′ W. lon., 3 days after, was surprised to find it wrong, when compared with the sun; was it too fast or too slow? how much, and why?

71. A building fell in Portland, Me., 70° 20′ W. lon., at 9 o'clock, A. M., and in 3 minutes the intelligence of the event reached St. Louis, Mo., 90° 15′ W. lon., by magnetic telegraph; when was it known at St. Louis? *Ans.* At 43 m. 20 sec. past 7 o'clock, A. M.

72. At the battle of Bunker Hill the roar of cannon was distinctly heard at Hanover, N. H, and business was suspended for a time; in what time did the sound pass, the distance being supposed 120 miles?

NOTE. — Sound moves 1142 feet in a second. *Ans.* 9 m. 14 sec. +.

73. Seeing the flash of a rifle in the evening, it was 8 seconds before I heard the report; what was the distance? *Ans.* 1 mi. 3856 ft.

74. A man in view on a hill opposite is chopping, at the rate of a blow in 2 seconds; I saw him strike 4 blows before I heard the first; what is his distance from me? *Ans.* 1 mi. 1572 ft.

75. A laborer dug a cellar, the length of which was 2 times the width, and the width 3 times the depth; he removed 144 cubic yards of earth; what was the length? *Ans.* 36 feet.

76. A owes B $750, due in 8 months; but receiving $300 ready money, he extends the time of paying the remainder, so that B shall lose nothing; when must it be paid? *Ans.* In 1 yr. 1 mo. 10 days.

77. The sum of two numbers is $266\frac{2}{3}$, and the product of the greater multiplied by 3, equals the product of the less multiplied by 5; what are the numbers? *Ans.* 100, and $166\frac{2}{3}$.

78. A park 10 rods square is surrounded by a walk which occupies $\frac{19}{100}$ of the whole park; what is its width? *Ans.* 8 ft. 3 in.

79. A, B and C commence trade with $3053'25, and gain $610'65; A's stock + B's, is to B's + C's, as 5 to 7; and C's stock — B's, is to C's + B's, as 1 to 7; what is each one's part of the gain?

Ans. A's gain $135'70, B's $203'55, C's $271'40.

MEASUREMENT OF SURFACES

¶ 235. *To find the area of a parallelogram,* multiply the length by the shortest distance between the sides.

An *Angle* is the space comprised between two lines that meet in a point. A *Right Angle* is formed by one line meeting another perpendicularly. An *Obtuse Angle* is greater, and an *Acute Angle* is less, than a right angle.

NOTE 1. — A parallelogram has its opposite sides equal, but its adjacent sides unequal, like the figure A B C D, or E F C D. The former is called a rectangle, see ¶ 48. The second is called a rhomboid, and is equal in size to the first.

1. What are the superficial contents of an oblique angled piece of ground, measuring 80 rods in length and 20 rods in a perpendicular line between its sides? *Ans.* 1600 sq. rods.

NOTE 2. — To find the contents of a rhombus, which, like the annexed figure, has its sides equal, but its angles not right angles; multiply the length of one side by the shortest distance to the side opposite.

To find the area of a trapezoid, multiply half the sum of the parallel sides by the shortest distance between them.

NOTE 3. — A trapezoid is a figure, like the one in the annexed diagram, bounded by four straight lines, only two of which are parallel.

2. What is the area of a piece of ground in the form of a trapezoid, one of whose parallel sides is 8 rods, the other 12 rods, and the perpendicular distance between them 16 rods?

$\frac{8+12}{2} \times 16 = 160$ sq. rods, *Ans.*

3. How many square feet in a board 16 feet long, 1'8 feet wide at one end, and 1'3 at the other? *Ans.* 24'8 feet.

To find the area of a triangle, multiply the base by half the altitude.

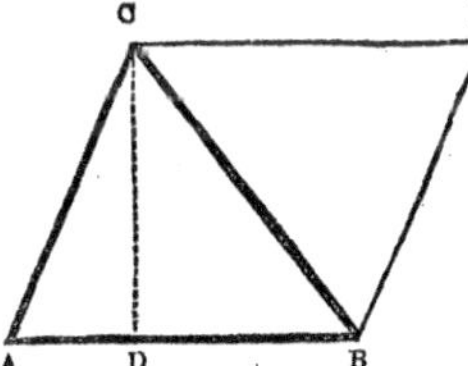

NOTE 4. — The figure A B C is a triangle, of which the side A B is the base, D C the altitude. The triangle is evidently half the parallelogram A B C F, the area of which equals A B $\times$ D C.

4. The base of a triangle is 30 rods, and the perpendicular 10 rods; what is the area? *Ans.* 150 rods.

5. If the contents are 600 rods, and the base 75 rods, what is the altitude? *Ans.* 16 rods.

6. Required the base, the area being 400, the altitude 40 rods. *Ans.* 20 rods.

7. How many square feet in a board 18 feet long, 1½ feet wide at one end, and running to a point at the other? *Ans.* 13½ feet.

To find the circumference of a circle when the diameter is known, multiply the diameter by $3\frac{1}{7}$, or accurately by 3'14159

To find the area, multiply the circumference by one fourth of the diameter; or multiply the square of the diameter by '7854.

8. What is the circumference of a circular pond, the diameter of which is 147 feet? What is the area?

Ans. to the last, $16971\frac{7}{10}$ feet.

9. If the circumference be 22 feet, what is the diameter?

Ans. 7 feet.

10. If the diameter of the earth is 7911 miles, what is the circumference? *Ans.* 24853 miles.

11. How many square inches of leather will cover a ball $3\frac{1}{2}$ inches in diameter?

NOTE 5. — The area of a ball is 4 times the area of a circle having the same diameter. *Ans.* $38\frac{1}{2}$ square inches.

12. How many square miles on the earth's surface?

Ans. 196,612,083.

Measurement of Solids.

¶ 236. NOTE 1. — The general principle for finding the contents of cubic bodies is to multiply the length by the breadth, and the product by the thickness, but the rule applies directly only to the cube or right prism, being subject to modifications as applied to solid figures of other forms. See ¶ 51.

1. How many solid inches in a globe 7 inches in diameter?

NOTE 2. — The solid contents of a globe are found by multiplying the area of its surface by $\frac{1}{6}$ part of its diameter, or the cube of its diameter by '5236.

Ans. $179\frac{2}{3}$ solid inches.

2. What number of cubic miles in the earth?

Ans. $259,233,031,435\frac{1}{2}$.

3. What are the solid contents of a log 20 feet long, of uniform size, the diameter of each end being 2 feet?

NOTE 3. — A figure like the above is called a cylinder. To find the solid contents, we find the area of one end by a foregoing rule, and multiply the area thus found by the length.

Ans. 62'83 + cu. ft.

4. A bushel measure is 18'5 inches in diameter, and 8 inches deep; how many cubic inches, does it contain? *Ans.* 2150'4 +.

NOTE 4. — Solids having bases bounded by straight lines, and decreasing uniformly till they come to a point, are called pyramids. Solids which thus decrease, with circular bases, are called cones. Pyramids and cones are just one third as large as cylinders, of which the area of each end is equal to the area of the bases of these solids. Hence, if we multiply the area of the base by the hight, and divide the product by 3, the quotient will be the solid contents.

5. What are the solid contents of a pyramid, the base of which is 4 feet square, and the perpendicular hight 9 feet?

Ans. 48 solid feet.

6. What are the solid contents of a cone, the hight of which is 27 feet, and the diameter of the base is 7 feet? *Ans.* 346½ solid feet.

7. What are the solid contents of a stick of timber 18 feet long one end of which is 9 inches square and the other end 4 inches square, uniformly diminishing throughout its whole length?

NOTE 5. — Such a figure is called the frustum of a pyramid, and the solid contents are found by adding to the areas of the ends the square root of their product, and multiplying the sum by one third of the hight. The pupil must notice that the diameters are expressed in inches, while the length is in feet

Ans. 5 solid feet, 936 solid inches.

8. What are the solid contents of a round log of wood, 36 feet long, 1,6 feet in diameter at one end, and diminishing gradually to a diameter of '9 of a foot at the other?

NOTE 6. — Such a figure is called the frustum of a cone, and the solid contents are found by adding to the squares of the two diameters the square root of the product of those squares, multiplying the sum by '7854, and the resulting product by one third of the length.

Ans. 45'333 + solid feet.

Gauging, or Measuring Casks.

¶ **237.** 1. How many gallons of wine will a cask contain the head diameter of which is 25 inches, and the bung diameter 31 inches, and the length 36 inches? How many beer gallons?

NOTE. — Add to the head diameter 2 thirds, or if the staves curve but slightly, 6 tenths of the difference between the head and bung diameters, the sum will be the average diameter. The cask will then be reduced to a cylinder, the contents of which may be found by a foregoing rule, in solid inches. The solid inches may be divided by 231, (¶ 114,) to find the number of wine gallons which the cask will contain, and by 282, (¶ 115,) to find the number of beer gallons.

Ans. 102'93+ wine gallons, 84'77+ beer gallons.

2. How many wine gallons in a cask, the bung diameter of which is 36 inches, the head diameter 27 inches, and the length 45 inches?

Ans. 166'617

Mechanical Powers.

¶ **238.** 1. A lever is 10 feet long, and the *fulcrum*, or prop, on which it turns is 2 feet from one end; how many pounds *weight* at the short end will be balanced by 42 pounds at the other end?

NOTE 1. — In turning round the prop, the long end will evidently pass over a *space* of 8 inches, while the short end passes over a *space* of 2 inches. Now, it is a fundamental principle in mechanics, that the *weight* and *power* will exactly *balance each other*, when they are *inversely* as the *spaces* they pass over. Hence, in this example, 2 pounds, 8 feet from the prop, will balance 8 pounds 2 feet from the prop; therefore, if we *divide the distance of the* POWER *from the prop by the distance of the* WEIGHT *from the prop, the quotient will always express the ratio of the* WEIGHT *to the* POWER; $\frac{8}{2} = 4$, that is, the *weight* will be 4 times as much as the *power*.

$42 \times 4 = 168.$ *Ans.* 168 lbs

2. Supposing the lever as above, what *power* would it require to raise 1000 pounds? *Ans.* $\frac{1000}{4} = 250$ pounds.

3. If the weight to be raised be 5 times as much as the power to be applied, and the distance of the weight from the prop be 4 feet, how far from the prop must the power be applied? *Ans.* 20 feet.

4. If the greater distance be 40 feet, and the less $\frac{1}{2}$ of a foot, and the power 175 pounds, what is the *weight?* *Ans.* 14000 pounds.

5. Two men carry a kettle, weighing 200 pounds; the kettle is suspended on a pole, the bale being 2 feet 6 inches from the hands of one, and 3 feet 4 inches from the hands of the other; how many pounds does each bear? *Ans.* $\begin{cases} 114\frac{2}{7} \text{ pounds.} \\ 85\frac{5}{7} \text{ pounds.} \end{cases}$

6. There is a windlass, the wheel of which is 60 inches in diameter, and the axis, around which the rope coils, is 6 inches in diameter; how many pounds on the axle will be balanced by 240 pounds at the wheel?

NOTE 2. — The *spaces* passed over are as the *diameters*, or the *circumferences;* therefore, $\frac{60}{6} = 10$, ratio.

Ans. 2400 pounds.

7. If the diameter of the wheel be 60 inches, what must be the diameter of the axle, that the *ratio* of the weight to the power may be 10 to 1? *Ans.* 6 inches.

NOTE 3. — This calculation is on the supposition, that there is no *friction*, for which it is usual to make allowances.

8. There is a screw, the threads of which are 1 inch asunder; if it is turned by a lever 5 feet, $= 60$ inches, long, what is the ratio of the weight to the power?

NOTE 4. — The power applied at the end of the lever will describe the circumference of a circle $60 \times 2 = 120$ inches in diameter, while the weight is raised 1 inch; therefore, the *ratio* will be found *by dividing the circumference of a circle, whose diameter is twice the length of the lever, by the distance between the threads of the screw.*

$$120 \times 3\tfrac{1}{7} = 377\tfrac{1}{7} \text{ circumference, and } \frac{377\frac{1}{7}}{1} = 377\tfrac{1}{7}, \text{ ratio, } Ans.$$

9. There is a screw, whose threads are $\frac{1}{4}$ of an inch asunder; if it be turned by a lever 10 feet long, what *weight* will be balanced by 120 pounds power? *Ans.* $362057\frac{1}{7}$ pounds.

10. There is a machine, in which the power moves over 10 feet, while the weight is raised 1 inch; what is the power of that machine, that is, what is the ratio of the *weight* to the *power?* *Ans.* 120.

11. A man put 20 apples into a wine gallon measure, which was afterwards filled by pouring in 1 quart of water; required the contents of the apples in cubic inches. *Ans.* $173\frac{1}{4}$ inches.

12. A rough stone was put into a vessel, whose capacity was 14 wine quarts, which was afterwards filled with $2\frac{1}{2}$ quarts of water; what was the cubic content of the stone? *Ans.* $664\frac{1}{8}$ inches.

NOTE 5. — For a more full consideration of the foregoing subjects, the pupil is referred to a more extended treatise on Mensuration in connection with the "series."

FORMS OF NOTES, RECEIPTS, AND ORDERS.

Notes.

No. I.

Keene, Sept. 17, 1846.

For value received, I promise to pay OLIVER BOUNTIFUL, or order, sixty-three dollars, fifty-four cents, on demand, with interest after three months.

WILLIAM TRUSTY.

No. II.

Ludlow, Sept. 17, 1846.

For value received, I promise to pay to O. R., or bearer, —— dollars —— cents, three months after date.

PETER PENCIL.

No. III.

By two persons.

Yates, Sept. 17, 1846.

For value received, we, jointly and severally, promise to pay C. D., or order, —— dollars —— cents, on demand, with interest.

Attest, PETER SAXE.

ALDEN FAITHFUL.
JAMES FAIRFACE.

Receipts.

Boston, Sept. 19, 1846.

Received from Mr. DURANCE ADLEY ten dollars in full of all accounts.

ORVAND CONSTANCE.

Newark, Sept. 19, 1846.

Received of Mr. ORVAND CONSTANCE five dollars in full of all accounts.

DURANCE ADLEY.

Receipt for Money received on a Note.

Rochester, Sept. 19, 1846.

Received of Mr. SIMPSON EASTEY (by the hand of TITUS TRUSTY) sixteen dollars twenty-five cents, which is endorsed on his note of June 3, 1846.

PETER CHEERFUL.

A Receipt for Money received on Account.

Hancock, Sept. 19, 1846

Received of Mr. ORLAND LANDIKE fifty dollars on account.

ELDRO SLACKLEY.

Receipt for Money received for another Person.

Salem, August 10, 1846.

Received from P. C. one hundred dollars for account of J. B.

ELI TRUMAN.

Receipt for Interest due on a Note.

Amherst, July 6, 1846.

Received of I. S. thirty dollars, in full of one year's interest of $500, due to me on the —— day of —— last, on note from the said I. S.

SOLOMON GRAY.

Receipt for Money paid before it becomes due.

Hillsborough, May 3, 1846.

Received of T. Z. ninety dollars, advanced in full for one year's rent of my farm, leased to the said T. Z., ending the first day of April next, 1847.

HONESTUS JAMES.

NOTE. — There is a distinction between receipts given in full of *all accounts*, and others in full of *all demands*. The former cut off accounts *only;* the latter cut off not only accounts, but all obligations and right of action.

Orders.

Utica, Sept. 9, 1846.

Mr. STEPHEN BURGESS. For value received, pay to A. B., or order, ten dollars, and place the same to my account. SAMUEL SKINNER.

Pittsburgh, Sept. 9, 1846.

Mr JAMES ROBOTTOM. Please to deliver to Mr. L. D. such goods as he may call for, not exceeding the sum of twenty-five dollars, and place the same to the account of your humble servant, NICHOLAS REUBENS.

FORMS OF BILLS.

"Before you build, sit down and count the cost."

Simeon Thrifty built a house for Thomas Paywell, according to a plan agreed upon between them, for the sum of $1500. The cellar of the house is 24 by 28 feet, and is dug 4 feet deep below the top of the ground. The cellar walls are 7 feet high. There is a wing at one end of the main building, 20 by 24 feet, which is underpinned with a wall 3 feet high. As one side of the wing joins the main building, for the underpinning of it but 3 walls are required, one 24, and two 20 feet long on the outside. The walls of the cellar, and the underpinning of the wing are 1½ feet thick. To the cellar there is a door 4 by 7 feet, and 2 windows, each 2 by 2½ feet. Simeon Thrifty, wishing to know how much it cost him to build the house, kept an accurate account of all the materials used, the labor employed, and the cost of each. The following are his bills: —

Bill of Timber.

6	sticks for posts to upright part, ...	each	14 ft. long, and	4 by	10 in.
6	" " wing, ...	"	11 "	4 "	10 "
2	" sills to upright part, ...	"	28 "	7 "	8 "
5	" " and sleepers to upright and wing, ...	"	24 "	7 "	8 "
4	" plates and side girts to upright,	"	28 "	6 "	7 "
7	" " and beams to upright and wing, ...	"	24 "	6 "	7 "
2	" eave gutters, ...	"	30 "	6 "	10 "
3	" "	"	20 "	6 "	10 "
52	rafters, ...	"	15 "	3 "	4 "
96	studs, ...	"	12 "	2 "	4 "
175	partition planks, ...	"	10 "	2 "	4 "
96	scantlings, ...	"	12 "	4 "	4 "
10	" for braces, ...	"	12 "	4 "	4 "
75	joists, ...	"	14 "	2 "	7 "
30	"	"	10 "	2 "	7 "

Bill of Lumber.

800 ft. best quality pine, for best doors, and other nice joiner work
10000 " common " " door and window casings, stairs, base boards common doors, &c., &c.
3850 " white wood siding.
2160 " bass " flooring.
1500 roof boards.
6000 " lath.
26 bunches shingles, 500 in each bunch.

Bill of Materials for Windows.

Sash and glass for 14 windows of 24 panes each, 7 by 9 inches.
" " 4 " " 20 " 7 " 9 "
" " 2 " " 6 " 7 " 9 "
" " 1 window, " 16 " 7 " 9 "
" " 1 " " 12 " 7 " 9 "
30 lb. putty.

Hardware Bill.

4 casks nails, 100 lbs. each.
22 pairs 3 inch door hinges, with screws.
2 " 4 " " "
20 door handles, "
2 outside door knobs and locks.
3 cupboard fastenings.
63 ft. tin eave conductors, including 4 elbows.
3 stove-pipe crocks.
3 " thimbles.
3 ft. tin pipe for sink-spout.
20 window springs.
4 papers 1 inch brads.

Bill of Materials for Chimneys and Plastering.

1600 bricks. 27 loads sand.
200 bush. lime. 10 bush. hair.

Bill of Prices of Materials.

Stone for cellar walls and underpinning, ...	$ '25	per perch.
All the timber except the eave gutters reduced to board measure, that is, 1 inch thick, . . .	10'	" M.
Eave gutters, ...	15'	" "
Pine lumber, best quality, ...	20'	" "
" " common, ...	10'	" "
Siding, flooring, and roof boards, ...	10'	" "
Lath, ...	5'	" "
Shingles, ...	1'50	" bunch.
Window sash, ...	'03	" pane.
" glass, ...	2'50	" box of 114 panes
Putty, ...	'07	" lb.
Nails, ...	'05½	" "
3 inch door butts, with screws, ...	'12½	" pair.
4 " " "	'15	" "
Door handles, ...	'12½	each.
Outside door knobs and locks, ...	1'50	"
Cupboard fastenings, ...	'12½	"
Eave conducters, ...	'12½	per ft.
Extra for elbows, ...	'06¼	each.
Stove-pipe crocks, ...	'37½	"
" thimbles, ...	'12½	"

Sink spout, ...	$ '44.
Window springs, ...	'06¼ each.
Brads, ...	'10 per paper
Bricks, ...	10' " M.
Lime, ...	'12½ " bush.
Sand, ...	1'00 " load.
Hair, ...	'25 " bush.

Bill of Prices of Labor.

Digging cellar, ...	$	'18 per cu. yd
5 stone masons, 6 days each, ...	"	2'00 " day.
2 carpenters, 12 "	"	1'50 " "
3 joiners, 40 "	"	1'75 " "
Painting and glazing, ...		100'00.
Furring ready for lathing, ...		12'00.
Lathing, ...		15'00.
2 plasterers, 7 days each, ...	"	2'50 per day.
3 brick-layers, 1 day "	"	2'75 "
Team and hired man, 4 months of 26 days each, ...	"	2'00 "

Simeon Thrifty commenced the house on the 1st day of May, and completed it on the 3d day of Sept.; allowing him $1'50 per week for his board, how much did he get for his own labor? *Ans.* $281'64$\frac{4}{11}$.

75$\frac{6}{11}$ perches stone, . .	$ 18'88$\frac{7}{11}$	Brads, ...	$ '40
600 ft. eave gutters, . .	9'00	1600 bricks, ...	9'60
8287 " other timber, . .	82'87	200 bush. lime, ...	25'00
800 " best pine lumber,	16'00	27 loads sand, ...	27'00
17510 " other lumber, . .	175'10	10 bush. hair, ...	2'50
6000 " lath, ...	30'00	Digging cellar, ...	4'48
26 bunches shingles, .	39'00	Laying stone work, . . .	60'00
Window sash, ...	13'68	Carpenters' work, ...	36'00
4 boxes window glass, .	10'00	Joiners' "	210'00
30 lbs. putty, ...	2'10	Painting and glazing, . .	100'00
400 " nails, ...	22'00	Furring, ...	12'00
Butt hinges, ...	3'05	Lathing, ...	15'00
Door handles, ...	2'50	Plastering, ...	35'00
Outside door knobs & locks,	3'00	Brick-laying, ...	7'50
Cupboard fastenings, . .	'37½	Team and hired man, . .	208'00
Tin eave conductors, . . .	7'87½	18 weeks' board, ...	27'00
Extra on elbows, ...	'25		
Stove-pipe crocks, ...	1'12½	Amount, . . .	$1218'35$\frac{7}{11}$
" thimbles, . . .	'37½		
Sink spout, ...	'44	Errors excepted,	
Window springs, . .	1'25		

www.ingramcontent.com/pod-product-compliance
Lightning Source LLC
LaVergne TN
LVHW020241110826
845151LV00003B/1000

* 9 7 8 1 4 2 5 5 2 9 4 0 6 *